IMAGING OF SURFACES AND INTERFACES

Frontiers of Electrochemistry

Series Editors

Jacek Lipkowski
Department of Chemistry and
   Biochemistry
University of Guelph
Guelph, Ontario N1G 2W1
Canada

Philip N. Ross
1 Cyclotron Road
Lawrence Berkeley Laboratory
University of California
Berkeley, CA 94720, USA

Advisory Board

D. M. Kolb, Ulm, Germany
M. Van Hove, Berkeley, CA
W. Schmickler, Ulm,
   Germany
R. Guidelli, Florence, Italy
A. Wieckowski, Urbana, IL

W. O'Grady, Washington, DC
M. J. Weaver, Lafayette, IN
W. R. Fawcett, Davis, CA
R. Parsons, Southhampton,
   United Kingdom
S. Trasatti, Milan, Italy

Series Listing

Adsorption of Molecules at Metal Electrodes,
   Jacek Lipkowski and Philip N. Ross, Eds.
Structure of Electrified Interfaces,
   Jacek Lipkowski and Philip N. Ross, Eds.
Electrochemistry of Novel Materials,
   Jacek Lipkowski and Philip N. Ross, Eds.
Electrocatalysis,
   Jacek Lipkowski and Philip N. Ross, Eds.

IMAGING OF SURFACES AND INTERFACES

Editors

JACEK LIPKOWSKI and **PHILIP N. ROSS**

WILEY-VCH

New York • Chichester • Weinheim • Brisbane • Singapore • Toronto

This book is printed on acid-free paper. ∞

Copyright © 1999 by Wiley-VCH, Inc. All rights reserved.

Published simultaneously in Canada.

No part of this publication may be reproduced, stored in a retrieval system or transmitted in any form or by any means, electronic, mechanical, photocopying, recording, scanning or otherwise, except as permitted under Sections 107 or 108 of the 1976 United States Copyright Act, without either the prior written permission of the Publisher, or authorization through payment of the appropriate per-copy fee to the Copyright Clearance Center, 222 Rosewood Drive, Danvers, MA 01923, (978) 750-8400, fax (978) 750-4744. Requests to the Publisher for permission should be addressed to the Permissions Department, John Wiley & Sons, Inc., 605 Third Avenue, New York, NY 10158-0012, (212) 850-6011, fax (212) 850-6008, E-Mail: PERMREQ@WILEY.COM.

For ordering and customer service, call 1-800-CALL WILEY.

Library of Congress Cataloging-in-Publication Data:
Imaging of surfaces and interfaces / editors, Jacek Lipkowski, Philip N. Ross.
    p.   cm. — (Frontiers of electrochemistry ; v. 5)
   Includes index.
   ISBN 0-471-24672-7 (cloth)
   1. Electrodes—Surfaces. 2. Interfaces (Physical sciences)
3. Spectroscopic imaging. 4. Surface chemistry.
5. Electrochemistry. I. Lipkowski, Jacek. II. Ross, P. N. (Philip N.) III. Series.
QD571.I43   1999
541.3'724—dc21   98-38181

Printed in the United States of America.

10 9 8 7 6 5 4 3 2 1

CONTENTS

Preface		vii
Contributors		ix
1.	**Low-Dimensional Metal Phases and Nanostructuring of Solid Surfaces** G. Staikov, W. J. Lorenz, and E. Budevski	1
2.	**Electron Diffraction and Electron Microscopy of Electrode Surfaces** G. Lehmpfuhl, Y. Uchida, M. S. Zei, and D. M. Kolb	57
3.	**Imaging Metal Electrocrystallization at High Resolution** R. Nichols	99
4.	**Imaging of Reaction Fronts at Surfaces and Interfaces** H. H. Rottermund, K. Krischner, and B. Pettinger	139
5.	**Potential Controlled Ordering in Organic Monolayers at Electrode–Electrolyte Interface** N. J. Tao	211
6.	**Scanning Probe Microscopy of Organic Thin Films at Electrode Surfaces** J.-B. Green, C. A. McDermott, M. T. McDermott, and M. D. Porter	249
7.	**Theoretical Aspects of the Scanning Tunneling Microscope Operating in an Electrolyte Solution** W. Schmickler	305
Index		339

PREFACE

The last two decades have witnessed an impressive development of new techniques for surface imaging. Surface and interfacial processes, which were investigated in the past by spectroscopic, diffraction, and electrochemical methods, can be visualized today by recording real-space images with atomic resolution and in real time. Structures of adlayers at surfaces and interfaces, which in the past were constructed from complex diffraction experiments, can now be observed directly on a display from a scanning tunneling or an atomic force microscope. More significantly, the distribution of atoms and molecules during a surface reaction and the dynamics of phase transitions in two dimensional adlayers can now be studied with the modern surface imaging techniques. The scanning probe techniques may also be used to manipulate atoms and molecules at surfaces in such a way as to create an information storage medium with a bit-size in the subnanometer regime. The objective of this volume is to review the impact of new surface imaging techniques in advancing the frontiers of modern electrochemistry. The technologically important field of electrocrystallization/electrodeposition was perhaps the first to benefit from the availability of modern surface imaging techniques. This point is illustrated by the first three chapters, which describe the formation of phases of low dimensionality and the early stages of the metal deposition process. The first chapter shows also how the newly acquired knowledge of the low dimensionality phases may be used to print nanopatterns at electrode surfaces. It has long been postulated that many heterogeneous reactions proceed chiefly on so-called active centers, but the true nature of these centers is often difficult to establish. The site specificity of electrode processes can now be elucidated with the help of these new surface imaging techniques. Numerous examples are shown in the chapters on metal electrocrystallization and on oscillating surface reactions where reaction fronts are imaged with high resolution. They demonstrate how much can be learned today about site specificity. The properties of organic films at surfaces and interfaces is a subject of significant applied and fundamental interest. Two chapters on this subject describe the advances of the scanning probe microscopies such as scanning tunneling microscopy (STM) and atomic force microscopy (AFM) to study organic films on metal electrodes and to image potential dependent phase transitions in organic monolayers. These advances in the application of STM have rekindled interest in the theory of electron tunneling at the metal/solution interface because this theory plays a central role in understanding STM images of electrode surfaces. Thus, progress in the theory of the operation of STM in an electrolyte solution is carefully reviewed in the last chapter.

This volume provides a comprehensive summary of progress in the imaging of surfaces in electrolyte and is addressed to a wide audience of scientists interested in the electrochemistry, surface science, materials science, and electrodeposition technologies. Each chapter provides sufficient background so that it can be read by a specialist and a nonspecialist alike.

JACEK LIPKOWSKI
PHILIP N. ROSS

CONTRIBUTORS

Evgeni Budevski, Central Laboratory of Electrochemical Power Sources, Bulgarian Academy of Sciences, Sofia 1113, Bulgaria

J.-B. Green, Naval Research Laboratory, Code 6170, 4555 Overlook Avenue, Washington, DC 20375-5342

Dieter M. Kolb, Department of Electrochemistry, University of Ulm, D-89069 Ulm, Germany

K. Krischner, Fritz-Haber-Institut der Max-Planck-Gesellschaft, Faradayweg 4-6, D-14196 Berlin, Germany

G. Lehmpfuhl, Fritz-Haber Institute, Max-Planck-Gesellschaft, Faradayweg 4-6, D-14195 Berlin, Germany

Wolfgang J. Lorenz, Institute of Physical Chemistry and Electrochemistry, University of Karlsruhe, Kaiserstrasse 12, D-76131 Karlsruhe, Germany

C. A. McDermott, Department of Chemistry, University of Alberta, Edmonton, Alberta T6G 2G2, Canada

M. T. McDermott, Department of Chemistry, University of Alberta, Edmonton, Alberta T6G 2G2, Canada

Richard J. Nichols, Chemistry Department, University of Liverpool, PO Box 147, Liverpool L69 7ZD, United Kingdom

Bruno Pettinger, Fritz-Haber-Institut der Max-Planck-Gesellschaft, Faradayweg 4-6, D-14196 Berlin, Germany

Marc Porter, Iowa State University, Ames, IA 50011

H. H. Rottermund, Fritz-Haber-Institut der Max-Planck-Gesellschaft, Faradayweg 4-6, D-14196 Berlin, Germany

Georgi Staikov, Institute of Physical Chemistry and Electrochemistry, University of Karlsruhe, Kaiserstrasse 12, D-76131 Karlsruhe, Germany

Wolfgang Schmickler, Department of Electrochemistry, University of Ulm, D-89069 Ulm, Germany

Nongjiang Tao, Department of Physics, Florida International University, Miami, FL

Y. Uchida, Fritz-Haber-Institute, Max-Planck-Gesellschaft, Faradayweg 4-6, D-14195 Berlin, Germany

M. S. Zei, Fritz-Haber-Institute, Max-Planck-Gesellschaft, Faradayweg 4-6, D-14195 Berlin, Germany

1

LOW-DIMENSIONAL METAL PHASES AND NANOSTRUCTURING OF SOLID SURFACES

GEORGI STAIKOV AND WOLFGANG J. LORENZ
Institute of Physical Chemistry and Electrochemistry, University of Karlsruhe, Kaiserstr. 12, D-76131 Karlsruhe, Germany

EVGENI BUDEVSKI
Central Laboratory of Electrochemical Power Sources, Bulgarian Academy of Sciences, Sofia 1113, Bulgaria

1.1 INTRODUCTION

The electrochemical formation of metal (Me) phases on native or foreign substrates (S) depends strongly on surface inhomogeneities (I) of the substrate. Surface inhomogeneities can be of different dimensionality (iD) denoted by I_{iD} with $i = 0,1,2$ affecting the thermodynamics, structure, and kinetics of the Me deposit.[1-5]

In the case of Me deposition on native substrates in the absence of surface inhomogeneities acting as growth sites (i.e., on an atomically smooth surface, or "quasi-perfect substrate"), the initial step of the substrate growth is 2D nucleation. The process of nucleation requires the surmounting of a considerable energy barrier and can only occur at a significant supersaturation.

The growth of a bulk (3D) Me phase in presence of inhomogeneities as sites of growth ("real substrates") can proceed continuously without nucleation phenomena, but, due to other kinetic hindrances, needs a supersaturation.

Imaging of Surfaces and Interfaces (Frontiers of Electrochemistry, Volume 5).
Edited by Jacek Lipkowski and Philip N. Ross.
ISBN 0-471-24672-7. © 1999 Wiley-VCH, Inc.

In both cases, the process of Me deposition occurs in the so-called supersaturation range and is denoted as Me overpotential deposition (OPD).[1] In Me electrocrystallization on real substrates with a relatively high density of crystal imperfections, the surface inhomogeneities play the most significant role in the process of deposition.

In the case of Me deposition on a foreign substrate, the formation of the new Me phase generally needs a preceding 3D nucleation. Therefore, the bulk deposition process of Me proceeds in the OPD range.

Me phases, usually with lower dimensionality, can also be formed, however, in the undersaturation range if the binding energy of Me adatoms on a foreign substrate is higher than that of Me adatoms on the native substrate.[1] This process is well known as Me underpotential deposition (UPD) and occurs at electrode potentials, E, more positive than the Nernst equilibrium potential of the 3D Me phase, $E_{Me/Me^{z+}}$:

$$\Delta E = E - E_{Me/Me^{z+}} = E - E^O_{Me/Me^{z+}} + \frac{RT}{zF} \ln \frac{a_{Me^{z+}}}{a_{3D\,Me}} > 0 \qquad (1)$$

where $E^O_{Me/Me^{z+}}$ denotes the standard potential of the 3D Me bulk phase, $a_{Me^{z+}}$ is the activity of Me^{z+} within the electrolyte, and $a_{3D\,Me}$ represents the activity of the pure, condensed 3D Me phase, which is a constant equal to unity by convention. The potential difference $\Delta E = E - E_{Me/Me^{z+}} > 0$ represents the underpotential range, which is directly related to the undersaturation range $\Delta\mu = -zF\Delta E < 0$ with respect to the Me phase.

Me UPD is an electrosorption process with a charge-covering stoichiometry given by the so-called electrosorption valency:[1]

$$\gamma = \frac{1}{F}\left(\frac{\partial q}{\partial \Gamma}\right)_E = \frac{1}{F}\left(\frac{\partial \mu}{\partial E}\right)_\Gamma \qquad (2)$$

where q is the relative specific ionic charge (charge density), Γ is the relative surface excess concentration of the adsorbed Me species, and μ denotes the chemical potential of Me^{z+} in the electrolyte. In absence of cosorption or competitive sorption phenomena or at a constant relative surface excess concentration of anions in the interphase within a selected UPD range, the electrosorption valency is equal to the ionic charge of Me^{z+} ($\gamma = z$) and the Me UPD process can be described by a quasi-Nernst equation

$$\Delta E = -\frac{RT}{zF} \ln f(\Gamma) \qquad (3)$$

where $f(\Gamma)$ is related to the adsorption isotherm depending on the vertical Me_{ads}-S and lateral Me_{ads}-Me_{ads} interactions and the crystallographic structure and homogeneity of S.

In the presence of substrate surface inhomogeneities, I_{iD} ($i = 0,1,2$), a stepwise formation of different low-dimensional Me phases becomes possible in the UPD range. However, cosorption or competitive sorption phenomena of electrolyte constituents different from Me^{z+} (e.g., anions, may significantly influence the Me UPD process).

The local formation of low-dimensional Me phases on foreign substrates, S, plays an important role in modern nanotechnology, because future aspects of science and technology in many fields such as physics, chemistry, materials science, electronics, sensors, biology, medicine, etc., are characterized by a miniaturization down to an atomic level.[1] Nanotechnology dealing with single atoms, molecules, or clusters will take the position of the micrometer technology, which has dominated the last 150 years. In surface nanotechnology, solid surfaces such as electron-conducting materials (metals, graphite, semiconductors, superconductors, electron-conducting polymers, etc.), ion-conducting materials (solid electrolytes, ion-conducting polymers, membranes, etc.), and insulators have to be analyzed (analytical aspect) in the nanometer range. On the other hand, well-defined nanostructuring and nanomodification of solid surfaces will play an increasing role in the future nanotechnology (preparative aspects). Both aspects can only be studied using local probe microscopy (SPM) techniques such as STM, AFM, and related methods that have been developed during the past fifteen years.[6-14]

The first topic of this paper deals with theoretical and experimental results of the formation of low-dimensional Me phases on foreign substrates. The second topic contains the current state on nanostructuring and nanomodification of electron-conducting solid surfaces under defined electrochemical conditions at solid/liquid interfaces using different in situ local probe methods. The role of surface inhomogeneities in Me phase formation and local structuring processes will be discussed.

1.2 FORMATION OF LOW-DIMENSIONAL METAL PHASES

1.2.1 Theoretical Considerations

Systems of different dimensionality i with $i = 3,2,1,0$ are well known in solid state physics. For example, diamond (3D), graphite (2D), monatomic carbon chains (1D) and fullerenes (0D) are treated as carbon "objects" with different dimensionality.[15] Fractal systems have non-integer dimensionalities, e.g., $1 < i < 2$. A special category are 0D systems. The "quantum dot" in semiconductor physics is the best example. It is a small disc compared to the Fermi wavelength cut out of a 2D electron gas.[15]

1.2.1.1 Thermodynamics of Low-Dimensional Me Phases By definition, thermodynamically stable phases are infinitely large. Therefore, a concept for the existence of Me phases of different dimensionality i can be derived only for

$i = 3,2,1$. 3D Me bulk phases, 2D Me overlays, and 2D Me-S surface alloys, as well as 1D Me chains can be considered as examples. 2D and 1D Me phases can be formed on atomically smooth terraces and at monatomic steps of a foreign substrate, respectively, as shown in Figure 1.1. The equilibrium between a condensed (solid-like) iD Me phase and its ambient phase is characterized by the equality of attachement-detachment frequencies of Me^{z+} ions at kinks as sites for crystal growth and dissolution.[1] The attachment-detachment of an Me^{z+} ion to or from a kink reproduces the kink site position as schematically shown in Figure 1.1. The adsorption of electrolyte constituents different from Me^{z+} does not affect the energetics of kink sites of 3D phases, but can significantly change the energetics of kink sites of low-dimensional phases. Consequently, the equilibrium potential of a 3D Me phase is independent of cosorption processes on which those of low-dimensional Me phases may depend.

All these considerations are obviously not valid in the case of 0D systems. However, in order to bridge between low-dimensional Me phases and 0D systems, not a single Me adatom on a certain substrate surface site, but a small localized and stable cluster of Me adatoms with specific energetics and structure different from those of the Me bulk phase, will be formally considered as a "0D Me cluster" in the following discussion.

Three questions arise. First, what are the stability conditions of iD Me phases? Second, how are the terms *super-* and *undersaturation* thermodynamically defined? Third, how does the formation of low-dimensional Me phases take place on real substrate surfaces under defined electrochemical conditions?

Generally, the stability ranges of expanded (gas-like) or condensed (liquid-like or solid-like) Me phases of different dimensionality, iD (with $i = 1,2,3$), are characterized by Nernst-type equations:

$$E = E^O_{3D\,Me} + \frac{RT}{zF} \ln\left(\frac{a_{Me^{z+}}}{a_{iD\,Me}}\right) \quad \text{with } i = 1,2,3 \qquad (4)$$

where E denotes the actual electrode potential, $E^O_{3D\,Me} \equiv E^O_{Me/Me^{z+}}$ is the standard potential of the 3D Me bulk phase, $a_{Me^{z+}}$ is the activity of Me^{z+} within the electrolyte, and $a_{iD\,Me}$ represents the activity of an iD Me phase.

In the case of condensed iD Me phases, $a_{iD\,Me}$ are constants and E represents the corresponding equilibrium potential $E_{iD\,Me}$. For 3D Me bulk phases, $a_{3D\,Me} = 1$ and $E_{3D\,Me}$ is usually denoted as $E_{Me/Me^{z+}}$ (cf. eq. (1)). However, for low-dimensional condensed Me phases ($i = 1,2$), the activities $a_{iD\,Me}$ are also constants, but less than unity. The activity $a_{iD\,Me}$ of condensed Me phases usually decreases with decreasing i so that the corresponding equilibrium potentials are shifted in the positive direction:

$$E_{3D\,Me} < E_{2D\,Me} < E_{1D\,Me} \qquad (5)$$

Condensed iD Me phases are stable in the potential ranges $E < E_{iD\,Me}$.

1.2 FORMATION OF LOW-DIMENSIONAL METAL PHASES

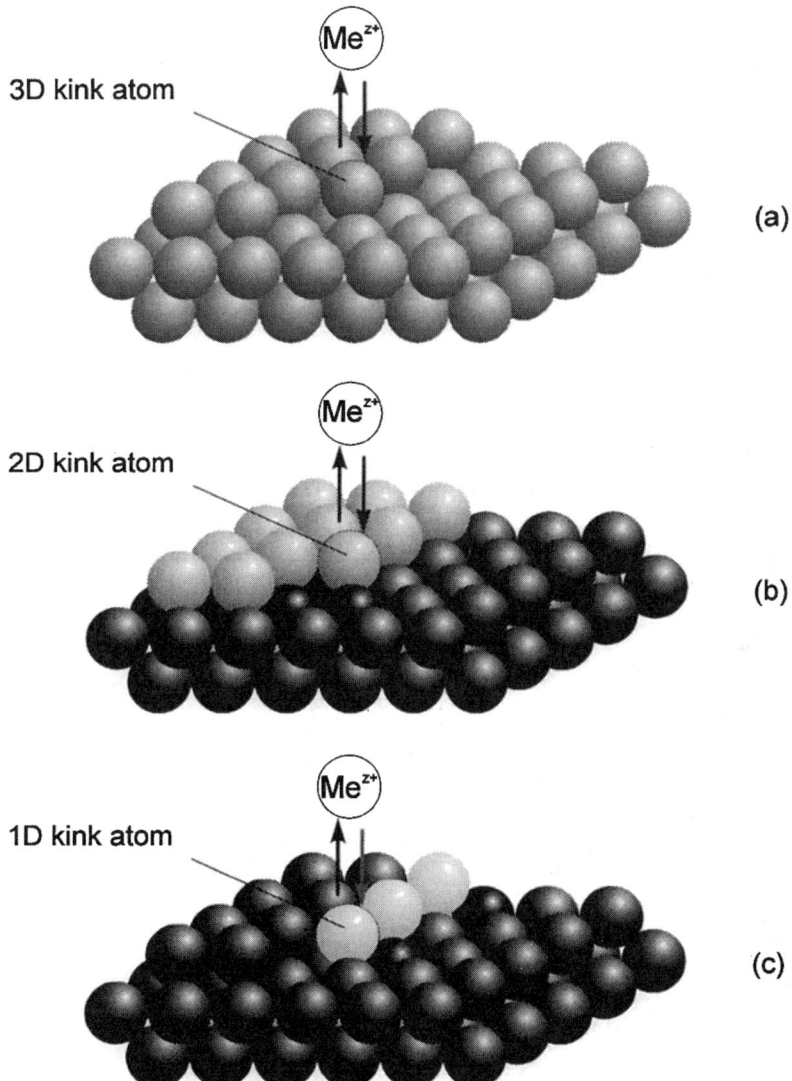

Figure 1.1 Schematic representation of iD kink atoms (i = 3,2,1) determining the equilibria of the corresponding condensed iD Me phases. (a) 3D Me phase; (b) 2D Me phase; (c) 1D Me phase.

In the case of expanded iD Me phases, $a_{i\text{D Me}}$ becomes a function of E and μ, expressed as the corresponding isotherm $a_{i\text{D Me}} = f(\Gamma)$ with $\Gamma = \Gamma(E, \mu)$ (cf. eq. (3)). Expanded iD Me phases with i = 1,2 are found in the potential ranges $E > E_{i\text{D Me}}$.

Phases of low dimensionality are only stable if the interaction energy between Me adatoms and the substrate is higher than that between Me adatoms and the native substrate. The higher the Me-S interaction energy is, the lower the dimensionality of the iD Me phase can be. The activity term $a_{i\text{D Me}}$ is related to this interaction energy, so that it is expected to decrease with decreasing dimensionality, giving higher (i.e., more positive), equilibrium potentials of phases of lower dimensionality i, as stated above in eq. (5).

Supersaturation or undersaturation with respect to the corresponding condensed iD Me phases, $\Delta\mu_{i\text{D}}$, in absence of all other kinetic hindrances are given by:

$$\Delta\mu_{i\text{D}} \stackrel{\text{def}}{=} \mu_{\text{Me}_{\text{ads}}}^{(\text{IP})}(E) - \mu_{\text{Me}_{\text{ads}}}^{(\text{IP})}(E_{i\text{D Me}}) = -zF(E - E_{i\text{D Me}}) \quad \text{with } i = 1,2,3 \quad (6)$$

where $\mu_{\text{Me}_{\text{ads}}}^{(\text{IP})}(E)$ and $\mu_{\text{Me}_{\text{ads}}}^{(\text{IP})}(E_{i\text{D Me}})$ represent the chemical potentials of Me adatoms in the interphase (IP) at E and $E_{i\text{D Me}}$, respectively. It should be mentioned that $\mu_{\text{Me}_{\text{ads}}}^{(\text{IP})}(E_{i\text{D Me}})$ is a constant equal to the chemical potential, $\mu_{i\text{D Me}}$, of the infinitely large condensed iD Me phase. Then, $\Delta\mu_{i\text{D}} > 0$ denotes a supersaturation and $\Delta\mu_{i\text{D}} < 0$ an undersaturation with respect to the corresponding condensed iD Me phase. Therefore, the potential differences $E - E_{i\text{D Me}}$ can be defined as:

$$E - E_{i\text{D Me}} \stackrel{\text{def}}{=} \begin{cases} \Delta E_{i\text{D}} \text{ (underpotential)} > 0 & \text{for } E > E_{i\text{D Me}} \\ \eta_{i\text{D}} \text{ (overpotential)} \quad\;\; < 0 & \text{for } E < E_{i\text{D Me}} \end{cases} \quad \text{with } i = 1,2,3 \quad (7)$$

representing the underpotential and the overpotential with respect to the equilibrium potential of the corresponding condensed iD Me phase. As seen, the underpotential $\Delta E_{i\text{D}}$ and the overpotential $\eta_{i\text{D}}$ are conditional and depend on the reference equilibrium potential $E_{i\text{D Me}}$.

Low-dimensional Me phases ($i = 2,1$) are expected to exist on an ideally polarizable and foreign substrate in the undersaturation range $E > E_{3\text{D Me}}$, if the binding energy of Me adatoms on S is stronger than that of Me on the native substrate:

$$\Psi_{\text{Me}_{\text{ads}}-\text{S}} \gg \Psi_{\text{Me}_{\text{ads}}-\text{Me}} \quad (8)$$

In this case, assuming an atomically flat and homogeneous substrate surface without any surface inhomogeneities, S becomes modified by an expanded and/or condensed 2D Me phase in the UPD range $\Delta E_{3\text{D}}$ with respect to the condensed 3D Me phase. A 3D Me phase is formed on top of the 2D Me phase of the UPD modified substrate surface in the OPD range following either a Frank-van der Merwe or a Stranski-Krastanov growth mode as schematically illustrated in Figures 1.1a and 1.2b.[1] The growth mode depends on the crystallographic misfit between Me and S, which is defined by

1.2 FORMATION OF LOW-DIMENSIONAL METAL PHASES

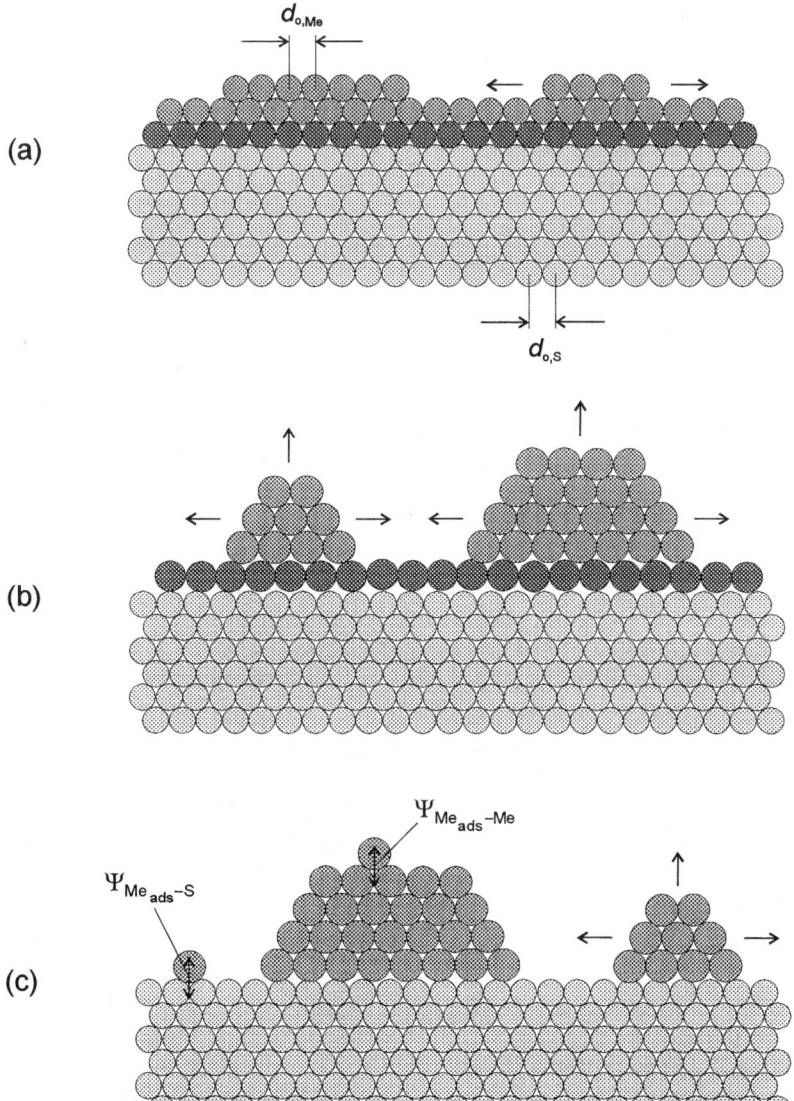

Figure 1.2 Schematic representation of different growth modes in metal (Me) deposition on foreign substrate (S) depending on the binding energy of Me_{ads} on S, $\Psi_{Me_{ads}-S}$, compared to that of Me_{ads} on native substrate Me, $\Psi_{Me_{ads}-Me}$, and on the crystallographic misfit, $f = (d_{O,Me} - d_{O,S})/d_{O,S}$, characterized by the interatomic distances $d_{O,Me}$ and $d_{O,S}$ of 3D Me and S phases, respectively. (a) "Frank–van der Merwe" growth mode (Me layer-by-layer formation) for $\Psi_{Me_{ads}-S} \gg \Psi_{Me_{ads}-Me}$ and $f \approx 0$; (b) "Stranski–Krastanov" growth mode (3D Me island formation on top of predeposited 2D Me overlayers on S) for $\Psi_{Me_{ads}-S} \gg \Psi_{Me_{ads}-Me}$ and $f \neq 0$; (c) "Volmer–Weber" growth mode (3D Me island formation) for $\Psi_{Me_{ads}-S} \ll \Psi_{Me_{ads}-Me}$ independent of f. The formation of the first Me UPD monolayer in (a) and (b) is indicated by a darker color.

$$f = \frac{d_{O,Me} - d_{O,S}}{d_{O,S}} \qquad (9)$$

where $d_{O,Me}$ and $d_{O,S}$ denote the atomic-nearest neighbor distances (atomic diameters) of 3D Me and S phases, respectively.

In the opposite case,

$$\Psi_{Me_{ads} - S} \ll \Psi_{Me_{ads} - Me} \qquad (10)$$

3D Me clusters are formed on UPD unmodified substrate surfaces at supersaturation η_{3D} with respect to the 3D Me phase according to a Volmer-Weber growth mode independent of f as demonstrated in Figure 1.2c.

However, real substrates with surface inhomogeneities of different dimensionality (I_{iD} with $i = 0,1,2$) can induce a stepwise formation of low-dimensional Me phases under the condition

$$\Psi_{Me_{ads} - I_{0D}} \gg \Psi_{Me_{ads} - I_{1D}} \gg \Psi_{Me_{ads} - I_{2D}} \gg \Psi_{Me_{ads} - Me} \qquad (11)$$

where $\Psi_{Me_{ads} - I_{iD}}$ ($i = 0,1,2$) denotes the binding energy of Me adatoms on an iD surface inhomogeneity of the foreign substrate S. $\Psi_{Me_{ads} - Me}$ represents the binding energy of a Me adatom on the native substrate. The stability ranges of iD Me phases are defined according to eqs. (4) through (7). The stepwise formation of 0D Me clusters as well as Me phases of different dimensionality is schematically illustrated in Figure 1.3. If the sequence in eq. (11) is not fulfilled, 0D Me clusters and/or certain low-dimensional Me phases may not exist. Additionally, coadsorption or competitive adsorption processes of electrolyte constituents different from Me^{z+} (e.g., anions) can significantly influence the stepwise formation process.

It should be noted, however, that the formation of iD Me phases is not restricted to the formation of only one 2D Me overlayer or one 1D Me row, but depends on the activity of the corresponding iD surface inhomogeneity. In the case of very strong Me$_{ads}$-I$_{iD}$ interactions, the formation of two or more 2D Me overlayers and two or more 1D Me rows becomes thermodynamically possible. In the presence of specifically active 0D substrate surface inhomogeneities, a formation of a small 0D Me cluster with defined energetics, size, and structure can be induced as previously mentioned.

Formation and dissolution of iD Me UPD phases can involve positive and negative iD nucleation and growth steps. Generally, a nucleation process represents a first order phase transition in which nucleation events lead from expanded (gas-like) phases to condensed (liquid-like or solid-like) phases. In the case of 2D nucleation, an expanded 2D Me overlayer can be transformed into a condensed state (or vice versa) by a discontinuous change of the corresponding extensive system variable, which is the relative surface excess concentration of Me^{z+} in the interphase denoted as Γ.[1] Higher order phase transitions, such as

1.2 FORMATION OF LOW-DIMENSIONAL METAL PHASES

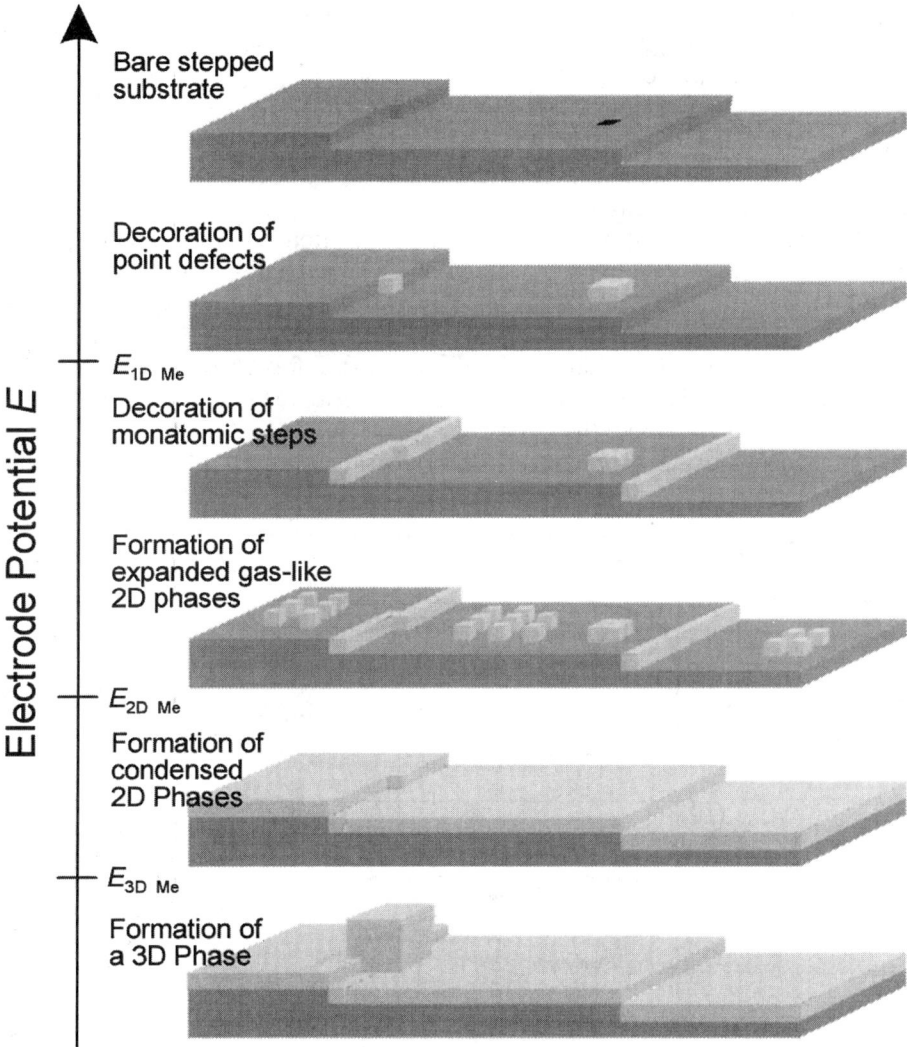

Figure 1.3 Schematic representation of a stepwise formation of Me phases of different dimensionality.

order-disorder transitions, are characterized by large fluctuations of a Landau order parameter at the critical point.[16,17] In 2D systems, Γ or lattice parameters can act as order parameters.

A first order phase transition is unequivocally characterized by a discontinuity of the $\Gamma(E)$ isotherm at μ = constant. At this special point in the $\Gamma(E)$ isotherm, expanded and condensed iD Me phases coexist. Therefore, nucleation and growth occur in the presence of a preformed expanded, but supersaturated low-dimensional Me adsorbate. The surface concentration of the expanded low-dimensional Me adsorbate is continuously changing according to the actual polarization state of the Me UPD system. It should be noted, however, that surface inhomogeneities of real substrates including surface contaminations will influence both the equilibrium potentials, $E_{iD\,Me}$, and the coverage isotherms, $\Gamma(E)$, and therefore can mask the discontinuity of Γ, which is characteristic for a first order phase transition.

However, the concept of first order phase transition is clearly not applicable to 0D Me clusters since the coexistence of expanded and condensed structures within a small 0D Me cluster has no physical sense. Also, the coexistence of expanded and condensed structures in 1D Me phases is questionable. From a theoretical point of view, a first order phase transition is absent in 1D systems, particularly when short range interactions are considered only.[18,19] This is illustrated in Figure 1.4 where the Gibbs energy of cluster formation, ΔG_{iD} for $i = 1,2,3$, is schematically plotted as a function of the number of atoms, N, using

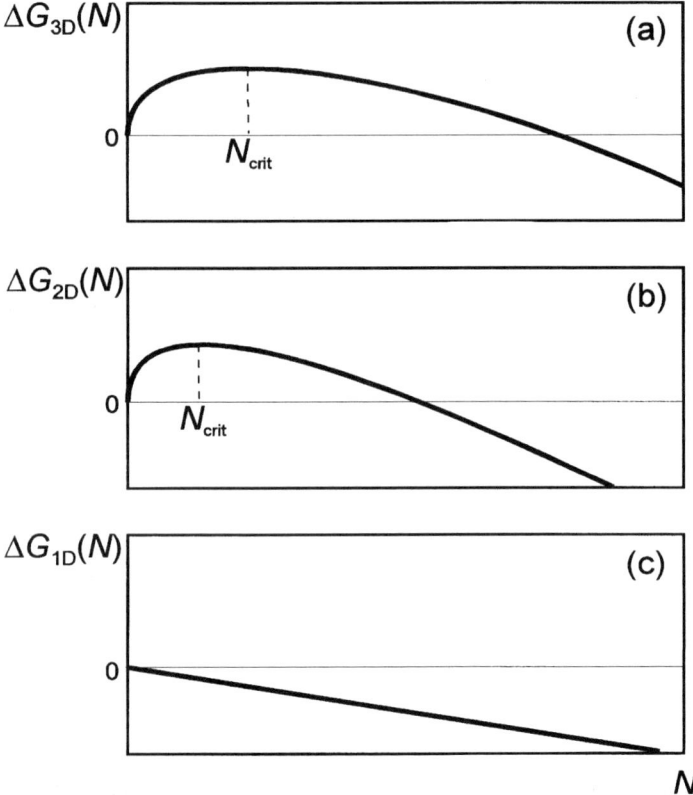

Figure 1.4 Gibbs energy of cluster formation, ΔG_{iD} ($i = 3,2,1$), as a function of the number of atoms, N. (a) 3D Me cluster formation; (b) 2D Me cluster formation; (c) 1D Me cluster formation.

1.2 FORMATION OF LOW-DIMENSIONAL METAL PHASES

the classical nucleation theory.[1] The functions exhibit maxima for the formation of 3D and 2D (Figs. 1.4a and 1.4b), but not for 1D clusters (Fig. 1.4c). This difference indicates that the formation of condensed 3D and 2D Me phases may involve nucleation, whereas the formation of condensed 1D Me phases occurs without nucleation phenomena. The maximum values of the ΔG_{iD} for $i = 2,3$ in Figures 1.4a and 1.4b are related to the critical 3D and 2D clusters (nuclei) involving N_{crit} particles.

Surface inhomogeneities of different dimensionality, I_{iD}, of real substrates significantly influence the energetics of 3D and 2D nucleation. In the following, the influence of monatomic steps and step-corners on the thermodynamics of 2D nucleation of condensed 2D Me phases is briefly discussed. According to the classical nucleation theory, the Gibbs energy of formation of a 2D nucleus on an atomically flat terrace (Terrace), $\Delta G_{\text{crit, Terrace}}$, is given by[1]

$$\Delta G_{\text{crit, Terrace}}^{2D} = \frac{N_A b \Omega \varepsilon^2}{zF|\eta_{2D}|} \tag{12}$$

where N_A is the Avogadro number, b is a geometric factor depending on the shape of the critical cluster, Ω represents the area occupied by one Me adatom in a condensed 2D Me overlayer, ε denotes the specific edge energy, and $zF|\eta_{2D}| = zF|E - E_{2D\,Me}|$ for $E \leq E_{2D\,Me}$ corresponds to a supersaturation $\Delta\mu_{2D}$ with respect to the condensed 2D Me phase.

However, the Gibbs energies for the formation of critical 2D Me clusters formed at monatomic steps of S, $\Delta G_{\text{crit, Step}}$, or at monatomic step-corners of S, $\Delta G_{\text{crit, Corner}}$, are reduced by additional terms including the lateral interaction energies between Me adatoms and the monatomic step, $\Psi_{\text{Me}_{\text{ads}} - \text{Step}}$ (Fig. 1.5). For a nucleus of a 2D Me phase with a square lattice structure and considering only lateral adatom interactions between first nearest Me_{ads} neighbors, $\Psi_{\text{Me}_{\text{ads}} - \text{Me}_{\text{ads}}}$, one obtains for $\Delta G_{\text{crit, Step}}$ and $\Delta G_{\text{crit, Corner}}$:[1,20]

$$\Delta G_{\text{crit, Step}}^{2D} = \Delta G_{\text{crit, Terrace}}^{2D} \left(1 - \frac{\Psi_{\text{Me}_{\text{ads}} - \text{Step}}}{\Psi_{\text{Me}_{\text{ads}} - \text{Me}_{\text{ads}}}}\right) \tag{13}$$

and

$$\Delta G_{\text{crit, Corner}}^{2D} = \Delta G_{\text{crit, Terrace}}^{2D} \left(1 - \frac{\Psi_{\text{Me}_{\text{ads}} - \text{Step}}}{\Psi_{\text{Me}_{\text{ads}} - \text{Me}_{\text{ads}}}}\right)^2 \tag{14}$$

For one case, $\Psi_{\text{Me}_{\text{ads}} - \text{Step}} < \Psi_{\text{Me}_{\text{ads}} - \text{Me}_{\text{ads}}}$ (no 1D Me phase formation at monatomic steps) follows $\Delta G_{\text{crit, Corner}} < \Delta G_{\text{crit, Step}} < \Delta G_{\text{crit, Terrace}}$. Consequently, 2D Me nucleation takes place preferentially at unmodified monatomic step-corners followed by that at monatomic steps. On the other

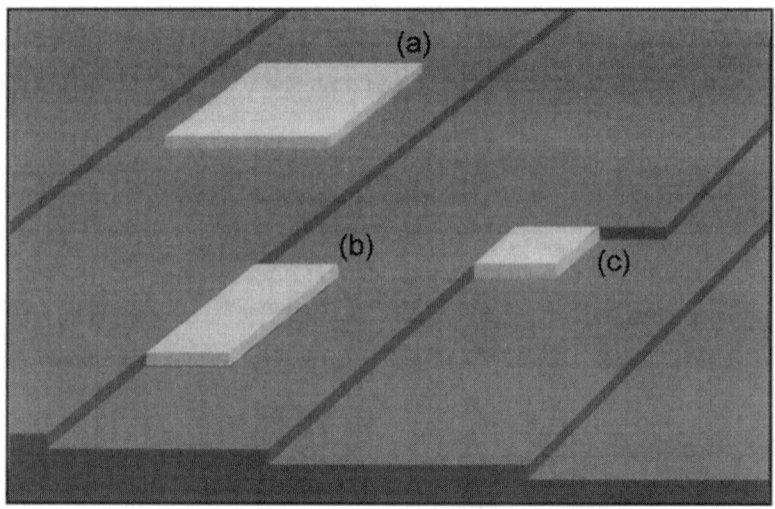

Figure 1.5 Nucleation of a condensed 2D Me phase on a stepped foreign substrate S. (a) 2D nucleation on atomically flat terraces of S; (b) 2D nucleation at monatomic steps of S; (c) 2D nucleation at monatomic step corners of S.

hand, $\Psi_{Me_{ads}-Step} \approx \Psi_{Me_{ads}-Me_{ads}}$ yields $\Delta G_{crit,Step} \approx 0$, and the formation of a condensed 2D Me_{ads} phase occurs without activation energy for 2D nucleation along the steps. If $\Psi_{Me_{ads}-Steps} \gg \Psi_{Me_{ads}-Me_{ads}}$, a formation of a 1D Me phase at monatomic steps becomes thermodynamically possible in the undersaturation range $E > E_{2D\,Me}$ (with respect to the condensed 2D Me phase). The width of such an 1D Me deposit formed at a monatomic step is not necessarily restricted to monatomic dimensions, but rather depends on the step activity as previously mentioned. These effects play an important role in step decoration processes in Me UPD systems at relatively high underpotentials.

1.2.1.2 Substrate Structure and Low-Dimensional Me Phases The assumption of atomically flat ideal substrate surfaces is rather unrealistic. Real crystals are always imperfect in some respect and exhibit various structural imperfections of different dimensionality:

- 0D or point imperfections (atomic disorder, chemical impurities, etc.)
- 1D or line imperfections (edge and screw dislocations, etc.)
- 2D or planar imperfections (grain or subgrain boundaries, stacking faults, etc.)
- 3D or volume imperfections (crystal domains with different chemical and/or physical properties).

The crystal surface, which can be considered as a 2D crystal imperfection, plays an important role in the process of phase formation and crystal growth.

1.2 FORMATION OF LOW-DIMENSIONAL METAL PHASES

The structure and homogeneity of a crystalline surface depend strongly on the structure and perfection of the corresponding 3D bulk crystal and on the applied surface treatment. Generally, the following surface inhomogeneities can be distinguished:

- 0D inhomogeneities (kinks, adatoms, vacancies, chemical impurities, emergence points of edge and screw dislocations, etc.)
- 1D inhomogeneities (monatomic steps, intersection lines of grain or subgrain boundaries and stacking faults, etc.)
- 2D inhomogeneities (reconstructed surface domains, 2D islands and pits, surface domains or terraces with different chemical and/or physical properties).

The heterogeneity of a real crystal surface can be well characterized on a nanometer scale by in situ local probe techniques. Emergence points of screw dislocations, monatomic steps separated by atomically flat terraces, 2D islands and pits, and so on, can be directly observed under defined electrochemical conditions as illustrated in Figure 1.6. The atomic structure of flat terraces and monatomic steps can be imaged by in situ local probe methods with lateral atomic resolution as shown in Figure 1.7.

The structure of low-dimensional Me phases depends on the following parameters: the vertical $Me_{ads} - I_{iD}$ interaction energy, $\Psi_{Me_{ads} - I_{iD}}$, the lateral $Me_{ads} - Me_{ads}$ interaction energy, $\Psi_{Me_{ads} - Me_{ads}}$, the degree of coverage depending on μ and E, the crystallographic structure of the underlying foreign substrate, and the crystallographic Me-S misfit defined in eq. (9).

Expanded iD Me phases (gas-like phases) are either randomly adsorbed or ordered as commensurate phases formed at ΔE_{iD}. Condensed iD Me phases with $i = 1,2$ formed at η_{iD} are either commensurate, higher order commensurate, or incommensurate depending on the Me-S crystallographic misfit. The transition between expanded 2D Me phases and condensed 2D Me phases usually takes place via a first order phase transition. In the case of higher order commensurate or incommensurate condensed 2D Me phases, the overlayer structure is usually rotated with respect to the underlying substrate structure. Due to the lattice interference, higher order commensurate or incommensurate 2D Me phases show Moiré patterns in STM images. In the presence of positive crystallographic Me-S misfit, condensed 2D Me phases are generally isotropically or anisotropically compressed. The internal strain is reflected in an underpotential-dependence of the nearest neighbor distance, d, of Me adatoms leading to a relative compression, which is given by:[1,21,22]

$$\frac{\Delta d}{d_O} = \frac{d_O - d}{d_O} = \frac{zF(\eta_{2D}^O - \eta_{2D})}{N_A 2(\lambda + \mu)v} \quad (15)$$

where d_O and v represent the nearest neighbor distance and the volume of a

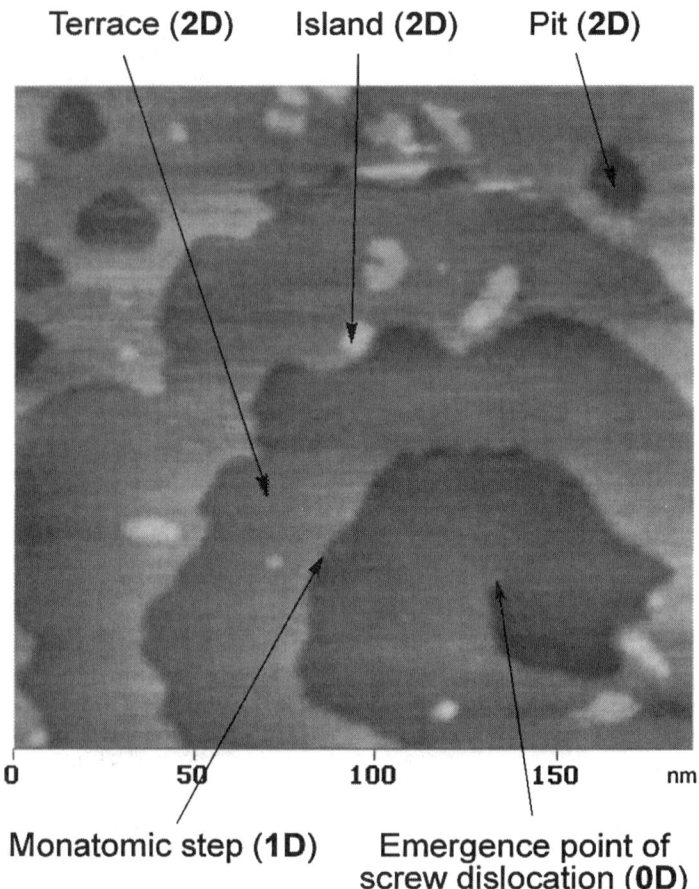

Figure 1.6 In situ STM image of an electrochemically polished Au(111) surface in the system Au(111)/5 × 10^{-3} M AgClO$_4$ + 5 × 10^{-1} M HClO$_4$ at T = 298 K and ΔE_{3D} = 200 mV.[1]

Me adatom in an uncompressed condensed 2D Me monolayer, respectively. η_{2D}^O and η_{2D} denote the overpotential values for the uncompressed and compressed (condensed) 2D Me monolayers, respectively. λ and μ are the so-called Lamé coefficients, or the compressibility and shear moduli, respectively.[23]

An iD Me phase can act as a precursor for an epitaxial formation of the corresponding $(i+1)$D Me phase.[1] If a significant difference in the structures of iD and $(i+1)$D Me phases exists, a finite transition range is necessary to adjust the stable lattice parameter of the $(i+1)$D Me phase via the formation of misfit dislocations.[1,24,25]

In many cases, cosorption of anions or other electrolyte constituents strongly influence the structure of iD Me phases. For example, expanded 2D Me phases can be stabilized by cosorption of anions.[1,26]

1.2 FORMATION OF LOW-DIMENSIONAL METAL PHASES

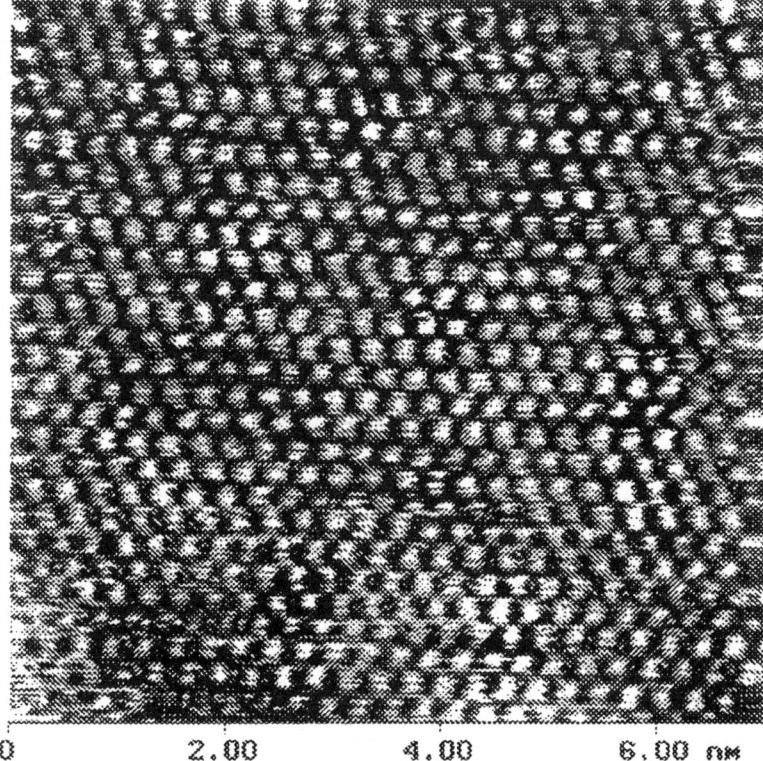

Figure 1.7 In situ STM image with lateral atomic resolution of an atomically flat Ag(111) terrace in the system Ag(111)/10^{-2} M HClO$_4$ at $T = 298$ K and $E_H = 350$ mV.[5]

1.2.1.3 Dynamics of Low-Dimensional Me Phase Formation The dynamics of iD Me phase formation and dissolution processes strongly depends on the binding energies of Me$_{ads}$ on I$_{iD}$ (cf. eq. (11)). Considering condensed (solid-like) iD Me phases ($i = 1,2,3$) and a direct charge transfer of Me^{z+} ions from the electrolyte to iD Me kink sites and vice versa (Fig. 1.1), the equilibrium potentials $E_{i\text{D Me}}$, thermodynamically defined in eqs. (4) and (5), are kinetically characterized by the equality of attachment and detachment frequencies of Me^{z+} at kinks of iD phases (abbreviated as "iD kinks"):[1]

$$\omega_{0, i\text{D kink}}(E_{i\text{D Me}}) \stackrel{\text{def}}{=} \omega_{\text{diss}, i\text{D kink}}(E_{i\text{D Me}}) = \omega_{\text{dep}, i\text{D kink}}(E_{i\text{D Me}}) \qquad (16)$$

where $\omega_{0, i\text{D kink}}(E_{i\text{D Me}})$ is the exchange frequency of iD kink atoms at the corresponding equilibrium potential $E_{i\text{D Me}}$. The surface concentration of iD kink sites, $n_{i\text{D kink}}$, does not influence the equilibrium potential, but affects significantly the partial exchange current density of iD kink atoms given by:

$$i_{O,iD\,\text{kink}}(E_{iD\,\text{Me}}) = ze\omega_{O,iD\,\text{kink}}(E_{iD\,\text{Me}})n_{iD\,\text{kink}}$$

$$= ze\omega_{O,iD\,\text{kink}}(E_{iD\,\text{Me}})\frac{L_S}{\delta_{iD\,\text{kink}}} \quad (17)$$

where L_S denotes the density of monatomic steps (i.e., the total step length per unit surface area), and $\delta_{iD\,\text{kink}}$ is the mean distance of iD kink atoms.

At any electrode potential in the Me UPD range $E > E_{3D\,\text{Me}}$, the experimentally measurable overall exchange current density, $i_{O,S/\text{Me}^{z+}}$, is the sum of different partial exchange current densities, $i_{O,j}$, of expanded (gas-like) and/or condensed (liquid-like or solid-like) iD Me phases contributing to the corresponding equilibrium:

$$i_{O,S/\text{Me}^{z+}}(E) = \sum_j i_{O,j}(E) \quad (18)$$

In selected Me UPD model systems, single terms in the sum of eq. (18) can be neglected as will be shown in the following experimental part.

Assuming that the lateral growth of 2D and 1D Me phases is controlled by a direct charge transfer (dt) of Me^{z+} to 2D and 1D kink sites only, the corresponding lateral growth rates, $v_{iD,\text{dt}}$, at low supersaturation are given by:

$$v_{2D,\text{dt}} = \frac{\omega_{O,2D\,\text{kink}}(ze)^2}{q_{2D\,\text{Me}}\delta_{2D\,\text{kink}}kT}|\eta_{2D}| \quad (19a)$$

$$= \frac{i_{O,2D\,\text{kink}}ze}{q_{2D\,\text{Me}}L_S kT}|\eta_{2D}| \quad (19b)$$

$$v_{1D,\text{dt}} = \frac{\omega_{O,1D\,\text{kink}}(ze)^2}{q_{1D\,\text{Me}}kT}|\eta_{1D}| \quad (20a)$$

$$= \frac{i_{O,1D\,\text{kink}}\delta_{1D\,\text{kink}}ze}{q_{1D\,\text{Me}}L_S kT}|\eta_{1D}| \quad (20b)$$

where $q_{2D\,\text{Me}}$ [As cm^{-2}] and $q_{1D\,\text{Me}}$ [As cm^{-1}] denote the charge amounts for the formation of a 2D Me monolayer per unit area and a 1D Me raw per unit length, respectively. Equations (19a) and (20a) reflect the physics of the process. In the 2D case, the propagation rate depends on the number of 2D kink sites per unit step length, $1/\delta_{2D\,\text{kink}}$. In contrast, in the 1D case, the propagation rate does not depend on the number of 1D kinks. In eqs. (19b) and (20b), experimentally measurable integral quantities (i.e., exchange current densities, $i_{O,iD\,\text{kink}}$) are introduced according to eq. (17).

In some cases, the propagation rate can be experimentally directly determined, for example, by optical or in situ SPM measurements. Moreover, the propagation rate can be also indirectly determined using integral information

1.2 FORMATION OF LOW-DIMENSIONAL METAL PHASES

from electrochemical measurements and the application of eqs. (19b) and (20b), which are, however, model-dependent.

Generally, excluding chemical reaction steps in the electrolyte, the following different steps must be considered in the kinetics of low-dimensional Me phase formation processes: bulk diffusion of Me^{z+} in the electrolyte, localized charge transfer of Me^{z+}, surface diffusion of Me adatoms, and first or higher order phase transitions. In a first model approach, the substrate surface was considered as quasi-homogeneous and the kinetics of the Me UPD process was assumed to be controlled by Me^{z+} ion charge transfer and/or semi-infinite bulk diffusion of Me^{z+} excluding surface diffusion of Me adatoms and phase transitions.[1] Later, a more sophisticated kinetic model including monatomic steps as 1D surface inhomogeneities was developed.[1,27] In this model, bulk diffusion of Me^{z+} was neglected, while three different fluxes were considered: localized charge transfer of Me^{z+} at monatomic steps and on atomically flat terraces as well as a superimposed surface diffusion leading to a leveling of local Me adatom gradients. In both models, however, the state and the properties of the interphase modified by the formation of different low-dimensional Me phases were not considered.

Assuming $\gamma = z$ (no cosorption and competitive sorption phenomena), the transfer function of the interfacial impedance of a Me UPD process controlled by Me^{z+} ion charge transfer and bulk diffusion of Me^{z+} ions is given by:

$$Z(s) = \frac{1}{sC_{dl} + \dfrac{1}{R_{ct} + \dfrac{1}{sC_{ads}} + Z_T(s)}} \qquad (21)$$

with

$s =$ Laplace variable

$C_{dl} = -\left(\dfrac{\partial q}{\partial E}\right)_{\Gamma}$ double layer capacitance

$R_{ct} = \dfrac{RT}{zF} \dfrac{1}{i_{0,S/Me^{z+}}}$ charge transfer resistance

$C_{ads} = -zF\left(\dfrac{\partial \Gamma}{\partial E}\right)_{\mu}$ adsorption capacitance

$Z_T(s) =$ transport impedance

In this model, the overall exchange current density, $i_{0,S/Me^{z+}}$, contains the contributions of different low-dimensional Me phases formed at different electrode potentials on surface inhomogeneities of different dimensionality, I_{iD}. The transport impedance in eq. (21), $Z_T(s)$, depends on the experimental conditions and, in the case of localized charge transfer, on the size and distribution of I_{iD} leading to non-uniform Me^{z+} fluxes within the diffusion layer.[28–30]

A great problem arises in all small-signal system perturbation measurements (transient technique in the time domain as well as Electrochemical Impedance Spectroscopy (EIS) in the frequency domain). These methods give only integral information on the overall electrode surface, whereas the charge transfer occurs localized as demonstrated for the formation of low-dimensional Me phases at I_{iD}. An exact interpretation of small-signal system perturbation data requires additional local information, which are only available from in situ local probe measurements. Consequently, only a combination of transient or EIS measurements with in situ SPM studies will lead to an exact interpretation of experimental data of the dynamics of low-dimensional Me phase formation processes.

A kinetic model for Me UPD systems involving formation of an expanded 2D Me phase and a first order phase transition leading to the formation of a condensed 2D Me phase on a quasi-homogeneous substrate surface was previously proposed.[1,4,31] It was demonstrated that sorption phenomena and nucleation and growth processes are coupled and that an exact separation (deconvolution) of both processes is rather complicated. As long as corresponding Γ amounts of the expanded Me phase formation contribute significantly to the overall measured Γ-values, an exact identification of nucleation and growth is relatively difficult by kinetic measurements. Usually, the fraction of the substrate surface covered by a condensed 2D Me monolayer, $S(t)$, is described by the Avrami equation:

$$S(t) = 1 - \exp[-S_{ex}(t)] \qquad (22)$$

which assumes multiple 2D Me nucleation on a quasi-homogeneous and atomically flat substrate surface with a sufficient density of randomly distributed nuclei.[1,4,32] In eq. (22), $S_{ex}(t)$ denotes the "extended" fractional coverage disregarding interference or overlapping of all growing 2D Me islands.

However, surface inhomogeneities of different dimensionality will influence not only the thermodynamics of 2D nucleation processes (cf. 1.2.1.1), but also the nucleation rate and the overlapping of growing islands.[1]

The influence of monatomic steps and step-corners on $\Delta G_{\text{crit},j}$ (j = Terrace,Step,Corner) as discussed in eqs. (12)–(14) is also reflected in the 2D nucleation rate per nucleation site, J_j

$$J_j = K_j \exp\left(-\frac{\Delta G_{\text{crit},j}}{kT}\right) \qquad \text{with } j = \text{Terrace, Step, Corner} \qquad (23)$$

where the preexponential factor K_j and $\Delta G_{\text{crit},j}$ depends on the supersaturation $zF|\eta_{2D}|$. However, the supersaturation dependence of K_j is relatively weak and can be disregarded. According to eq. (23), a detectable 2D nucleation rate J_j is

1.2 FORMATION OF LOW-DIMENSIONAL METAL PHASES

obtained after passing a critical value, $zF|\eta_{2D}|_{crit}$, which is significantly reduced for a 2D nucleation process at corners and monatomic steps in comparison to that on flat terraces as schematically shown in Figure 1.8.

Monatomic steps at real surfaces do influence significantly not only the 2D nucleation act, but also the spreading and overlapping of 2D islands, which determines the shape of potentiostatic current transients. Thus, the presumptions of the Avrami theory (cf. eq. (22)) are not more valid under real conditions. There are different problems related to edge effects arising from the presence of monatomic steps.

In the simplest case, a condensed 2D Me phase is formed on a foreign substrate exhibiting parallel monatomic steps with a regular pattern characterized by a constant step spacing $2\delta_{Step}$. Neglecting any adsorptive contributions and assuming a barrierless nucleation and growth of the condensed 2D Me phase only at monatomic steps, the current density transient is given by:[1,4]

$$-i(t) = \left| \begin{array}{ll} \dfrac{q_{2D\,Me}v_{2D,dt}}{2\delta_{step}} = q_{2D\,Me}L_S v_{2D,dt} & \text{for } 0 \leq t \leq \dfrac{2\delta_{step}}{v_{2D,dt}} \\ 0 & \text{for } t = \dfrac{2\delta_{step}}{v_{2D,dt}} \end{array} \right. \quad (24)$$

where $v_{2D,dt}$ is the propagation rate of the growing condensed 2D Me phase by

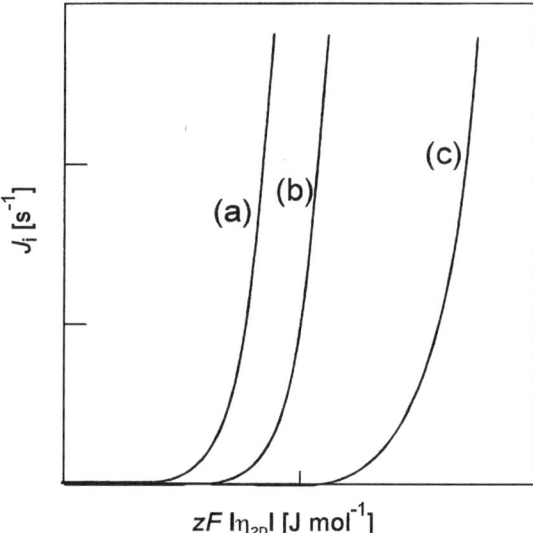

Figure 1.8 Rates of 2D nucleation, J_j, on a stepped foreign substrate S as a function of supersaturation, $zF|\eta_{2D}|$. (a) 2D nucleation at monatomic step corners of S; (c) 2D nucleation at monatomic steps of S; (b) 2D nucleation on atomically flat terraces of S.

direct charge transfer of Me^{z+} to the step edges. The propagation rate depends on the overpotential, $|\eta_{2D}|$, and the partial exchange current density, $i_{O, 2D\,kink}$, according to eq. (19b). The rectangular current transient according to eq. (24) is schematically shown in Figure 1.9a. The current describes the lateral growth of the condensed 2D Me monolayer starting at the monatomic steps and implies the assumption that the steps are instantaneously covered by the condensed 2D Me phase at $t = 0$. The initial current density $i(t = 0)$ is related to the step density L_S.

However, real surfaces exhibit monatomic steps with irregular patterns (i.e., variable δ_{step}). In this case, the initial current $i(t = 0)$ is also related to the step density, L_S, whereas the shape of the transient depends on the step distribution function as schematically shown in Figure 1.9b for an arbitrary step distribution. Consequently, a quantitative analysis of current transients or EIS data requires highly precise charge measurement and the knowledge of the charge densities of the condensed ($q_{2D\,Me}$) and expanded ($q_{2D\,ep}$) 2D Me phases at $t = 0$, which can be obtained from $q(E)$ or $\Gamma(E)$ isotherms. Additionally, an exact knowledge of the substrate surface morphology and the local mechanism of 2D Me phase formation processes depending on $|\eta_{2D}|$ is necessary. Information on the defect structure (step distribution) of a substrate surface on an atomic level can only be obtained by in situ SPM.[1]

1.2.1.4 Me-S Surface Alloy Formation In Sections 1.2.1.1 through 1.2.1.3, the formation of low-dimensional Me phases on foreign substrates was considered disregarding the formation of Me-S alloys. However, the formation of *i*D Me-S alloy phases in the underpotential range must be taken into account if the solubility of Me in S is different from zero and/or when single place exchange processes between Me adatoms and S atoms can take place. These processes are usually kinetically strongly hindered at room temperature.

The formation of low-dimensional Me-S surface alloy phases and 3D Me-S bulk alloy phases also strongly depends on surface inhomogeneities. Therefore,

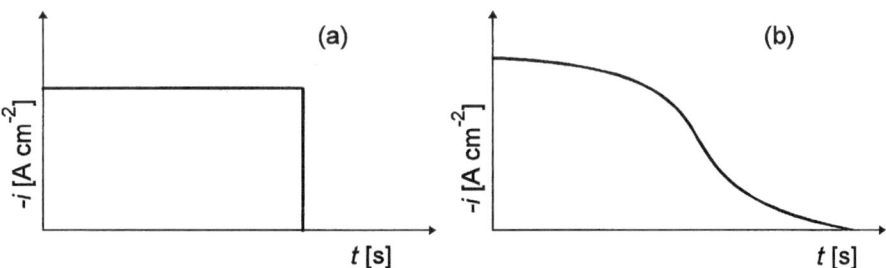

Figure 1.9 Schematic representation of current transients for a lateral growth of a condensed monolayer starting at monatomic steps of a foreign substrate. (a) foreign substrate exhibiting parallel steps with a regular step spacing; (b) foreign substrate exhibiting parallel steps with an irregular step spacing (arbitrary step distribution).

1.2 FORMATION OF LOW-DIMENSIONAL METAL PHASES

the role of surface inhomogeneities has to be taken into account in thermodynamic, structural, and kinetic aspects of iD Me-S alloy formation processes.

The thermodynamics of a Me UPD system forming low-dimensional Me-S surface alloy phases are similar to the thermodynamics of Me UPD systems without alloy formation. However, both iD Me phases and iD Me-S alloy phases belong to the solid/liquid interphase and can be thermodynamically treated according to Guggenheim's interphase concept.[1]

Generally, the binding energy of Me in an iD Me phase is weaker than that of Me in an iD Me-S surface alloy phase. Therefore, the dissolution of an iD Me phase will take place at more negative electrode potentials than the dissolution of a corresponding iD Me-S alloy phase.

The surface structure and morphology is drastically changed due to iD Me-S surface alloy formation. In many systems, iD Me-S alloy formation is connected with considerable changes of the substrate lattice type, unit cell, and/or interatomic distances producing significant internal mechanical stress.

Formation and dissolution of low-dimensional Me-S surface alloys and/or of 3D Me-S bulk alloys are strongly irreversible processes. The formation and dissolution of iD Me-S surface alloys and/or 3D Me-S bulk alloys of a binary system can be described by the reaction

$$x \text{ Me}_{ads} \text{ (IP)} + y \text{ S(IP)} \Leftrightarrow \text{Me}_x - \text{S}_y \text{ (IP)} \quad (25)$$

where Me in $\text{Me}_x\text{-S}_y$ (IP) denotes an incorporated Me atom within an iD Me-S alloy. The Me incorporation process is usually coupled with transport of S in opposite direction. These transport processes are affected by holes and interstitials.

iD Me-S alloy formation and dissolution are considered as either a heterogeneous chemical reaction (site exchange) or a mass transport process (solid state mutual diffusion of Me and S). In site exchange models, usual rate equations for the kinetics of heterogeneous reactions of first order (with respect to the species Me in Me_{ads} and $\text{Me}_x\text{-S}_y$) are applied. In solid state diffusion models, Fick's second law at defined boundary conditions must be solved using Laplace transformation. It is plausible that the formation of low-dimensional Me-S alloy phases is better described by a site exchange model than by a diffusion model. In addition, the kinetics of the initial stage of Me-S alloy formation processes can also involve nucleation and growth phenomena.

1.2.1.5 Summary The theoretical considerations are comprehensively illustrated in Figure 1.10. In systems with weak Me-S interactions, low-dimensional Me phases do not exist. In the supersaturation range, 3D Me phase formation starts as $E < E_{\text{3D Me}}$ by 3D nucleation and growth. The size of the nucleus (critical Me cluster) depends on the overpotential η_{3D}. In systems with strong Me-S interaction, low-dimensional Me phases can be formed. Equilibrium potentials $E_{i\text{D Me}}$ with $i = 2,1$ can be attributed to the corresponding low-dimensional con-

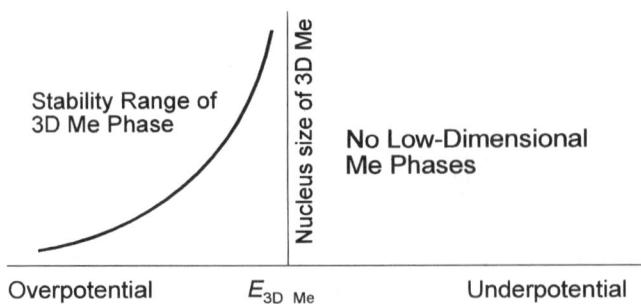
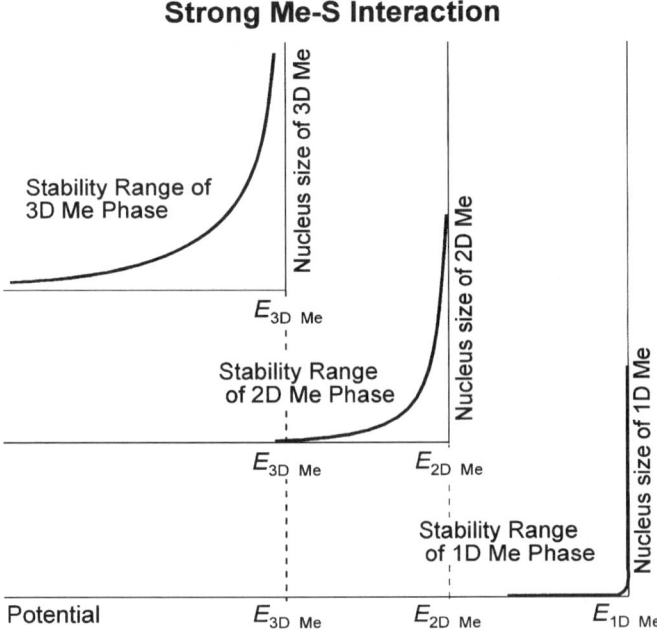

Figure 1.10 Formation and stability of Me phases and clusters of different dimensionality.

densed iD Me phases in analogy to the 3D Me phase. The stability ranges of condensed iD Me phases (infinitely large) and small iD Me clusters (finite size) are characterized by the corresponding equilibrium potentials, $E_{i\text{D Me}}$, and the nucleus size at the actual overpotential, $\eta_{i\text{D}}$, respectively. The sizes of critical iD clusters depend more strongly on the overpotentials $\eta_{i\text{D}}$ compared to that of systems with low Me-S interaction. For low-dimensional Me phases, this overpotential dependence increases with decreasing dimensionality. Expanded low-dimensional Me phases exist at $E > E_{iD\,Me}$ in the corresponding underpotential ranges.

1.2.2 Experimental Results

The theoretical concept of low-dimensional Me phases discussed in Section 1.2.1 has been developed on the basis of experimental results obtained in selected Me UPD model systems. The main progress was achieved by the in situ application of local probe methods with atomic resolution of well-defined single crystal surfaces under electrochemical conditions (Figs. 1.6 and 1.7). In situ SPM techniques open a window for a better understanding of the role of surface inhomogeneities in the stepwise formation of Me phases. An important precondition for these investigations was the development of surface preparation techniques such as growth methods, cutting techniques, and flame-annealing in order to produce substrate surfaces with defined surface inhomogeneities, I_{iD}. Another precondition was the development of an independent bipotentiostatic control of both probe and substrate for in situ SPM measurements. The following discussion of experimental results is generally restricted to Me UPD systems in which the electrosorption valency is equal to the ionic charge of Me^{z+} ($\gamma = z$ in eq. (2)), indicating that competitive sorption phenomena and, in most systems, cosorption phenomena of electrolyte constituents different from Me^{z+} can be neglected.

Extensive theoretical and experimental investigations on nucleation and crystal growth in the past 100 years led to a comprehensive understanding of the thermodynamic, structural, and kinetic aspects of the formation of 3D crystal phases.[1] Theoretical studies are connected with the names of Gibbs, Volmer, Kossel, Stranski, Kaischev, Becker, Döring, among others. The important role of surface inhomogeneities in heterogeneous nucleation and growth processes was taken into account only in the last 50 years starting with the fundamental work of Burton, Cabrera, and Frank. They have shown that crystals can grow without nucleation in the presence of monatomic steps generated by screw dislocations or other crystal defects. The general concept of crystal phase formation has been successfully applied to the electrochemical phase formation and electrocrystallization by Fischer, Kaischev, Fleischmann, Thirsk, Harrison, Gorbunova, and Bockris.

First experimental evidence for an electrochemical formation of 2D Me phases in the UPD range was obtained in the 1960s by different authors, especially Schmidt et al., using polycrystalline substrates.[1] In the 1970s, Me UPD experiments combined with ex situ UHV studies were started on well-defined single crystal faces. In these studies, important influences of the crystallographic substrate orientation and of surface inhomogeneities, on thermodynamic, structural, and kinetic aspects of the 2D Me phase formation processes were demonstrated and theoretically interpreted in first simple approaches.[1] The model systems discussed in this paper for low-dimension Me UPD phase formation are listed in Table 1.1.

TABLE 1.1 Me UPD Model Systems

Substrate S	Metal Me	Orientation (S) (hkl)	Misfit f	Electrolyte	$E^0_{3D\,Me}$ mV	E_N (S)* mV	γ
Au	Ag	(111)	0.002	H^+, ClO_4^-	799	470	?
		(100)	0.002	H^+, ClO_4^-	799	240	?
		(111)	0.002	H^+, SO_4^{2-}	799	470	1
		(100)	0.002	H^+, SO_4^{2-}	799	240	1
Ag	Pb	(111)	0.211	H^+, ClO_4^-	−125	−478	2
		(100)	0.211	H^+, ClO_4^-	−125	−648	2
Au	Pb	(111)	0.214	H^+, ClO_4^-	−125	470	2
		(100)	0.214	H^+, ClO_4^-	−125	240	2
Au	Cu	(111)	−0.114	H^+, SO_4^{2-}	340	470	≤2
		(100)	−0.114	H^+, SO_4^{2-}	340	240	≤2
Ag	Cu	(111)	−0.116	H^+, SO_4^{2-}	340	−478	?
		(100)	−0.116	H^+, SO_4^{2-}	340	−648	?

*E_N (S) denotes the zero charge potential of S.

As examples for the electrochemical behavior of Me UPD model systems, Figures 1.11 and 1.12 show cyclic voltammograms and $\Gamma(\Delta E_{3D})$ isotherms, measured by the so-called twin-electrode thin-layer (TTL) technique, in the systems Au(100)/Ag$^+$ and Ag(100)/Pb^{2+}, respectively.[1,33,34] Different adsorption (A_n) and desorption (D_n) peaks in the cyclic voltammograms and corresponding steps in the $\Gamma(\Delta E_{3D})$ isotherms indicate a stepwise formation and dissolution of low-dimensional Me UPD phases. A detailed interpretation of the different adsorption and desorption peaks became only possible using in situ SPM as noted in the following discussion.

The formation of different *expanded* 2D Me phases takes place at ΔE_{2D} and is indirectly measured by electrochemical methods and directly observed by in situ SPM. The first example in Figure 1.13b, shows an expanded, "quasi-hex" Me overlayer with an Au(100)-c($\sqrt{2}\times5\sqrt{2}$R 45° Ag superstructure in the system Au(100)/Ag$^+$, which is a characteristic system for negligible crystallographic Me-S misfit (Table 1.1).[1,33,35] Expanded 2D Me structures in the model systems Ag(100)/Pb^{2+} and Au(100)/Pb^{2+} are shown in Figure 1.14.[1,35,36] Both systems have nearly the same, relatively high positive Me-S misfit (Table 1.1). Even though, they show remarkable differences in the UPD behavior. The Pb UPD range in the system Au(100)/Pb^{2+} extends to much more positive potentials than that in the system Ag(100)/Pb^{2+} reflecting large interaction differences: $\Psi_{Pb_{ads}-Au} \gg \Psi_{Pb_{ads}-Ag}$. At ΔE_{2D}, domains of expanded 2D phases with a S(100)-c(2×2) Pb structure could be imagined by in situ STM in both systems. However, another expanded, but more dense 2D Me overlayer with an Au(100)-2C(3$\sqrt{2}\times\sqrt{2}$)R 45° Pb superstructure can be observed in the Au(100)/Pb^{2+} system at lower ΔE_{2D}, reflecting the strong Pb-Au interaction.[1,35,36] The system Au(hkl)/Cu^{2+} is characterized by a strong Me-S interaction and a significant

1.2 FORMATION OF LOW-DIMENSIONAL METAL PHASES

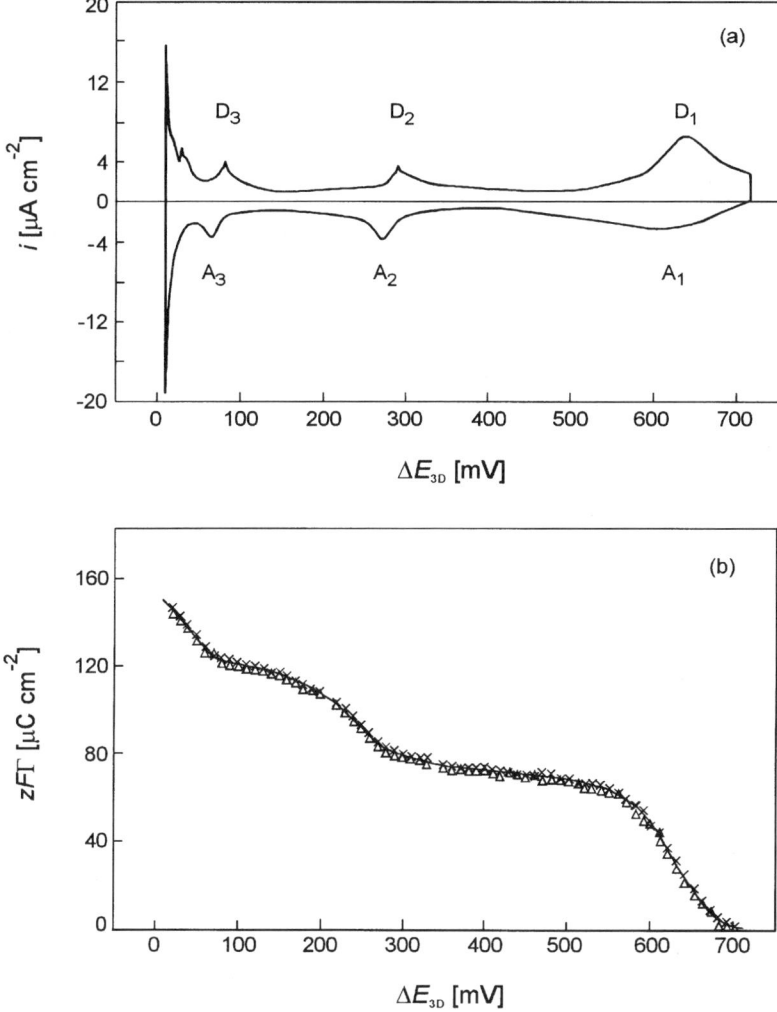

Figure 1.11 UPD of Ag on Au(100). (a) cyclic voltammogram measured with a scan rate of $|dE/dt| = 7$ mV s^{-1} in the system Au(100)/5×10^{-3} M Ag$_2$SO$_4$ + 5×10^{-1} M H$_2$SO$_4$ at $T = 298$ K. A$_n$ and D$_n$ with $n = 1,2,3$ denote cathodic adsorption and anodic desorption peaks, respectively; (b) $\Gamma(E)$ isotherm measured by the "twin-electrode thin-layer" (TTL) technique in the system Au(100)/4.2×10^{-4} M Ag$_2$SO$_4$ + 5×10^{-1} M H$_2$SO$_4$ at $T = 298$ K.[1,33]

negative Me-S lattice misfit (Table 1.1). Moreover, in this system a coadsorption of sulphate ions in a restricted underpotential range is observed. Therefore, the Au(111) substrate is modified at ΔE_{2D} by a commensurate 2D Cu overlayer with an Au(111)-2($\sqrt{3}\times\sqrt{3}$)R 30° Cu honeycomb structure, which is stabilized by HSO$_4^-$ ions coadsorbed at the honeycomb centers. In situ STM

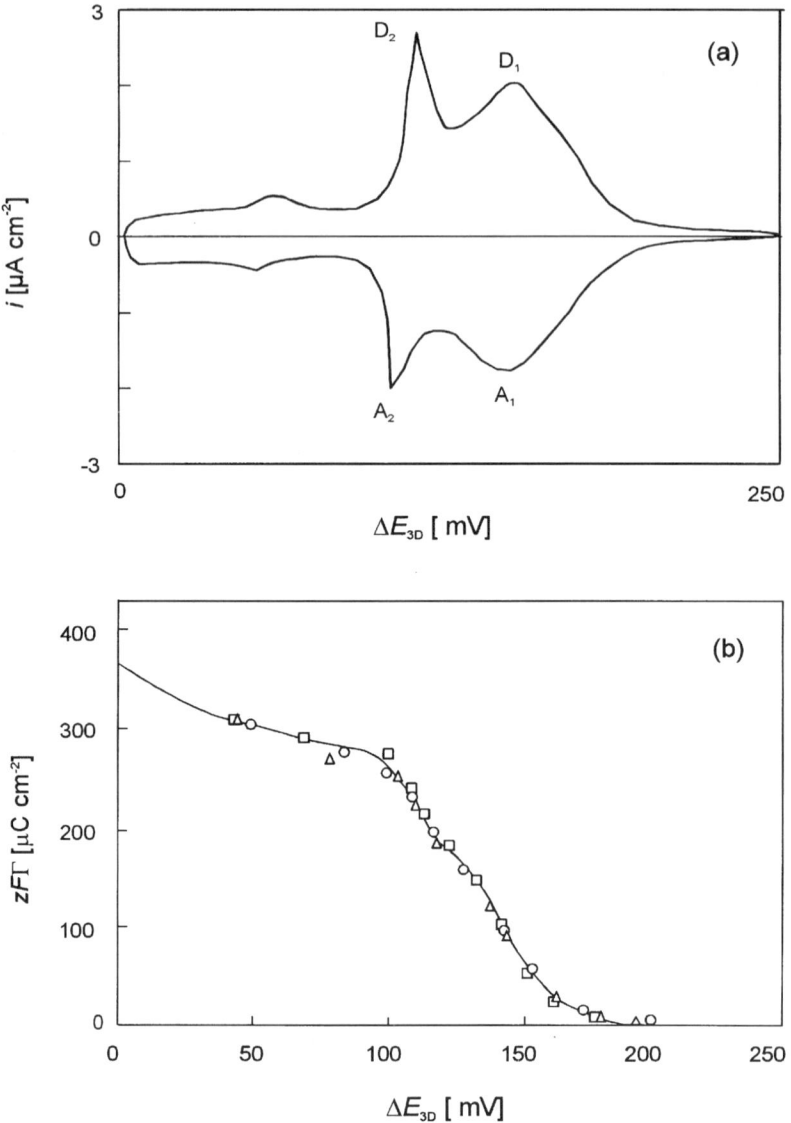

Figure 1.12 UPD of Pb on Au(100). (a) cyclic voltammogram measured with a scan rate of $|dE/dt| = 0.42$ mV s^{-1} in the system Au(100)/5×10^{-4} M Pb(ClO$_4$)$_2$ + 5×10^{-1} M NaClO$_4$ + 5×10^{-3} M HClO$_4$ at $T = 298$ K. A$_n$ and D$_n$ with $n = 1,2$ denote cathodic adsorption and anodic desorption peaks, respectively; (b) $\Gamma(E)$ isotherm measured by the "twin-electrode thin-layer" (TTL) technique in the system Ag(100)/x M Pb(ClO$_4$)$_2$ + 5×10^{-1} M NaClO$_4$ + 5×10^{-3} M HClO$_4$ with $10^{-4} \leq x \leq 10^{-3}$ at $T = 298$ K. Different symbols denote experimental values obtained at different x.[1,34]

1.2 FORMATION OF LOW-DIMENSIONAL METAL PHASES

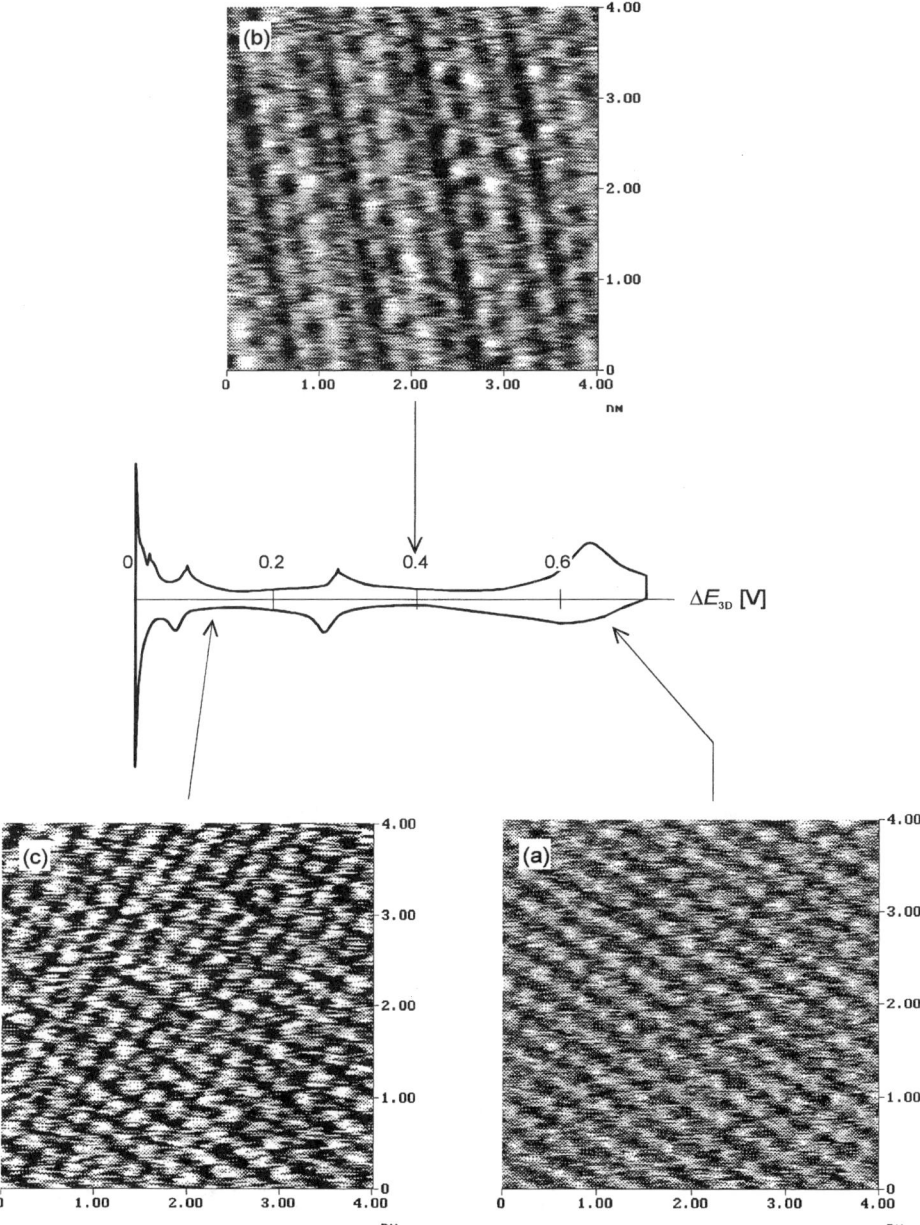

Figure 1.13 Cyclic voltammogram and in situ STM images in the system Au(100)/5 × 10^{-3} M Ag_2SO_4 + 5 × 10^{-1} M H_2SO_4 at T = 298 K. (a) atomic structure of the bare Au(100) substrate at 650 mV ≤ ΔE_{3D} ≤ 700 mV; (b) "quasi-hex" Ag overlayer with an expanded Au(100)-c($\sqrt{2}$ × 5$\sqrt{2}$)R 45° Ag superstructure at 200 mV ≤ ΔE_{3D} ≤ 550 mV; (c) condensed commensurate Ag overlayer with an Au(100)-(1 × 1) Ag quadratic structure at ΔE_{3D} ≤ 200 mV.[1,33,35]

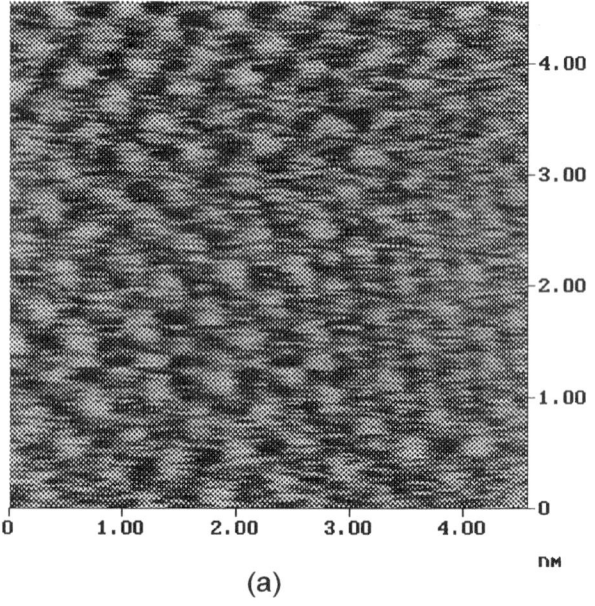

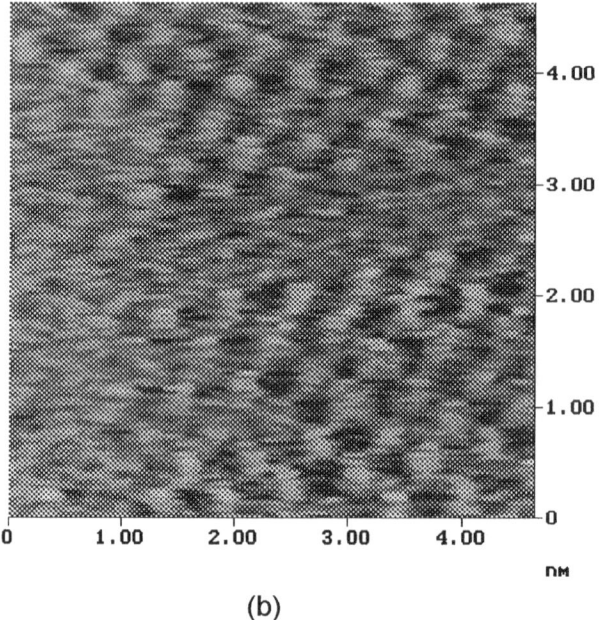

Figure 1.14 In situ STM images of 2D Pb domains with expanded Ag(100)-c(2 × 2) Pb and Au(100)-c(2 × 2) Pb structures. (a) system Ag(100)/5 × 10^{-3} M Pb(ClO$_4$)$_2$ + 10^{-2} M HClO$_4$ at T = 298 K and ΔE_{3D} = 175 mV; (b) system Ag(100)/5 × 10^{-3} M Pb(ClO$_4$)$_2$ + 10^{-2} M HClO$_4$ at T = 298 K and ΔE_{3D} = 550 mV.[1,36]

1.2 FORMATION OF LOW-DIMENSIONAL METAL PHASES

images show only the Au(111)-($\sqrt{3}\times\sqrt{3}$)R 30° HSO_4^- sublattice as illustrated in Figure 1.15b.[26]

The formation of different *condensed* 2D Me phases takes place at η_{2D} via a first order phase transition and can be indirectly detected by electrochemical methods and directly observed by in situ SPM. As expected, condensed

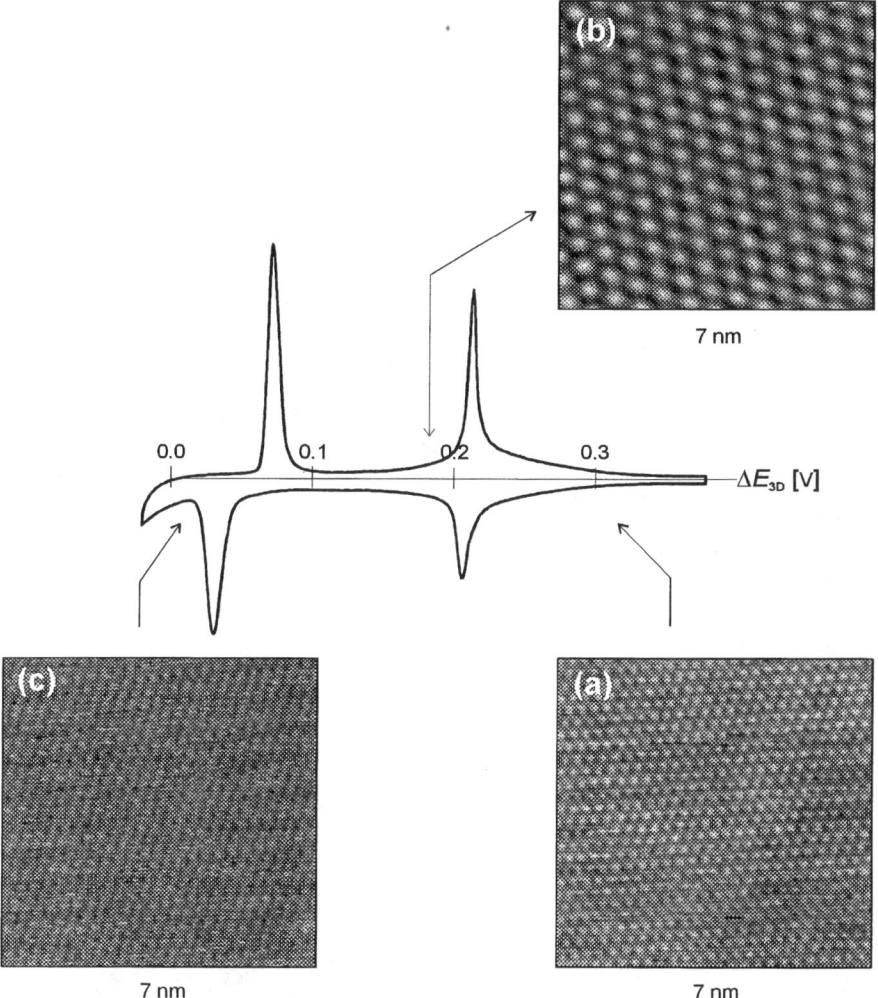

Figure 1.15 Cyclic voltammogram and in situ STM images in the system Au(111)/10^{-3} M $CuSO_4$ + 5 × 10^{-2} M H_2SO_4 at T = 298 K. (a) atomic structure of the bare Au(111) substrate at 300 mV ≤ ΔE_{3D} ≤ 500 mV; (b) Au(111)-($\sqrt{3} \times \sqrt{3}$)R 30° HSO_4^- overlayer structure at 100 mV ≤ ΔE_{3D} ≤ 220 mV stabilizing the Au(111)-2($\sqrt{3} \times \sqrt{3}$)R 30° Pb honeycomb structure, which is not detectable by in situ STM; (c) condensed commensurate Cu overlayer with an Au(111)-(1 × 1) Cu structure at ΔE_{3D} ≤ 20 mV. By courtesy of D. M. Kolb.[26]

2D Me phases with commensurate S(hkl)/-(1×1) Me structures are formed in systems with either negligible or negative crystallographic misfit. As examples, the Au(100)-(1×1) Ag and Au(111)-(1×1) Cu structures are shown in Figures 1.13c and 1.15c, respectively.[1,26,33,35] According to theory, condensed and isotropically and anisotropically compressed incommensurate 2D Me phases with closed packed 2D hcp structures are formed in systems with a significant positive crystallographic misfit.[1] As examples, the isotropically compressed Ag(111)-hcp Pb and the anisotropically compressed Ag(100)-hcp Pb structures are shown in Figures 1.16 and 1.17, respectively.[1,21,36,37] The linear dependence of the interatomic distance of Me atoms in condensed and compressed overlayers on η_{2D} according to eq. (15) is demonstrated in Figure 1.18 for the Ag(111)-hcp R 4.5° Pb structure.[1,38] In this model system, $E_{2D\,Pb}$ was determined from cyclic voltammetry and in situ STM measurements yielding a value of about $E_{2D\,Pb} = E_{3D\,Pb} + 150$ mV as noted in the following discussion.

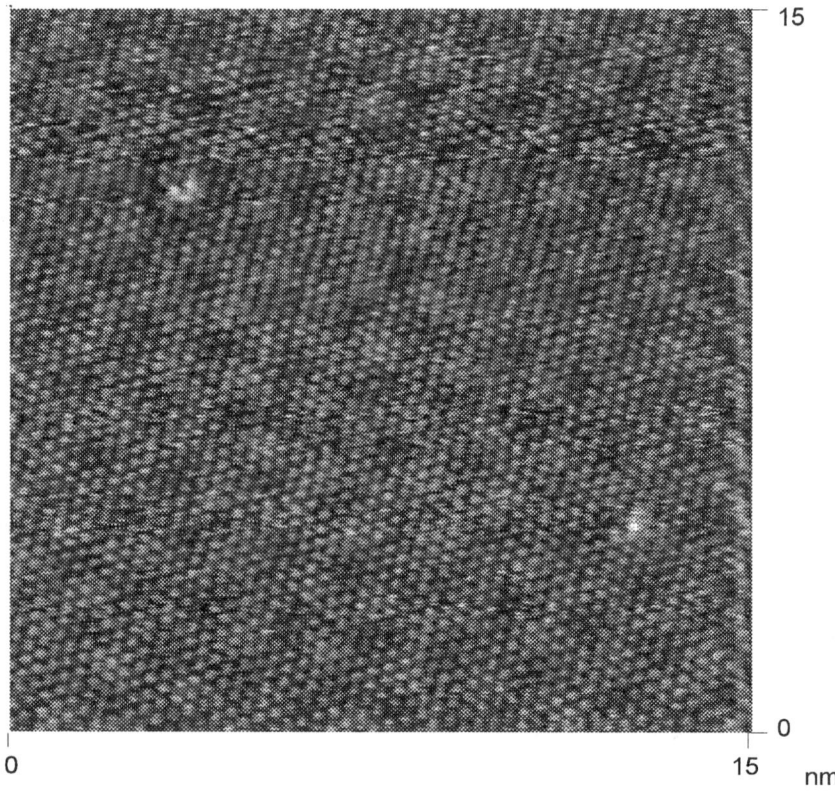

Figure 1.16 In situ STM image of a compressed 2D hcp Pb overlayer on Ag(111) showing a superstructure with moiré pattern observed in the system Ag(111)/5 × 10^{-3} M Pb(ClO$_4$)$_2$ + 10^{-2} M HClO$_4$ at $T = 298$ K and $\Delta E_{3D} = 30$ mV.[1,37]

1.2 FORMATION OF LOW-DIMENSIONAL METAL PHASES

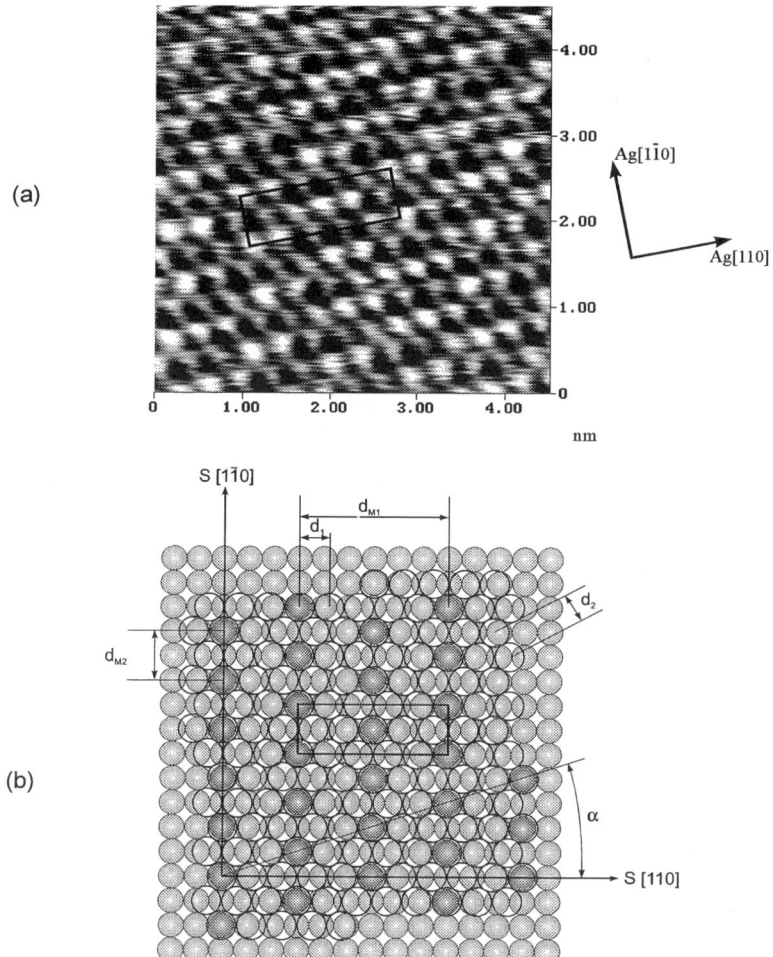

Figure 1.17 Anisotropically compressed 2D hcp Pb overlayer on Ag(100) observed in the system Ag(100)/5 × 10^{-3} M Pb(ClO$_4$)$_2$ + 10^{-2} M HClO$_4$ at T = 298 K and ΔE_{3D} = 80 mV. (a) in situ STM image of the 2D hcp Pb overlayer showing a superstructure with moiré pattern; (b) schematic representation of the S(100)-c(6 × 2) Me moiré superstructure.[1,36]

An example for a first order phase transition from a condensed 2D Me phase into an expanded one is shown in Figure 1.19 for the system Au(100)/Pb^{2+}.[1,35,36] The anisotropically compressed hcp Pb overlayer with an Au(100)-c(6×2)Pb superstructure formed at η_{2D} is transformed to an expanded 2D Pb phase with an Au(100)-2c(3√2×√2)R 45° Pb superstructure at ΔE_{2D}. In this system, $E_{2D\,Pb}$ was roughly evaluated to about $E_{2D\,Pb} = E_{3D\,Pb}$ + 145 mV. A first order 2D phase transition involves nucleation and growth processes, which are expected to start preferentially at monatomic steps.

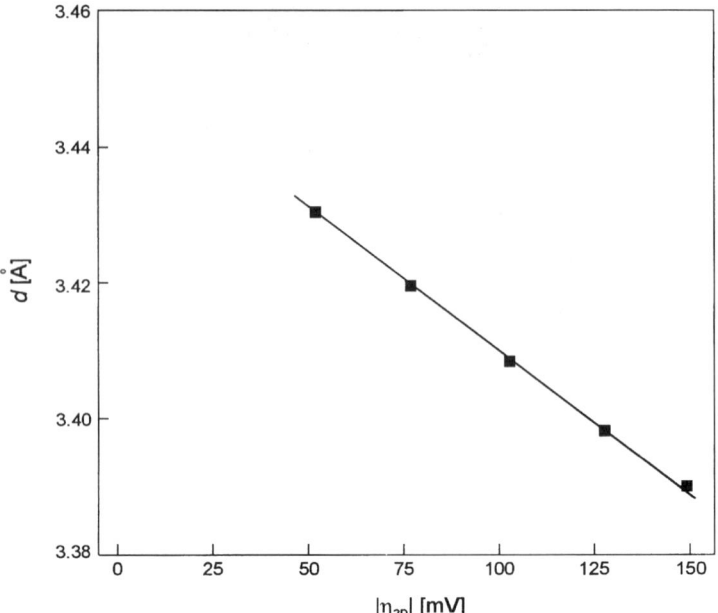

Figure 1.18 Interatomic distance, d, in a condensed 2D Pb phase as a function of η_{2D} obtained from in situ grazing incident X-ray scattering (GIXS) measurements in the system Ag(100)/5×10^{-3} M Pb(ClO$_4$)$_2$ + 10^{-1} M NaClO$_4$ + 10^{-2} M HClO$_4$ at T = 298 K.[38]

The model system Ag(111)/Pb^{2+} is appropriate to study the role of monatomic steps in the stepwise low-dimensional Me phase formation by combined electrochemical and in situ SPM measurements.[1,4,39–43] The cyclic voltammogram of this system is shown in Figure 1.20. The different adsorption and desorption peaks A$_n$/D$_n$ with n = 1,2,3 in Figure 1.20 reflect different steps and stages in the formation and dissolution of low-dimensional Me phases depending on the surface morphology as schematically shown in Figure 1.21 for the different deposition and dissolution steps.

As schematically illustrated in Figure 1.21, the peak A$_1$ is connected with the formation of a 1D Me phase decorating the bottom part of monatomic step edges. The width of this decoration extends to about 2 nm. From the cyclic voltammogram in Figure 1.20 and in situ STM measurements, the equilibrium potential $E_{1D\,Pb}$ can be roughly estimated to about $E_{1D\,Pb} = E_{3D\,Pb} + 170$ mV. The peak A$_2$ is related to the formation of a condensed 2D Me phase in substrate pits as well as on flat terraces, where a restricted region around the top part of the step edges, having a width of about 2 nm, remains still uncovered.*

---

*Extended (long-time) polarization at this stage leads to the formation of a 2D Pb-Ag surface alloy.[1,39–41] This process was found to start at step edges of the incomplete condensed 2D Pb monolayer.

1.2 FORMATION OF LOW-DIMENSIONAL METAL PHASES

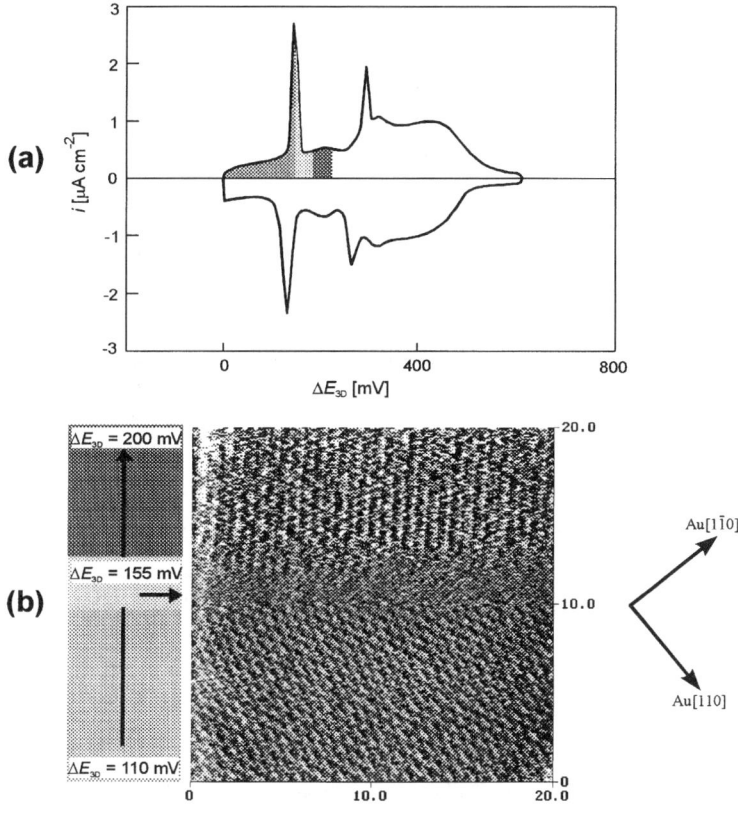

Figure 1.19 First order phase transformation in the system Au(100)/5 × 10^{-3} M Pb(ClO$_4$)$_2$ + 10^{-2} M HClO$_4$ at T = 298 K. (a) cyclic voltammogram at a scan rate of $|dE/dt|$ = 10 mV s^{-1}; (b) in situ STM image during an anodic potential sweep from ΔE_{3D} = 110 mV to ΔE_{3D} = 200 mV. A condensed 2D hcp Pb phase showing an Au(100)-c(6 × 2) Pb moiré superstructure is imaged at ΔE_{3D} < 155 mV in the lower part of this image. At ΔE_{3D} > 170 mV the upper part of the STM image shows an expanded 2D Pb phase with an Au(100)-2c(3√2×√2)R 45° Pb superstructure. In the intermediate part of this STM image, the Au(100) substrate appears indicating a very mobile expanded Pb UPD adlayer.[1,35,36]

The formation of the condensed 2D Pb phase is a first order phase transition process that is reflected in an hysteresis of the appearance of the adsorption and desorption peaks A$_2$ and D$_2$ in the cyclic voltammogram. The corresponding equilibrium potential $E_{2D\,Pb}$ is located between these two peaks at $E_{2D\,Pb}$ = $E_{3D\,Pb}$ + 150 mV as seen in Figures 1.20 and 1.21. The peak A$_3$ corresponds to the completion of the top parts of step edge regions, as well as to a kinetically hindered nucleation process of a 2D Pb phase and its growth on top of isolated substrate islands. The completed 2D Pb monolayer is compressed and internally

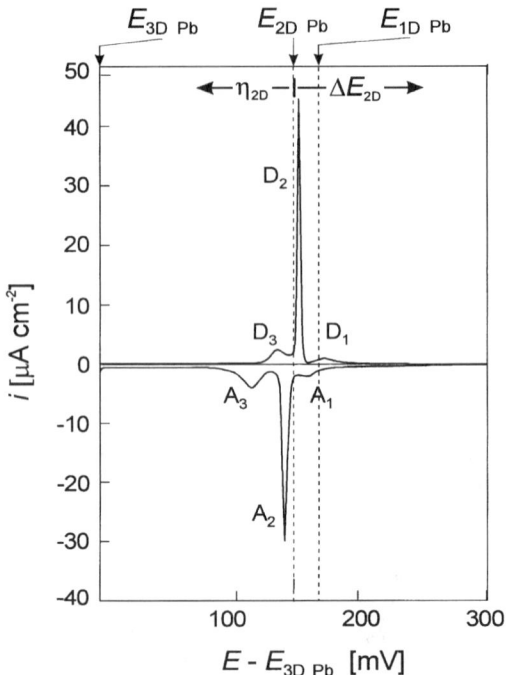

Figure 1.20 Cyclic voltammogram measured with a scan rate of $|dE/dt| = 1$ mV s^{-1} in the system Ag(111)/5×10^{-3} M Pb(ClO$_4$)$_2$ + 5×10^{-1} M NaClO$_4$ + 5×10^{-3} M HClO$_4$ at $T = 298$ K. A$_n$ and D$_n$ with $n = 1,2,3$ denote cathodic adsorption and anodic desorption peaks, respectively. Equilibrium potentials, $E_{i\text{D Pb}}$ ($i = 1,2,3$), of the corresponding iD Pb phases as well as the overpotential, η_{2D}, and the underpotential, ΔE_{2D}, with respect to the condensed 2D Pb phase, are indicated in the upper abscissa.

strained in the potential range -150 mV $\leq \eta_{2D} \leq -50$ mV as demonstrated in Figure 1.18. The peak D$_3$ is related to the depletion of the top parts of step edge regions only. The dissolution of the condensed 2D Pb phase on top of isolated islands and on flat terraces occurs in peak D$_2$. The peak D$_1$ reflects the dissolution of the step decoration. This interpretation was possible only recently on the basis of in situ SPM results.[40,41]

The formation of the 1D Pb phase (decoration of monatomic steps) and the initial stage of the formation of the condensed 2D Pb phase starting at monatomic steps is shown by the in situ STM line scan in Figure 1.22 corresponding to the stages (a$_2$) or (d$_3$) in Figure 1.21.

The nucleation and growth of the condensed 2D Pb phase strongly depends not only on the supersaturation, but also on the substrate surface morphology as illustrated by in situ STM line scans combined with step polarization experiments in Figure 1.23. The step polarization routine used in these experiments starts from an adsorbate-free initial state at ΔE_{1D} to low and high supersaturations with respect to the 2D Pb phase at $\eta_{2D}^{(1)}$ and at $\eta_{2D}^{(2)}$, respectively, and vice versa.

1.2 FORMATION OF LOW-DIMENSIONAL METAL PHASES

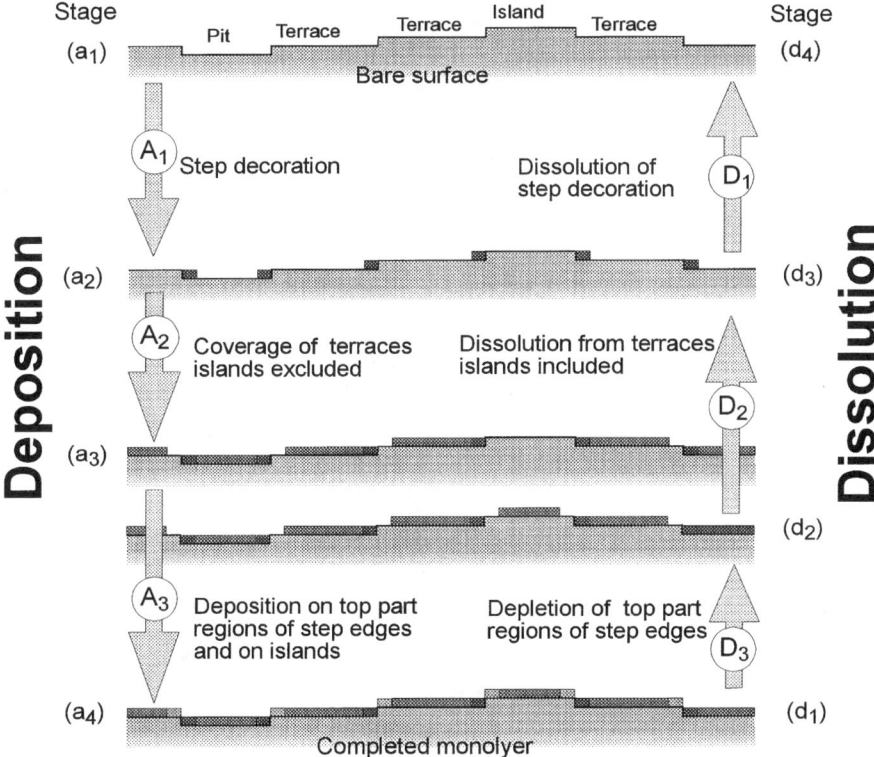

Figure 1.21 Schematic representation of the stepwise deposition and dissolution of low-dimensional Pb phases in the system $Ag(111)/Pb^{2+}$, ClO_4^-.

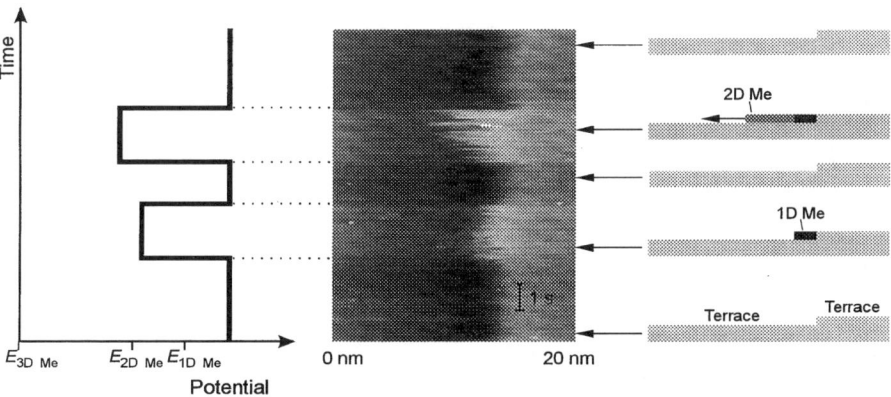

Figure 1.22 In situ STM line scan plot showing the step decoration by a 1D Pb phase and the initial stage of 2D phase formation on a stepped Ag(111) substrate in the system $Ag(111)/4 \times 10^{-3}$ M $Pb(ClO_4)_2 + 10^{-2}$ M $HClO_4$ at $T = 298$ K. By courtesy of H. Siegenthaler.[40,41]

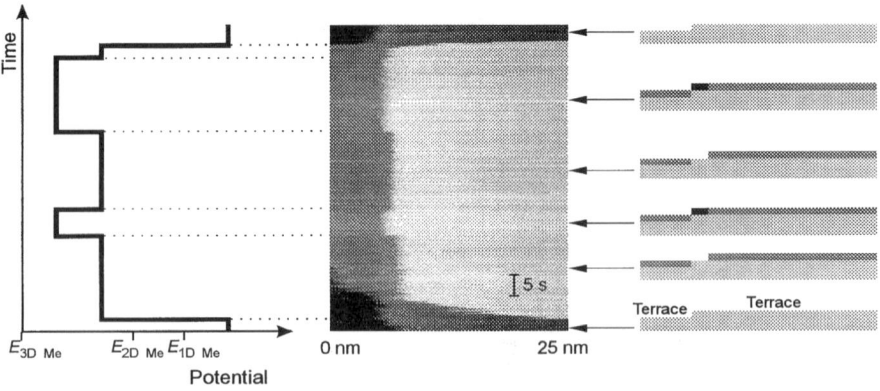
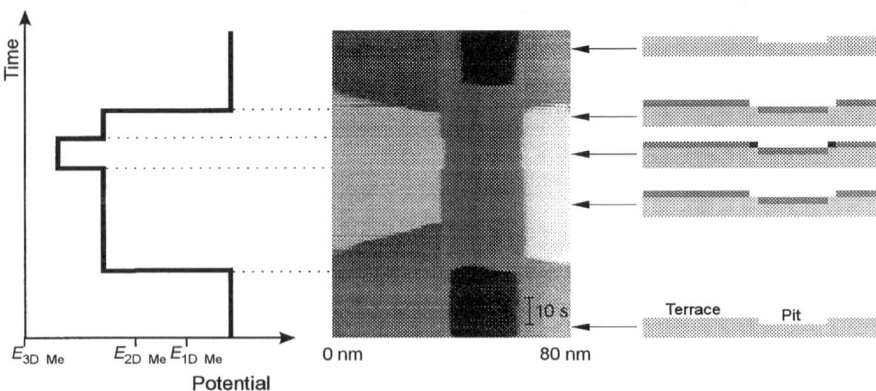
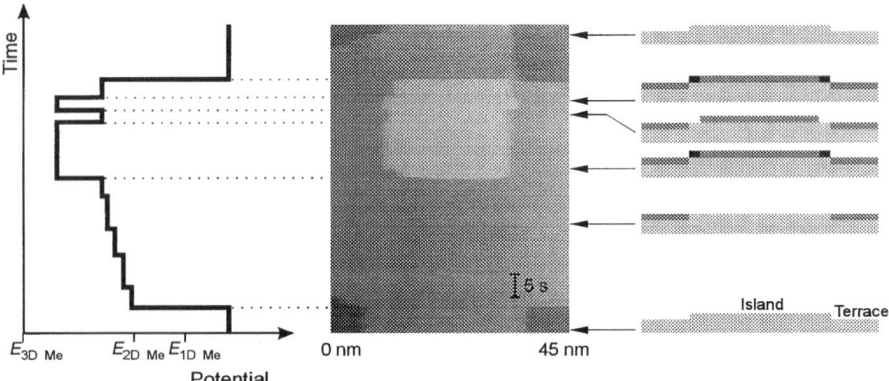

Figure 1.23 In situ STM linescan plots showing the formation of a condensed Pb monolayer on a terrace, at a pit, and on an island of an Ag(111) substrate in the system Ag(111)/4 × 10^{-3} M Pb(ClO$_4$)$_2$ + 10^{-2} M HClO$_4$ at T = 298 K. By courtesy of H. Siegenthaler.[40,41]

1.2 FORMATION OF LOW-DIMENSIONAL METAL PHASES

The covering of atomically smooth terraces of a stepped surface at $\eta_{2D}^{(1)}$ and $\eta_{2D}^{(2)}$ is shown in the lower part of Figure 1.23a corresponding to the stages (a_3) and (a_4) in Figure 1.21. The reverse desorption process is seen in the upper part of Figure 1.23a and corresponds to the stages (d_1)–(d_4) in Figure 1.21.

The lower part of Figure 1.23b shows the filling up of a pit and the covering of adjacent terraces at $\eta_{2D}^{(1)}$ and at $\eta_{2D}^{(2)}$ corresponding to the stages (a_3) and (a_4) in Figure 1.21, respectively. The stepwise dissolution of the condensed 2D Pb monolayer is seen in the upper part of Figure 1.23b corresponding to the different stages (d_1)–(d_4) in Figure 1.21.

The irreversible covering on top of separated substrate islands is clearly seen in the lower part of Figure 1.23c. The 2D nucleation and growth process in this case starts only after exceeding a certain supersaturation corresponding to $\eta_{2D}^{(2)}$ and to the stage (a_4) in Figure 1.21. As expected, the reverse dissolution process starts with the depletion of the top parts around step edges at $\eta_{2D}^{(1)}$ corresponding to the stage (d_2) in Figure 1.21.

The condensed 2D Pb phase including the covered islands dissolves immediately after exceeding the reversible potential $E_{2D\,Pb}$ corresponding to the stage (d_3). The dissolution begins at the top part of step edges as seen in Figure 1.23b. The dissolution of the 2D condensed Pb phase in pits is a 2D negative nucleation hindered process, which is not shown in Figure 1.21 but clearly seen in the upper part of Figure 1.23b.

The dissolution of the 1D Pb phase occurring in peak D_1 leads finally to the bare substrate denoted as (d_4) in Figure 1.21.

Figures 1.22 and 1.23 clearly demonstrate that formation and dissolution of low-dimensional Me phases are strongly affected by monatomic steps. Particularly, the formation of a condensed 2D Me phase in the absence of monatomic steps (i.e., on extended terraces or islands), needs a considerable overpotential in order to initiate 2D nucleation. In contrast, in the presence of monatomic steps the formation of the condensed 2D Me phase proceeds at much lower overpotentials and starts from the bottom part of step edges, already modified by a 1D Me phase (Fig. 1.23b). These experimental results are in good agreement with the theoretical concepts implemented in eq. (23) and Figure 1.8.

The results discussed previously show that two low-dimensional Pb phases can exist at the equilibrium potential $E_{2D\,Pb}$. Depending on the polarization routine to adjust this potential the substrate is modified either by 1D or 2D Pb phases.

If $E_{2D\,Pb}$ is adjusted cathodically starting from ΔE_{1D} (bare substrate), a formation of the condensed 2D Pb phase cannot occur and only a 1D Pb phase (decoration of monatomic steps) is formed. If, however, $E_{2D\,Pb}$ is adjusted anodically starting from η_{2D}, the surface is still covered by the condensed 2D Pb phase. EIS measurements of both Pb phases at $E_{2D\,Pb}$ are shown in Figure 1.24.[42,43] The impedance data were analyzed using the transfer function given in eq. (21) with a Warburg impedance for $Z_T(s)$ as a first approximation and using a special nonlinear fit program.[42–44] Overall exchange current densities were calculated from charge transfer resistances as already described. The over-

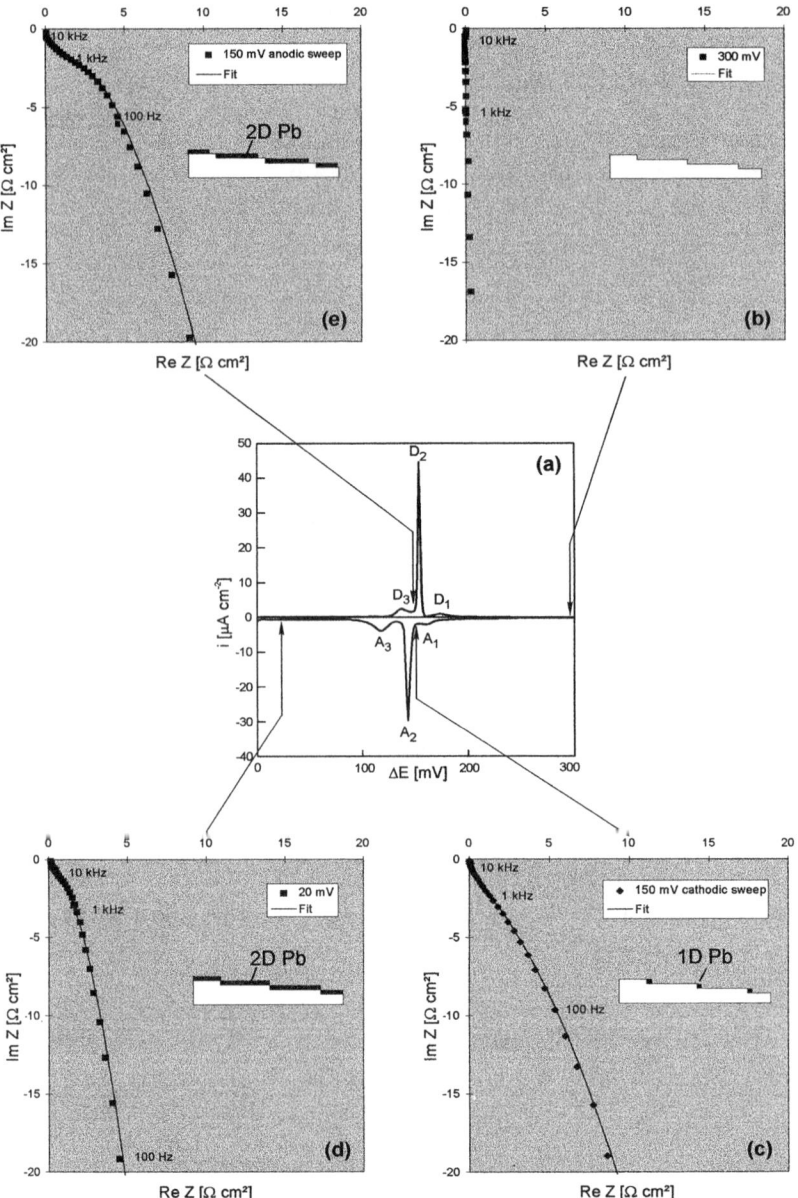

Figure 1.24 Interface impedance behavior in the system Ag(111)/5 × 10^{-3} M Pb(ClO$_4$)$_2$ + 5 × 10^{-1} M NaClO$_4$ + 5 × 10^{-3} M HClO$_4$ at T = 298 K. (a) cyclic voltammogram measured with a scan rate 1 mV s^{-1}; (b) impedance spectrum at ΔE_{3D} = 300 mV; (c) impedance spectrum at ΔE_{3D} = 150 mV measured after a cathodic potential sweep; (d) impedance spectrum at ΔE_{3D} = 20 mV; (e) impedance spectrum at ΔE_{3D} = 150 mV measured after an anodic potential sweep. The inserts in the impedance spectra schematically represent the condition of the substrate.[42,43]

1.2 FORMATION OF LOW-DIMENSIONAL METAL PHASES

all exchange current density in eq. (18) is assumed to be mainly determined by the partial exchange current densities of the condensed iD Me phases only. As a main result, the partial exchange current density of 2D Pb kinks, $i_{0,2D\,kink}$ = 3 mA cm^{-2} was found to be about two times higher than that of 1D Pb kinks, $i_{0,1D\,kink}$ = 1.6 mA cm^{-2}. As mentioned, $i_{0,iD\,kink}$ is proportional to the surface concentration of iD kink sites, $n_{iD\,kink}$ (cf. eq. (17)), but depends also on the activation energies of the Me^{z+} ion charge transfer from the electrolyte to iD kinks and vice versa.[1] The activation energies are strongly influenced by the interaction energies $\Psi_{Me_{ads}-I_{i}D}$ (i = 1,2).

If comparable kink site densities are assumed for both 1D and 2D Me phases, the partial exchange current density $i_{0,1D\,kink}$ should be lower than $i_{0,2D\,kink}$ as found experimentally in the system discussed. The assumption of comparable kink site densities in the 1D and 2D Pb phases is, however, rather speculative.

The experimental value $i_{0,2D\,kink}$ can be used for an estimation of the propagation rate of a condensed 2D Pb monolayer on Ag(111) from eq. (19b), assuming a direct charge transfer mechanism. The average monatomic step density, L_S, was directly determined from in situ SPM measurements. With $i_{0,2D\,kink}$ = 3 mA cm^{-2}, $q_{2D\,Me}$ = 0.3 mC cm^{-2}, L_S = 2.10^6 cm^{-1}, and η_{2D} = −2 mV a value of $v_{2D,dt}$ = 7 nm s^{-1} is obtained in a good agreement with the propagation rate 5 nm s^{-1} < $v_{2D,dt}$ < 10 nm s^{-1} measured directly by in situ STM under the same conditions.[40–43] This result indicates that the lateral growth of the condensed 2D Pb phase on Ag(111) preferentially occurs via a direct transfer to the growing monatomic step edges of the Pb monolayer. A similar estimation of the propagation rate of a condensed 1D Pb phase according to eq. (20b) is not possible due to the unknown average value of $\delta_{1D\,kink}$.

The results in Figures 1.22 and 1.23 demonstrate that the formation of the condensed 2D Me phase starts exclusively at monatomic steps at low supersaturation. Potentiostatic current transients in the system Ag(111)/Pb^{2+}, starting from ΔE_{1D} (bare substrate) to different η_{2D} are shown in Figure 1.25.[4] The shape of the experimental current transients is in good agreement with the theoretical prediction in Figure 1.9b for a non-uniform distribution of monatomic steps.

It should be noted that a stepwise formation and dissolution of low-dimensional Me phases takes place in many other Me UPD systems. However, the different Me phase formation stages are often not so well separated as in the Ag(111)/Pb^{2+} model system.

Concerning the existence, formation, and behavior of 0D Me clusters (phases) on 0D surface inhomogeneities, I$_{0D}$, at high undersaturation, we are in a very early stage of experimental knowledge. Up to now, in situ STM images of small 0D Me clusters were not reported in the literature. However, from an historical point of view, we have been in a similar position in the 1940s concerning 2D nucleation and phase formation as well as spiral growth of crystals. Apart from theoretical predictions, no experimental results indicated 2D UPD and OPD phases. It took roughly thirty years to confirm experimentally the theoretically predicted 2D nucleation mechanism. An analogous situation appeared in the case of 1D phases. The important role of 1D surface inho-

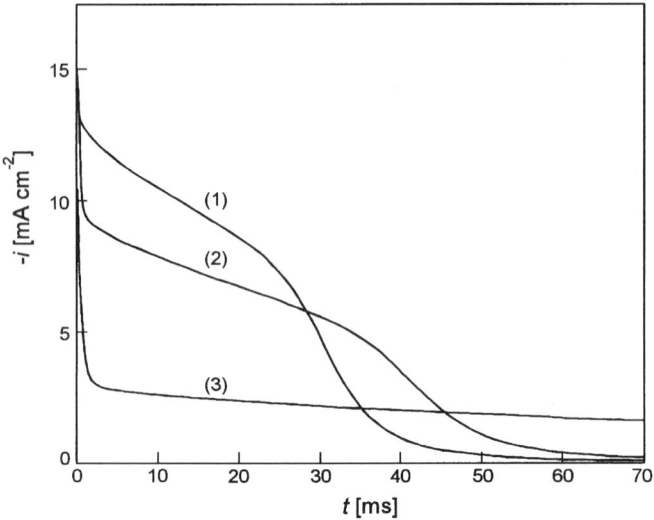

Figure 1.25 Current transients for the formation of a condensed Pb monolayer on a stepped Ag(111) substrate in the system Ag(111)/5×10^{-3} M Pb(ClO$_4$)$_2$ + 5×10^{-1} M NaClO$_4$ + 5×10^{-3} M HClO$_4$ at T = 298 K. Potential pulses from ΔE_{2D} = 150 mV to η_{2D} = -130 mV (1); -100 mV(2); -40 mV(3).[4,5]

mogeneities, such as monatomic steps in 1D phase formation processes, was largely disregarded up to recent in situ SPM studies of Me UPD. Although no experimental evidence shows 0D Me clusters formation at high undersaturation, the so-called deactivating coverage of interface inhibitors and the poisoning of point-like active centers are well accepted concepts in corrosion science and electrocatalysis, respectively. Therefore, the concept of 0D Me clusters may also be confirmed in future by in situ SPM measurements in selected Me UPD systems involving alloys as foreign substrates. For example, the Cu–Ag interaction is relatively weak and does not lead to pronounced Cu UPD phenomena in the system Ag(*hkl*)/Cu^{2+}.[1,45] On the contrary, the system Au(*hkl*)/Cu^{2+} shows significant Cu UPD features due to a strong Cu–Au interaction.[1,26] Using as a substrate a single crystal of an Ag–Au solid solution (bulk alloy) with a small content of gold, the surface will exhibit single gold atoms as 0D surface inhomogeneities (atomic 0D surface disorder) in a silver matrix. Then, 0D Cu UPD clusters might be formed at these 0D surface inhomogeneities. Similarly, the adsorption of a few organic molecules at such 0D structured substrate surfaces can be expected. An open question focuses on the time-stability and the surface-immobility of 0D clusters for in situ SPM imaging by a moving SPM probe. This problem is particularly of great importance for electrochemical nanostructuring of solid substrates. The different methods and techniques for a local nanostructuring of electron-conducting substrates will be discussed in the next section.

1.3 NANOSTRUCTURING OF SOLID SURFACES

1.3.1 General Remarks

Structuring and modification of solid surfaces, such as metals, graphite, superconducting films, semiconductors, and conducting polymers, in the nanometer range (from Ångstrøms to micrometers) is important for modern nanotechnology. In many cases, nanostructuring and nanomodification are closely connected with the formation or dissolution of low-dimensional Me phases. Nanostructures are, for example, very important to produce quantum confinements on semiconductor surfaces for single electron devices. In the preceding section, the analytical aspects of nanotechnology were discussed, taking into account the important influences of surface inhomogeneities on the formation of low-dimensional Me phases. In this section, the preparative aspects of electrochemical nanostructuring will be discussed.

Again, three questions arise. First, is it possible to deposit metal nanoclusters with a sufficient reproducibility at a defined location of an atomically flat terrace without surface inhomogeneities using in situ SPM techniques under electrochemical conditions? Second, is it possible to produce metal nanostructures of defined geometry on real solid surfaces independent of surface inhomogeneities? Third, are small Me clusters and defined nanostructures stable in time, or which procedures must be applied to stabilize such nanoobjects?

Local probe techniques can be carried out ex situ, non situ, or in situ with respect to applied environmental conditions. Ex situ local probe investigations are performed under UHV conditions on well-defined substrates (e.g., single crystal surfaces). Such ex situ measurements are often far from real conditions that involve adsorption and film formation phenomena of contaminants. Therefore, ex situ UHV techniques are usually equipped with appropriate transfer devices to switch substrates from the real environment to UHV and vice versa. Non situ local probe measurements are also started under defined UHV conditions to characterize the bare substrate surface, but they are continued under a finite vapor pressure in order to form adsorbates or monatomic or multiatomic (-molecular) films. This procedure is used to model real environmental conditions. In situ local probe measurements are carried out at solid/gas or solid/liquid interfaces under defined real conditions involving adsorption as well as phase formation and growth processes.

In situ local probe investigations at solid/liquid interfaces can be performed under defined electrochemical conditions when the solid phase is electron-conducting and the liquid phase is ion-conducting. In this case, electrochemistry offers a great advantage since the Fermi levels of both substrate and probe (STM-tip or metallized AFM-cantilever) can be adjusted precisely and independently of each other using bipotentiostatic control. This corresponds to a four-probe technique with substrate as working electrode, tip, or conducting cantilever as local probe, reference, and counter electrodes.[1] The Fermi level is defined by

$$E_F^{(j)} = \tilde{\mu}_e^{(j)} = \mu_e^{(j)} - e\phi^{(j)} \tag{26}$$

where $\tilde{\mu}_e^{(j)}$ and $\mu_e^{(j)}$ (in eV) denote the electrochemical and chemical potentials of electrons in phase j, respectively, and $\phi^{(j)}$ is the inner or Galvani potential of phase j, which can be measured as electrode potential E versus a reference electrode. In STM studies, Fermi level control leads to defined surface properties of tip and substrate and, therefore, to defined tunneling conditions for distance tunneling spectroscopy (DTS) and voltage tunneling spectroscopy (VTS).[13,35,46,47] Without bipotentiostatic control, only the potential difference between tip and substrate (i.e., the tunneling voltage $U_T = E_{Tip} - E$) can be held constant without control of the surface properties and, therefore, of the tunneling conditions. Consequently, the influence of the electrode potential on electrochemical interface processes is undefined. By contrast, in situ local probe studies under defined electrochemical conditions allow accurate control of the supersaturation or undersaturation in iD Me phase formation processes, as shown in Section 2.

The advantages and disadvantages of studies at solid/liquid interfaces in comparison with those at solid/gas interfaces for local nanostructuring and nanomodification of solid surfaces will be discussed in the following. At present, defined local structuring and modification of electron-conducting solid state surfaces (metals, superconductors, semiconductors) is mainly performed indirectly at solid/gas interfaces using physical (PVD) or chemical (CVD) vapor deposition techniques or electrodeposition of metals in combination with surface lithography. Usual photolithographic techniques operate in the micrometer range of lateral resolution. A well-known example is the LIGA technique to form structures and objects in the micrometer scale. A higher lateral resolution in the range 10 to 10^3 nm can be obtained using electron or molecular beam lithography. The lateral resolution is limited by the excitation source wavelength. Therefore, nanotechnology in the range of a few nanometers or even Angstrøms seems to be difficult using these conventional techniques. A "direct writing" with electron or molecular beam excitation is only possible when metal deposition processes are started indirectly. A great disadvantage using PVD, CVD, and electron or molecular beam techniques is the less controllable and only very slowly changeable supersaturation. Additionally, pronounced local fluctuation phenomena in the gas phase lead to a local distribution of the supersaturation and, therefore, to a distribution of critical nuclei sizes, which are denoted as "magic numbers." Thus, controlled epitaxial growth is difficult to be obtained by metal deposition from the gas phase. Usually, polycrystalline deposits of different microstructures are formed. This problem is, for example, well known for laser-deposited epitaxial thin films of high-T_c-superconductors. Furthermore, PVD, CVD, and electron or molecular beam processes increase the temperature of the substrate surface. Dry-etching processes of illuminated lithographic domains often lead to damage of the substrate surface, which are difficult to be controlled. Finally, the indirect structuring and modification processes require multiple environmental changes under ultrapure external conditions. The multiple changes of the environment represent a key factor for contamination and resulting degradation mechanisms thus affecting the quality of micro- and nano-devices in electronics.

On the other hand, the supersaturation is more or less homogeneously distributed at solid/liquid interfaces under electrochemical conditions leading to a unique value of the critical nucleus size at a constant overpotential (cf. Fig. 1.10). Consequently, local nanostructuring and nanomodification of solid state surfaces by in situ local probe techniques in electrochemical systems provides a number of great advantages. As already mentioned, supersaturation or undersaturation can be precisely controlled and rapidly changed via the electrode potential in electrochemical systems. Additionally, the Fermi levels of the substrate and the probe of in situ local probe techniques can be adjusted and independently changed. In such a way, local cathodic metal deposition and anodic dissolution (electrochemical etching) within the same system can be applied, depending on the electrochemical conditions, thus avoiding multiple environmental changes. Furthermore, thermodynamics and kinetics of the formation of metal phases by nucleation and growth processes are well defined. Epitaxial metal deposition as well as selective etching are rather easily obtained. The elimination of beam-based lithographic procedures in electrochemical processes avoids a localized increase of temperature via dissipative energy transfer. Finally, metal phases of different dimensionality in nanostructuring processes can be studied more exactly within defined supersaturation or undersaturation ranges in electrochemical systems than at solid/gas interfaces (cf. Section 1.2).

An important aim of modern nanotechnology is the structuring and modification of solid state surfaces in the nanometer or subnanometer range. An implementation can only be realized if metal deposition or dissolution processes can be locally focused. This can be achieved either from the gas phase or from the electrolyte phase using in situ local probe techniques as a powerful tool. In this case, the tip and the cantilever of STM and AFM instruments, respectively, operate as miniaturized electrodes. Again, the execution of localized metal deposition or dissolution reactions in the vapor phase environment features the same disadvantages discussed previously. In particular, localized metal deposition occurs only using high energy differences between the tip or the (metallized) cantilever and the substrate. By contrast, localized metal deposition or dissolution processes that are electrochemically induced involve energy differences orders of magnitude lower. Furthermore, the electrochemical potentials (Fermi levels) of both the tip or the metallized cantilever as well as the substrate can be separately adjusted and controlled allowing the development of appropriate techniques for localized Me deposition or dissolution.

At present, different techniques are applied in electrochemical systems for local nanostructuring and nanomodification of solid surfaces. The techniques applied can be characterized by different external perturbations of the system substrate/electrolyte/SPM probe, which lead to a local nanostructuring of the substrate surface as a system response. An attempt is made to classify these nanostructuring methods on the basis of three general aspects as schematically illustrated in Figure 1.26.

i) *Probe-induced techniques* (Fig. 1.26) are characterized by an initial Me deposition at the SPM probe and the subsequent transfer of a small Me cluster to the substrate surface below the probe. Me predeposition at the probe (tip or metallized cantilever) is usually produced by an appropriate negative prepolarization of the probe. The transfer of a Me cluster from the probe to the substrate surface is usually achieved by either approaching (not contacting) the probe to the substrate surface by a perturbation of the z-piezo voltage or applying an appropriate polarization routine on the system substrate/electrolyte/tip (cf. ii). Probe-induced methods usually do not damage the substrate surface.

ii) *Field-induced techniques* (Fig. 1.26b) are characterized by an inhomogeneous electric field distribution between the SPM probe and the substrate, locally changing the structure of the STM tunneling gap, the structure of the electrochemical double layers at the STM-tip/electrolyte and substrate/electrolyte interfaces, the Me^{z+} concentration within the gap, and/or the charge carrier density and distribution of the underlying substrate domain. The last effect is especially noteworthy in the case of semiconductor substrates. Relatively high electric fields may also cause surface defects by a substrate modification. In this case, field induced methods damage the substrate surface.

iii) *Defect-induced techniques* (Fig. 1.26c) are characterized by a mechanical production of small surface defects of different dimensionality, which can act as nucleation centers in the subsequent Me phase formation process. Defect-induced techniques always more or less damage the substrate surface.

Depending on the properties of real systems, appropriate combinations of these three different techniques are necessary for a successful nanostructuring of solid surfaces.

However, it should be mentioned that the stability of small Me clusters and defined Me nanostructures on foreign substrate surfaces is still an open problem. Generally, small clusters of any dimensionality have a more negative equilibrium potential than the corresponding infinitely large phases. This effect is related to the Gibbs–Thomson or Kelvin equation and leads to a dissolution of small iD phases at the corresponding equilibrium potentials, $E_{\text{1D Me}}$. This behavior of small clusters of iD phases in systems with different Me-S interaction is illustrated in Figure 1.10 showing the corresponding stability ranges. It is evident that a stabilization of small clusters of iD phases at open circuit conditions needed for long-time stable nanostructures can only be achieved using stabilizing effects such as adsorption of interface inhibitors, passivation, and inclusion of contaminants.

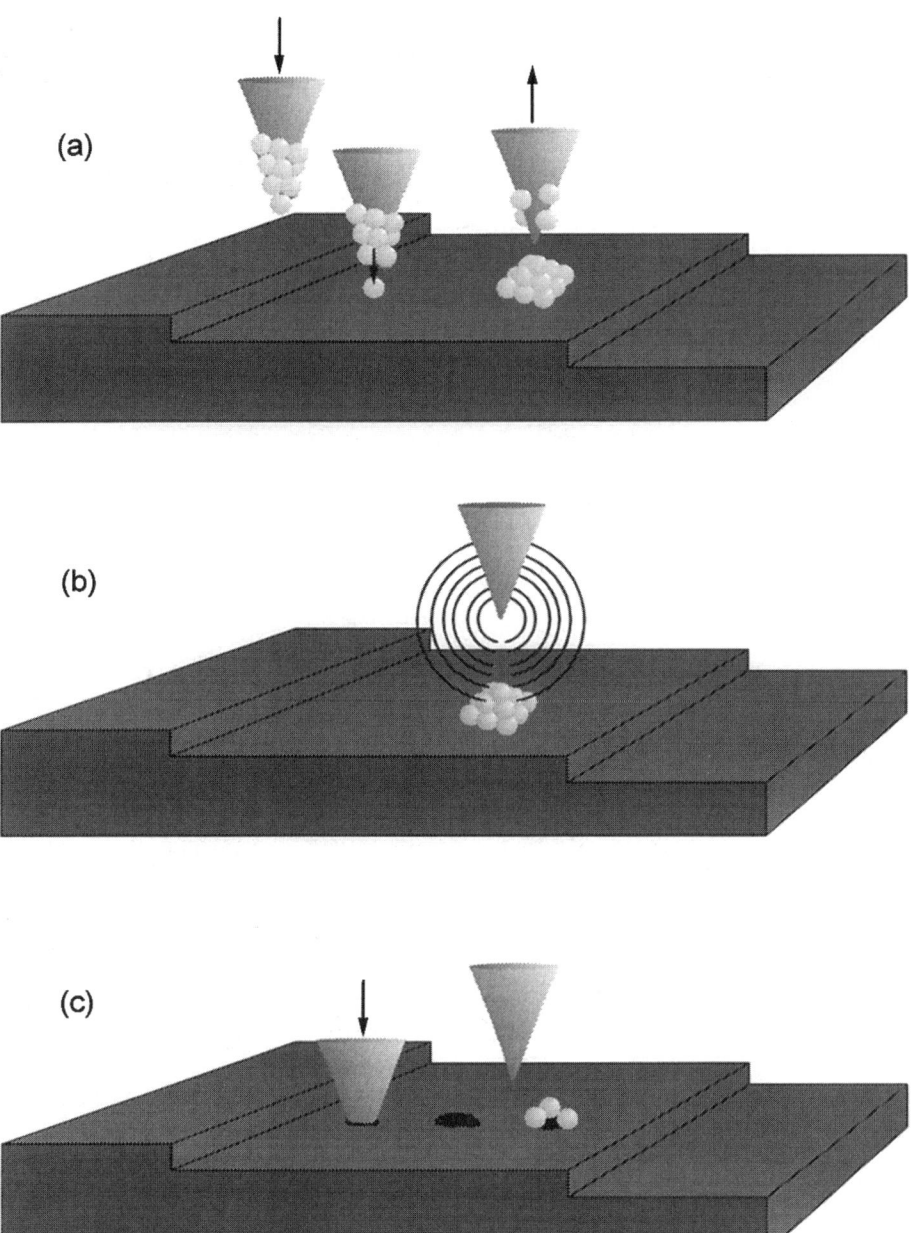

Figure 1.26 Schematic representation of different techniques for local nanostructuring and nanomodification of solid surfaces by Me electrodeposition. (a) probe-induced; (b) field-induced; (c) defect-induced.[1]

Investigations on local structuring and modification of electron-conducting solid state surfaces using in situ SPM techniques under electrochemical conditions are just in the beginning. Recently, interesting experiments were carried out on surfaces of metal single crystals, highly oriented pyrrolyic graphite (HOPG), epitaxial thin films of high-T_c-superconductors (HTSC), and semiconductor single crystal surfaces, which will be discussed in the following.

1.3.2 Experimental Results

A tip-induced local metal deposition was first achieved by Kolb et al.[48–52] in the systems Au($hkll$)/Cu^{2+}, Ag(hkl)/Cu^{2+}, and Au(hkl)Pd^{2+} with (hkl) = (100), (111) using in situ STM technique. Me is initially predeposited on the tip. Subsequently, the z-piezo voltage is changed and the tip approaches the surface, which leads to a transfer of Me clusters from the tip of the substrate surface forming a so-called "connective neck." By lateral movement of the tip and a simultaneous modulation of the z-piezo voltage by a sinusoidal perturbation with a small amplitude leading to a vertical oscillation of the tip, Me cluster trains or regular Me cluster arrays can be produced as shown in Figure 1.27.[50] The size and height of single Me clusters can be controlled by the polarization routine. It should be noted, however, that this tip-induced method only operates in systems with relatively high Me-S adhesion energy, that is, in systems with relatively strong Me-S interaction (cf. eq. (8)). Thus, it has to be expected that such systems are characterized by the formation of a 2D Me phase in the underpotential range and the 3D Me nanostructuring occurs on top of the Me UPD modified substrate.

First field-induced studies for nanostructuring of solid surfaces using in situ STM were performed in the system HOPG (0001)/Ag$^+$ by Penner et al.[53–55] using a potentiostatic two-probe pulse technique with high tunneling voltage amplitudes ($U_T \approx 6V$). A preferred 3D Ag cluster formation under the tip was achieved in the time of the pulse duration as shown in Figure 1.28. The 3D Ag clusters had a diameter of 20 nm $\leq d_{Cluster} \leq$ 40 nm and a height of about 5 nm. The mechanism of this local Me deposition process can be interpreted by a field-induced initial formation of surface defects (shallow pits) at the HOPG surface. In this way, different metals such as Ag and Cu could be successively deposited by changing the aqueous electrolyte solutions containing Me^{z+} forming nanometer-scale Ag-Cu galvanic cells as shown in Figure 1.29. Similar experiments were carried out in the system HOPG(0001)/Ag$^+$ by Pötzschke et al.[42,56,57] It was found that the 3D Ag cluster size strongly depends on the charge amount transferred during the pulse. Minimum cluster sizes with a diameter less than 10 nm and a height less than 2 nm could be formed as shown in Figure 1.30.

Defect-induced local nanostructuring of Au(100) terraces with small Cu clusters was independently obtained by Froese et al.[57,58] and Gewirth et al.[59,60] as illustrated in Figure 1.31. The influence of the unmetallized cantilever on the electrochemical copper deposition was investigated in the contact mode with a

1.3 NANOSTRUCTURING OF SOLID SURFACES

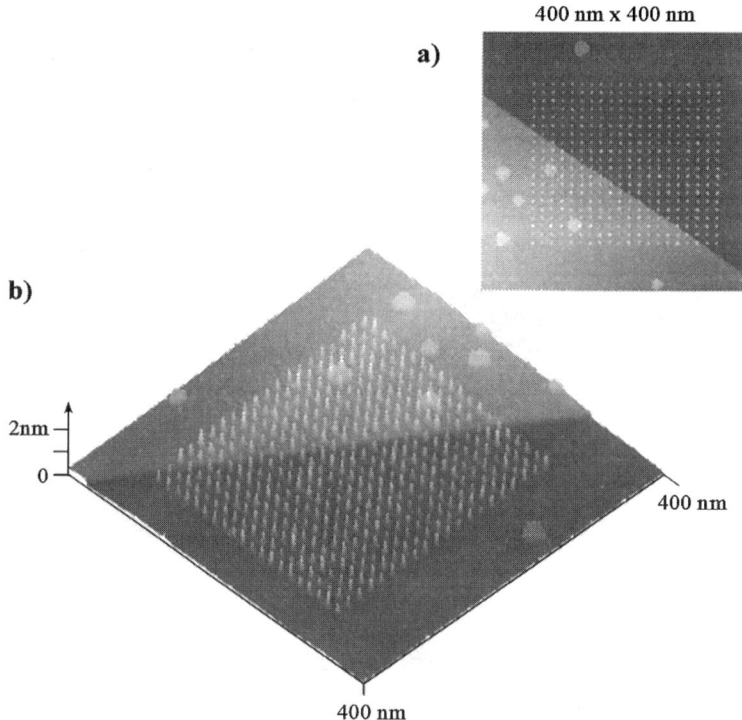

Figure 1.27 Array of 400 Cu clusters on Au(111) substrate deposited using a tip-induced technique in the system Au(111)/10^{-3} M CuSO$_4$ + 5 × 10^{-2} M H$_2$SO$_4$ at T = 298 K. (a) in situ STM image (top view); (b) 3D presentation of (a). By courtesy of D. M. Kolb.[50]

Figure 1.28 In situ STM image showing an Ag cluster on HOPG deposited using a field-induced technique in the system HOPG(0001)/5 × 10^{-4} M AgF at T = 298 K.[55] Reprinted by permission of Kluwer Academic Publishers.

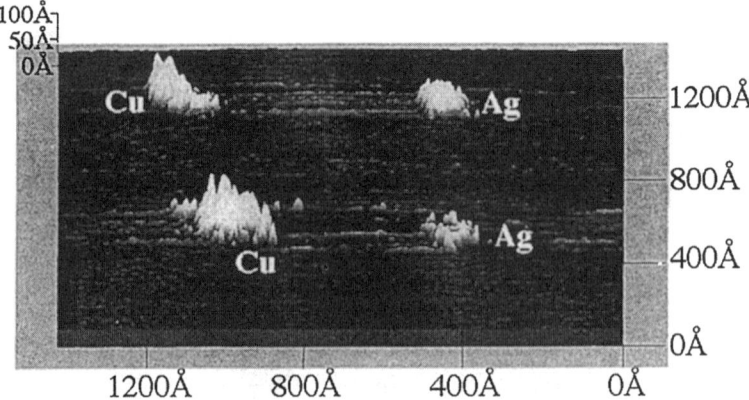

Figure 1.29 In situ STM image showing Cu and Ag clusters on HOPG deposited using a field-induced technique in the systems HOPG(0001)/5×10^{-4} M $CuSO_4$ and HOPG(0001)/5×10^{-4} M AgF at $T = 298$ K.[55] Reprinted by permission of Kluwer Academic Publishers.

constant and relatively low cantilever force of $F = 1.10^{-9}$ N. Under this condition, localized Cu deposition is induced by the cantilever in the overpotential range -30 mV $\leq \eta_{3D} \leq -20$ mV. These overpotentials are sufficient to locally deposit Cu clusters in the scan region only, but not at the remaining surface. As the Au(100) surface has been found undamaged after lifting the Cu clusters, one has to assume that the nucleation is induced by either an elastic deformation of the surface or by a creation of fresh surface areas free of adsorbed contaminants.

Another defect-induced local nanostructuring using the AFM-cantilever was studied in the system Au(100)/Cu^{2+} by Froese et al.[57,58] Figure 1.32a shows

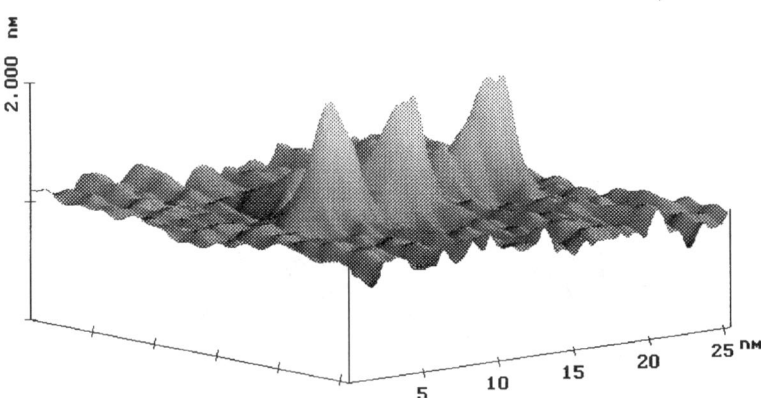

Figure 1.30 In situ STM image Ag clusters subsequently deposited on a flat HOPG terrace using a field-induced technique in the system HOPG(0001)/10^{-2} M $AgClO_4$ + 1M $HClO_4$ at $T = 298$ K.[1,42]

1.3 NANOSTRUCTURING OF SOLID SURFACES

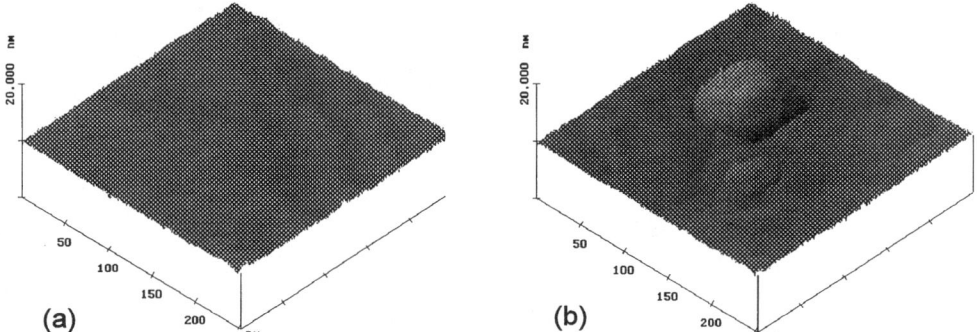

Figure 1.31 Defect-induced Cu deposition on Au(100) in the system Au(100)/5 × 10^{-3} M $CuSO_4$ + 10^{-2} M H_2SO_4 at $T = 298$ K. (a) in situ AFM image of the bare Au(100) substrate at $\eta_{3D} = 0$ mV; (b) in situ AFM image at $\eta_{3D} = 0$ mV after scanning distinct surface area of 50 nm × 50 nm in AFM contact mode at $\eta_{3D} = -25$ mV.[57,58]

the gold single crystal face after forming five surface defects by a relatively high cantilever force of F > 100 nN. Subsequently, Cu clusters were preferentially formed at these surface inhomogeneities at a relatively low overvoltage in the range -20 mV $\leq \eta_{3D} \leq -10$ mV as illustrated in Figure 1.32b. A similar technique was successfully applied for the nanostructuring of high-temperature superconducting (HTSC) thin films with c-axis orientation.[57,58] Epitaxial HTSC films were prepared by laser deposition or $SrTiO_3$ single crystal faces. Figure 1.33 shows an in situ AFM experiment on a $YBa_2Cu_3O_{7-\delta}$ ($\delta > 0.5$) surface in

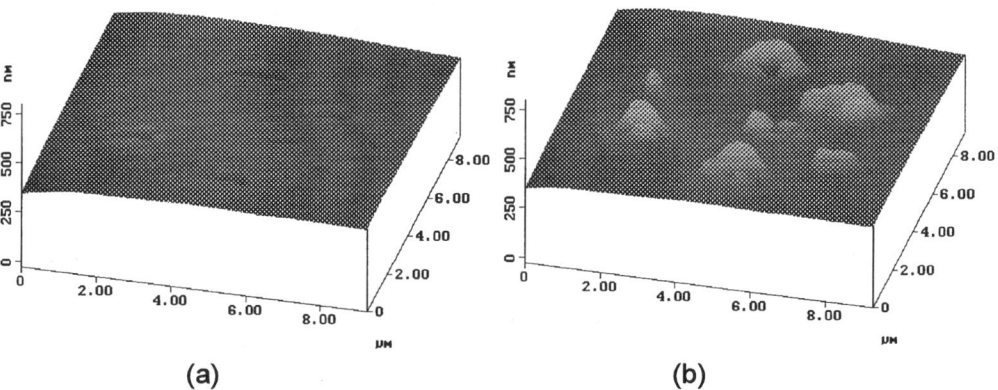

Figure 1.32 Defect-induced Cu deposition on Au(100) in the system Au(100)/5 × 10^{-3} M $CuSO_4$ + 10^{-2} M H_2SO_4 at $T = 298$ K. (a) in situ AFM image of the bare Au(100) surface featuring surface defects created by the AFM cantilever at $\eta_{3D} = 0$ mV; (b) in situ AFM image at $\eta_{3D} = 0$ mV after scanning the surface in AFM contact mode at $\eta_{3D} = -18$ mV for $t = 10$ s.[57,58]

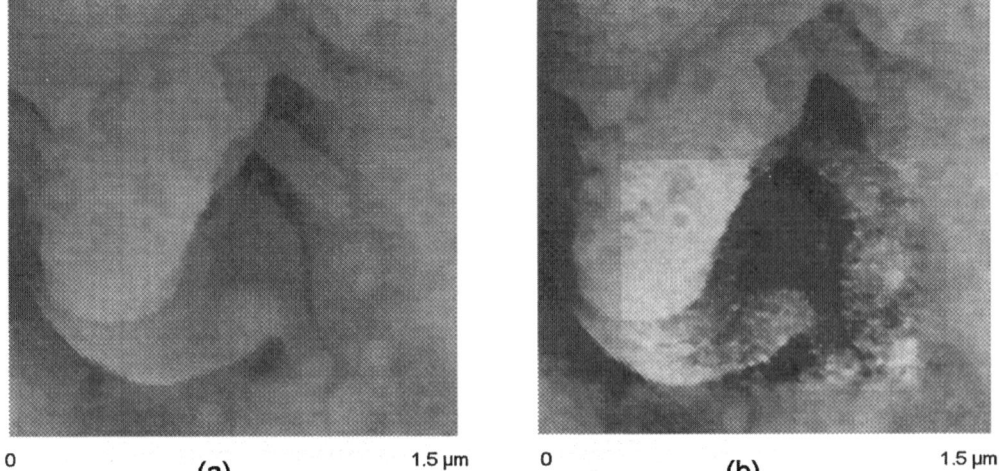

Figure 1.33 Defect-induced Cu deposition on an HTSC thin film epitaxially grown on SrTiO$_3$ in the system YBa$_2$Cu$_3$O$_{7-\delta}$/CH$_3$CN + 10^{-1} M C$_{16}$H$_{36}$ClNO$_4$ + 3.3 × 10^{-3} M [CH$_3$COCH=C(O—)CH$_3$]$_2$Cu at T = 298 K. (a) in situ AFM image of a YBa$_2$Cu$_3$O$_{7-\delta}$ substrate without Cu at η_{3D} = 0 mV; (b) in situ AFM image at η_{3D} = 0 mV after scanning a distinct surface area of 1 μm × 1 μm in the AFM contact mode at η_{3D} = −100 mV.[57,58]

nonaqueous and deaerated solution before and after local Cu deposition in the scanned region with a cantilever force of F = 200 nN. Note, that the deposition of 3D bulk copper on this substrate needs an overvoltage of $\eta_{3D} \leq -110$ mV, whereas the defect-induced Cu deposition starts already at lower values of the overpotential.

The disadvantage of defect-induced nanostructuring methods is that Me phase formation preferentially takes place not only at surface inhomogeneities produced by the SPM probe, but at all surface defects existing on a real substrate surface. Consequently, an application of this nanostructuring method requires a masking (inhibition) of undesirable surface inhomogeneities.

In complicated systems, a combination of different nanostructuring techniques has to be applied. For example, nanostructuring of semiconductor surfaces by electrochemical Me phase formation is complicated due to the band gap and the participation of surface states in semiconductor electrode processes.[61–64] In addition, the formation of 2D Me phases in the undersaturation range has experimentally not been observed in various semiconductor/electrolyte systems due to a relatively weak Me-S interaction. Therefore, it can be expected that a simple probe-induced nanostructuring process will not operate in such systems. However, very recently a nanostructuring by electrochemical metal deposition has been achieved in the system n-Si(111)/Pb^{2+} using a combination of STM tip-induced and field-induced techniques as demonstrated in Figure 1.34.[42]

1.3 NANOSTRUCTURING OF SOLID SURFACES

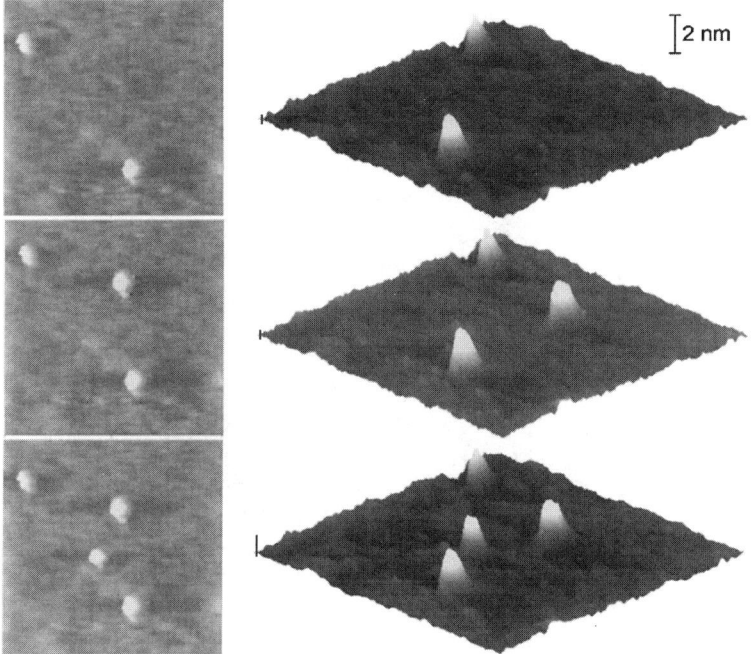

Figure 1.34 In situ STM images showing four Pb clusters subsequently deposited on n-Si(111) using a combined tip- and field-induced technique in the system n-Si(111)/5 × 10^{-3} M Pb(ClO$_4$)$_2$ + 10^{-2} M HClO$_4$ at T = 298 K. Scan size 50 nm × 50 nm.[42]

In these experiments, lead was primarily deposited on the tip by a cathodic polarization. Subsequent applications of an anodic pulse on the tip leads to a dissolution of lead. Thus, the Pb^{2+} concentration within the tunneling gap is locally increased producing a shift of the local equilibrium potential, $E_{3D\,Pb}$, in the positive direction and leading to a significant increase of the supersaturation to initiate 3D Pb nucleation.* Simultaneously, the electrochemical Pb deposition process at the semiconductor/electrolyte interface is enhanced by the inhomogeneous electric field distribution. A similar combined tip-induced and field-induced technique was recently successfully used by Kirschner and Schindler[66] for nanostructuring of metal surfaces in the system Au(111)/Cu^{2+} as shown in Figure 1.35.

The results in these model systems confirm the introductory expectations that a defined local surface structuring and modification in the nanometer range can be realized by electrochemical means using different in situ SPM techniques and polarization routines.

---

*A similar principle is used by Siegenthaler et al. in the recently developed "thin-layer STM probe" technique.[65]

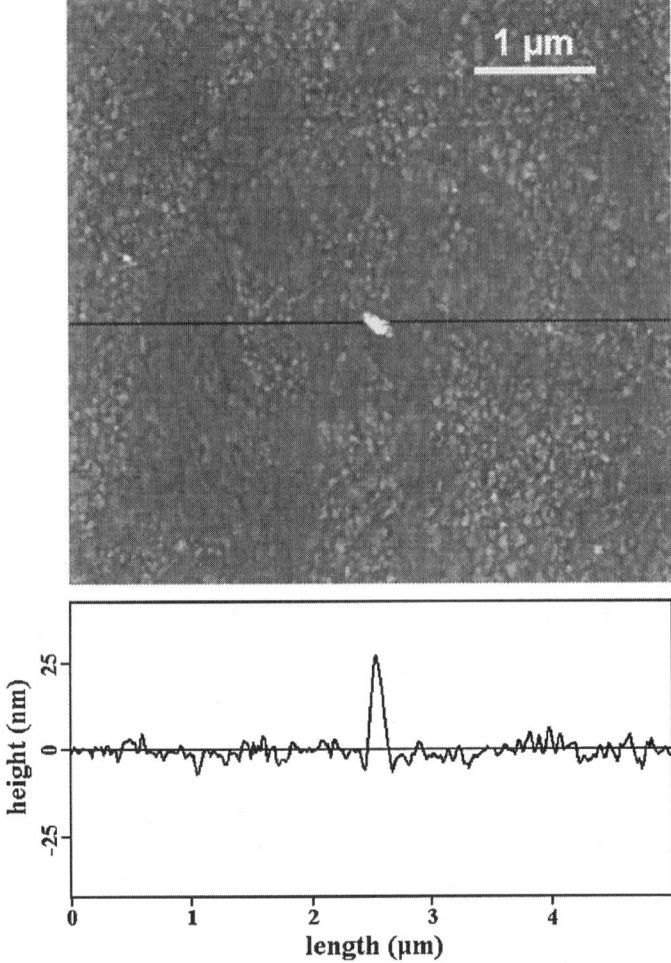

Figure 1.35 In situ STM image showing a Cu cluster deposited on a polycrystalline Au substrate using a combined tip- and field-induced technique in the system Au/10^{-3} M CuSO$_4$ + 5 × 10^{-2} M H$_2$SO$_4$ at T = 298 K. By courtesy of J. Kirschner and W. Schindler.[66]

1.4 CONCLUSIONS

The existence and properties of low-dimensional Me phases are clearly demonstrated by different electrochemical techniques and in situ SPM measurements. Thermodynamics, structure, and kinetics of 3D, 2D, and 1D Me phases as well as the corresponding phase transitions seem to be well understood. However, experimental evidence of higher order phase transitions (at constant surface coverage Γ) in 2D and 1D Me phases does not exist up to now. In addition, some open questions still exist for 1D Me phases concerning the coexistence of expanded and condensed 1D Me phases, a problem that is related to a first order phase transition. Furthermore, the average distances of 2D and 1D kinks in the corresponding phases are experimentally not determined. For 3D and 2D Me phases, this value can be obtained from a theoretical consideration; however, such a theoretical basis does not exist for 1D Me phases. Finally, 0D Me clusters are more or less hypothetical in this stage of knowledge.

There is no doubt that nanostructuring and nanomodification of solid surfaces can be performed by in situ SPM under defined electrochemical conditions. The elementary steps of phase formation and dissolution at solid/liquid interfaces under electrochemical conditions are well understood and a transfer of this knowledge into nanostructuring techniques can certainly be of great advantage. However, major problems are still unsolved (e.g., the stabilization of nanoclusters). Therefore, extensive theoretical and experimental investigations are needed before electrochemical nanotechniques can take a major role in the forthcoming nanotechnology of the next century.

ACKNOWLEDGMENTS

The authors thankfully acknowledge financial support of the Deutsche Forschungsgemeinschaft (DFG), Bundesministerium für Wissenschaft, Bildung, Forschung und Technologie (BMBF), Bundesministerium für Wirtschaft (BMW), Arbeitsgemeinschaft Industrieller Forschungsvereinigungen (AIF), Fonds der Chemie, and the Bulgarian Academy of Sciences (BAW) for long-term research on these topics and cooperation between BAW and the University of Karlsruhe. Morever, the authors gratefully acknowledge their colleagues Prof. Dr. H. Siegenthaler (Bern), Prof. Dr. D. M. Kolb (Ulm), and Prof. Dr. J. Kirschner (Halle) for giving the scientific material presented in Figs. 2.15, 2.22, 2.23, 3.2, and 3.10. Finally, the authors thank their long-time coworker, Dr. R. T. Pötzschke, for many discussions and assistance in the preparation of this article.

REFERENCES

1. E. Budevski, G. Staikov, W. J. Lorenz. *Electrochemical Phase Formation and Growth—An Introduction in the Initial Stages of Metal Deposition* (VCH, Weinheim, 1996).
2. W. K. Burton, N. Cabrera, F. C. Frank, *Phil. Trans. Roy. Soc.* **A243,** 299 (1951).
3. W. J. Lorenz, G. Staikov, in *Proceedings of the Third Symposium on Electrochemically Deposited Thin Films*, Vol. 96–19, M. Paunovich and D. A. Scherson, eds., (The Electrochemical Society, Pennington, N.J., 1997), p. 171.
4. G. Staikov, W. J. Lorenz, in *Proceedings of the Third Symposium on Electrochemically Deposited Thin Films*, Vol. 96–19, M. Paunovich and D. A. Scherson, eds., (The Electrochemical Society, Pennington, N.J., 1997), p. 3.

5. G. Staikov, W. J. Lorenz, *Can. J. Chem.* **75,** 1624 (1997), in press.
6. G. Binnig, H. Rohrer, C. Gerber, E. Weibel, *Phys. Rev. Lett.* **49,** 57 (1982).
7. G. Binnig, H. Rohrer, *IBM J. Res. Develop.* **30,** 355 (1986).
8. G. Binnig, C. F. Quate, C. Gerber, *Phys. Rev. Lett.* **56,** 930 (1986).
9. *Scanning Tunneling Microscopy and Related Methods*, NATO ASI Series E: Applied Sciences, Vol. 184, R. J. Behm, N. Garcia, and H. Rohrer, eds. (Kluwer Acad. Publ., Dordrecht, 1990).
10. *10 Years of STM*, Proc. 6th Int. Conf. on STM, Interlaken, Switzerland, August 12–16, 1991, P. Descouts and H. Siegenthaler, eds., *Ultramicroscopy* **42–44** (1992).
11. *Scanning Tunneling Microscopy I*, Springer Ser. Surf. Sci., Vol. 20, H.-J. Güntherodt and R. Wiesendanger, eds. (Springer, Berlin, Heidelberg, 1992).
12. *Scanning Tunneling Microscopy II*, Springer Ser. Surf. Sci., Vol. 28, R. Wiesendanger and H.-J. Güntherodt, eds. (Springer, Berlin, Heidelberg, 1992).
13. *Nanoscale Probes of the Solid/Liquid Interface*, NATO ASI Series E: Applied Sciences, Vol. 288, A. A. Gewirth and H. Siegenthaler, eds. (Kluwer Acad. Publ., Dordrecht, 1995).
14. S. Magonov, M.-H. Whangbo. *Surface Analysis with STM and AFM—Experimental and Theoretical Aspects* (VCH, Weinheim, 1996).
15. S. Roth. *One-Dimensional Metals: Physics and Materials Science* (VCH, Weinheim, 1995).
16. L. D. Landau, E. M. Lifshitz. *Statistical Physics* (MIT Press, Cambridge, Mass. and London, 1966).
17. H. B. Callen. *Thermodynamics and an introduction to thermostatistics* (John Wiley & Sons Inc., New York, 1985).
18. R. B. Griffits, in *Phase Transitions and Critical Phenomena*, Vol. 1, C. Domb and M. S. Green, eds. (Academic Press, London, New York, 1972), p. 7.
19. V. Tsakova, J. W. Schultze, *Bulg. Chem. Comm.* **27,** 138 (1994).
20. R. Kaischew, *Comm. Bulg. Acad. Sci. Phys. Ser.* **1,** 100 (1950).
21. W. Obretenov, U. Schmidt, W. J. Lorenz, G. Staikov, E. Budevski, D. Carnal, U. Müller, H. Siegenthaler, E. Schmidt, *J. Electrochem. Soc.* **140,** 692 (1993).
22. G. Staikov, E. Budevski, W. Obretenov, W. J. Lorenz, *J. Electroanal. Chem.* **349,** 355 (1993).
23. L. D. Landau, E. M. Lifshitz *Theory of Elasticity* (Pergamon, London, 1959).
24. J. H. van der Merwe, in *Single Crystal Films*, M. H. Frankombe and H. Sato, eds. (Pergamon Press, Oxford, 1964), p. 139.
25. J. W. Mathews, D. C. Jackson, A. Chambers, *Thin Solid Films* **26,** 129 (1975).
26. T. Will, M. Dietterle, D. M. Kolb, in *Nanoscale Probes of the Solid/Liquid Interface*, NATO ASI Series E: Applied Sciences, Vol. 288, A. A. Gewirth and H. Siegenthaler, eds. (Kluwer Acad. Publ., Dordrecht, 1995), p. 137.
27. K. Engelsmann, W. J. Lorenz, E. Schmidt, *J. Electroanal. Chem.* **114,** 11 (1980).
28. J. Hitzig, J. Titz, K. Jüttner, W. J. Lorenz, E. Schmidt, *Electrochim. Acta* **29,** 287 (1984).
29. E. Schmidt, J. Hitzig, J. Titz, K. Jüttner, W. J. Lorenz, *Electrochim. Acta* **31,** 1041 (1986).

REFERENCES

30. F. Mansfeld, W. J. Lorenz, in *Techniques for Characterization of Electrodes and Electrochemical Processes*, R. Varma and J. R. Selman, eds. (John Wiley & Sons, New York, 1991), p. 581.
31. G. Staikov, K. Jüttner, W. J. Lorenz, E. Schmidt, *Electrochim. Acta* **23**, 305 (1978).
32. M. Avrami, *J. Chem. Phys.* **7**, 1103 (1939); **8**, 212 (1940); **9**, 177 (1941).
33. S. Garcia, D. Salinas, C. Mayer, E. Schmidt, G. Staikov, W. J. Lorenz, *Electrochim. Acta*, **43**, 3007 (1998).
34. H. Bort, K. Jüttner, W. J. Lorenz, E. Schmidt, *J. Electroanal. Chem.* **90**, 413 (1978).
35. S. Vinzelberg, Ph.D. dissertation (University of Karlsruhe, 1995).
36. U. Schmidt, S. Vinzelberg, G. Staikov, *Surf. Sci.* **348**, 261 (1996).
37. U. Müller, D. Carnal, H. Siegenthaler, E. Schmidt, W. J. Lorenz, W. Obretenov, U. Schmidt, G. Staikov, E. Budevski, *Phys. Rev. B* **46**, 12899 (1992).
38. M. F. Toney, J. G. Gordon, G. L. Borges, O. R. Melroy, D. Yee, L. B. Sorensen, *Phys. Rev. B*, **49**, 7793 (1994).
39. D. Carnal, P. I. Oden, U. Müller, E. Schmidt, H. Siegenthaler, *Electrochim. Acta* **40**, 1223 (1995).
40. H. Siegenthaler, E. Ammann, *189th Meeting of The Electrochemical Society* (The Electrochemical Society, Los Angeles, 1996), Abs. 1072.
41. E. Ammann, M.S. thesis (University of Bern, 1995).
42. R. T. Pötzschke, Ph.D. dissertation (University of Karlsruhe, 1998).
43. J. Sackmann, A. Bunk, R. T. Pötzschke, G. Staikov, W. J. Lorenz, *Electrochim. Acta*, **43**, 2863 (1998).
44. M. Ebert, M.S. thesis (University of Karlsruhe, 1986).
45. M. Dietterle, T. Will, and D. M. Kolb, *Surf. Sci.*, **342**, 29 (1995).
46. J. Halbritter, G. Repphun, S. Vinzelberg, G. Staikov, W. J. Lorenz, *Electrochim. Acta* **40**, 1385 (1995).
47. G. Staikov, W. J. Lorenz, *Z. Phys. Chem.*, **208**, 17 (1999).
48. R. Ullmann, T. Will, D. M. Kolb, *Chem. Phys. Lett.* **209**, 238 (1993).
49. R. Ullmann, T. Will, D. M. Kolb, *Ber. Bunsenges. Phys. Chem.* **99**, 1414 (1995).
50. D. M. Kolb, R. Ullmann, T. Will, *Science* **275**, 1097 (1997).
51. G. E. Engelmann, J. C. Ziegler, D. M. Kolb, *J. Electrochem. Soc.*, **145**, L33 (1998).
52. D. M. Kolb, R. Ullmann, J. C. Ziegler, *Electrochim. Acta*, **43**, 2751 (1998).
53. W. Li, J. A. Virtanen, R. M. Penner, *Appl. Phys. Lett.* **60**, 1181 (1992).
54. W. Li, J. A. Virtanen, R. M. Penner, *J. Phys. Chem.* **96**, 6529 (1992).
55. W. Li, T. Duong, J. A. Virtanen, R. M. Penner, in *Nanoscale Probes of the Solid/Liquid Interface*, NATO ASI Series E: Applied Sciences, Vol. 288, A. A. Gewirth and H. Siegenthaler, eds. (Kluwer Acad. Publ., Dordrecht, 1995), p. 183.
56. R. T. Pötzschke, C. A. Gervasi, S. Vinzelberg, G. Staikov, W. J. Lorenz, *Electrochim. Acta* **40**, 1469 (1995).
57. R. T. Pötzschke, A. Froese, W. Wiesbeck, G. Staikov, W. J. Lorenz, in *Proceedings of the Third Symposium on Electrochemically Deposited Thin Films*, Vol. 96–19, M. Paunovich and D. A. Scherson, eds. (The Electrochemical Society, Pennington, N.J., 1997), p. 21.

58. A. Froese, Ph.D. dissertation (University of Karlsruhe, 1996).
59. J. R. LaGraff, A. A. Gewirth, *J. Phys. Chem.* **99,** 10009 (1995).
60. J. R. LaGraff, A. A. Gewirth, in *Nanoscale Probes of the Solid/Liquid Interface*, NATO ASI Series E: Applied Sciences, Vol. 288, A. A. Gewirth and H. Siegenthaler, eds. (Kluwer Acad. Publ., Dordrecht, 1995), p. 83.
61. P. Bindra, H. Gerischer, D. M. Kolb, *J. Electrochem. Soc.* **124,** 1012 (1977).
62. P. Allongue, in *Modern Aspects of Electrochemistry*, Vol. 23, B. I. Conway, J. O'Bockris, and R. E. White, eds. (Plenum Press, New York, 1992), p. 239.
63. P. Allongue, in *Advances in Electrochemical Science and Engineering*, Vol. 4, H. Gerischer and Ch. Tobias, eds. (VCH, Weinheim, 1995), p. 1.
64. B. Rashkova, B. Guel, R. T. Pötzschke, G. Staikov, W. J. Lorenz, *Electrochim. Acta* **43,** 3021 (1998).
65. E. Ammann, P. Haering, P.-F. Indermuehle, R. Koetz, N. F. de Rooij, H. Siegenthaler, *The 1997 Joint Meeting of The Electrochemical Society and The International Society of Electrochemistry* (The Electrochemical Society, Paris, 1997), Abs. 830.
66. J. Kirschner, W. Schindler, private communication.

2

ELECTRON DIFFRACTION AND ELECTRON MICROSCOPY OF ELECTRODE SURFACES

G. LEHMPFUHL, Y. UCHIDA, M. S. ZEI, AND D. M. KOLB*

Fritz-Haber-Institut der Max-Planck-Gesellschaft, Faradayweg 4-6, D-14195 Berlin, Germany

1.1 INTRODUCTION

It is well-documented in the surface science literature that the structure of solid surfaces plays a key role in many heterogeneous reactions [1]. Naturally the same is true for electrochemical reactions occuring at solid electrodes. Stimulated by the overwhelming success of surface scientists in describing surfaces and surface reactions at an atomic level by using structurally well-defined single crystal surfaces, electrochemists began to follow that route in the 1970s. However, with no techniques at hand, which allowed them to characterize at an atomic level surface structures in situ (i.e., in contact with an aqueous electrolyte), electrochemists started to adopt UHV techniques, particularly low energy electron diffraction (LEED) to check the quality of UHV-prepared single crystal surfaces before and after electrochemical treatment. Even though, at the beginning, studies in UHV were restricted to bare surfaces to test their single-crystallinity [2], they were soon extended to investigations of strongly bound adsorbates, such as iodide [3], metal [4] or oxide [5, 6] overlayers. Such studies were pioneered by Hubbard [7] and followed by the groups of Wino-

---

Imaging of Surfaces and Interfaces (Frontiers of Electrochemistry, Volume 5).
Edited by Jacek Lipkowski and Philip N. Ross.
ISBN 0-471-24672-7. © 1999 Wiley-VCH, Inc.

*Current address: Department of Electrochemistry, University of Ulm, D-89069 Ulm, Germany.

grad [4, 6], Yeager [2, 8], Ross [9–11], and Kolb [12, 13]. A major breakthrough was achieved by Hansen [14–16], who explored systematically the possibility of emersing an electrode with its electric double layer intact.

The study of electrode surfaces ex situ by electron diffraction had far-reaching consequences for the whole area of physical electrochemistry. It set the newly emerging single-crystal electrochemistry on a sound basis. It provided for the first time detailed surface structural data, which some 15–20 years later were often confirmed by in situ techniques such as surface X-ray scattering (SXS) or scanning tunneling microscopy (STM), a fact frequently ignored by the users of the latter methods. It also opened the door to a modern, interdisciplinary branch of electrochemistry, frequently termed as electrochemical surface science [17], which has attracted surface physicists. After its important role as midwife for modern physical electrochemistry, electron diffraction lost its prime role to in situ structure techniques such as SXS and STM, both of which have shortcomings with respect to availability and areal limitations. One of the purposes of this review is to demonstrate that there is still room for ex situ techniques in electrochemistry.

The structure of surfaces can be investigated by the scattering of electrons [12, 18–20]. Elastic scattering of monoenergetic electrons from an ordered array of atoms is called diffraction. Diffraction gives information of the arrangement of atoms in the surface, while inelastic scattering processes, leading to the production of Auger electrons, can give information on the chemical nature of atoms in and on the surface. Diffraction studies require single crystals with a very flat surface. Since electron scattering methods need vacuum conditions, the study of electrode surfaces cannot be carried out in situ. The electrodes have to be transferred from the electrochemical cell into a vacuum chamber, preferably by a closed transfer system that minimizes the danger of contamination in air. Therefore, electrode emersion under potential control is a crucial step; a successful ex situ study requires (i) structure conservation, and (ii) dry emersion (i.e., emersion of the electrode with its double layer, but without noticeable amounts of bulk electrolyte [21, 22].

In the following we shall give a brief introduction to electron diffraction theory before we discuss some experimental details of ex situ studies. In the main body of our review we present case studies for electron diffraction and microscopy of electrode surfaces, and we finish with a brief outlook for the ex situ route.

2.2 SOME BASIC CONSIDERATIONS OF DIFFRACTION

The following is a short and simple introduction to electron diffraction. For a more thorough treatment we refer to the standard literature [23]. Electron diffraction is based upon the wave nature of the electron and on the ordered arrangement of the atoms in a crystal. For X-rays the phenomenon of diffraction was first shown by v. Laue and it was described by the Bragg equation

$$2d \sin \theta = n\lambda \tag{1}$$

2.2 SOME BASIC CONSIDERATIONS OF DIFFRACTION

This equation describes diffraction as a reflection from two parallel semitransparent planes (Fig. 2.1). It represents the condition for directional intensity enhancement of a reflected plane wave with wave length λ from a sequence of reflecting planes with distance d. Θ is the angle between the direction of the incident wave and the reflecting plane. This equation for X-rays can be applied to electrons as well, because according to De Broglie the latter can be considered as plane waves. Electron energy E and wave length λ are related to each other approximately by

$$\lambda[\text{Å}] = \sqrt{150/E\,[eV]}^{**} \quad (2)$$

Only for such angles of incidence, which fulfill the Bragg condition, is the plane wave reflected from the next lattice plane in phase with the first one (see Fig. 2.1), leading to the intensity enhancement. In the Bragg equation d represents the distance between the reflecting lattice planes. The lattice planes are characterized by the Miller indices, which correspond to the reciprocal intercepts of a plane with the crystal axes.

The Miller indices can be considered as the coordinates of a vector in the reciprocal space of the crystal lattice, which is perpendicular to the lattice plane,

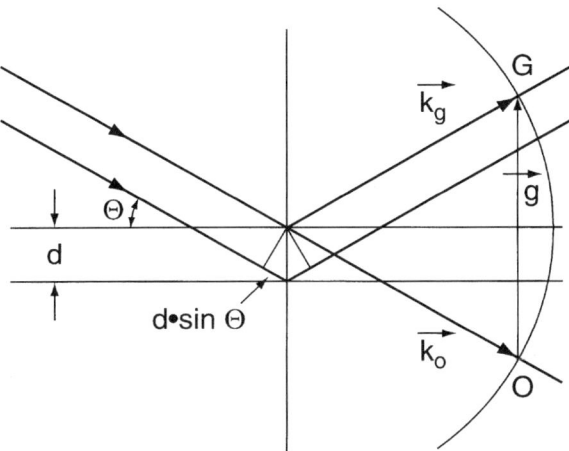

Figure 2.1. Sketch of a plane wave being reflected from a set of lattice planes with distance d, in order to explain the Bragg-equation. The wave vectors of incident and reflected beam, $\vec{k}_0$ and $\vec{k}_g$, are also shown together with the vector $\vec{g}$ of the reciprocal lattice.

**More accurately the wave length is given by the relativistically corrected formula [24]–[26]:

$$\lambda[\text{Å}] = 12.264(\sqrt{E[eV]}\,\sqrt{1 + 0.9785\,10^{-6}E[eV]})^{-1} \quad (2a)$$

For low energies, the order of 100 eV, the correction can be neglected.

with a magnitude that is reciprocal to the separation of the planes. Under this assumption all possible planes of a crystal lattice form the reciprocal lattice. Each set of lattice planes corresponds to a set of points in the reciprocal space, forming the reciprocal lattice. Now, the diffraction pattern can be understood as an intersection of a sphere with radius $1/\lambda$, the so-called Ewald sphere, with the reciprocal lattice. This is also shown in Figure 2.1 with vectors in the reciprocal space, the wave vector $\vec{k}_0$ in the direction of the incident plane wave with length $1/\lambda$, the wave vector $\vec{k}_g$ in the direction of the reflected wave with the same length and the reciprocal lattice vector $\vec{g}$, perpendicular to the reflecting lattice planes with a length equal $1/d$, the reciprocal lattice distance. The vector equation

$$\vec{k}_g = \vec{k}_0 + \vec{g} \qquad (3)$$

as indicated in Figure 2.1 is equivalent to the Bragg condition (1) with $|\vec{g}| = 1/d$ and

$$|\vec{k}_0| = |\vec{k}_g| = 1/\lambda$$

since

$$\sin \Theta = 1/2 \, |\vec{g}|/|\vec{k}_g| \qquad (4)$$

demonstrating the Ewald sphere construction. The center of the Ewald sphere is the origin of the wave vector $\vec{k}_0$ of the incident wave, which points to the zero point of the reciprocal lattice. The intersection circle of the Ewald sphere passing through the origin and the reciprocal lattice point G is indicated. All intersection points of the Ewald sphere with the reciprocal lattice indicate possible diffraction directions.

If the reciprocal lattice points would be only mathematical points as for an infinite perfect crystal, the diffraction condition would be very sharp and only very few points would be intersected by the Ewald sphere. However, in practice the reciprocal lattice points are broadened by the shape transform of the crystal as was pointed out by v. Laue [27]. The broadening is also reciprocal to the dimensions of the crystal. For a thin crystal plate the reciprocal lattice point is elongated in the direction perpendicular to the plate. This has consequences for the shape of the reciprocal lattice of an infinitely thin crystal lattice (i.e., a surface lattice). The reciprocal lattice of this two-dimensional lattice consists of periodically arranged rods perpendicular to the surface. The intersection of the Ewald sphere with such an extended area of reciprocal lattice points indicates the possible reflections. Here, it should be mentioned that the Ewald sphere itself is also not sharp because of the energy spread of the electrons and the finite illuminating aperture of the electron beam, which corresponds to slightly

2.2 SOME BASIC CONSIDERATIONS OF DIFFRACTION

different lengths and directions of the wave vector $\vec{k}_0$ of the incident beam. Both energy spread and finite aperture would result in a set of spheres passing through the origin of the reciprocal lattice. Furthermore, the reciprocal lattice points are also broadened by imperfections of the crystal.

This qualitative description of diffraction should help in understanding the principle of diffraction from a surface. In geometrical terms it means, as already stated, that the diffraction pattern is understood as a section through the reciprocal lattice. It is true only for the position of the diffraction spots, not for their intensities. The latter have to be estimated by more sophisticated considerations (e.g., by the application of the kinematical theory or more exactly by dynamical theories using the potentials of the scatterers). This process is described in standard literature [25].

2.2.1 LEED and RHEED

Depending on the energy of the electrons, we consider low energy electron diffraction (LEED) to be that performed with electrons of several tens of eV and high energy electron diffraction (HEED) to be that where the electrons have several thousand eV energy. While LEED is a reflection technique with nearly perpendicular incidence of the electrons to the surface, we have to distinguish in HEED between transmission and reflection. For transmission high energy electron diffraction (THEED) the crystals have to be thin, on the order of several hundreds of Ångstroms. The investigation of bulk material, however, as an electrode, can only be done by the reflection high energy electron diffraction (RHEED) technique. In LEED the direction of the incident beam is usually perpendicular to the surface, while for RHEED the incidence is almost grazing, only one or a few degrees to the surface. This distinction has consequences for the Ewald construction of the diffraction pattern because of the different sizes of the Ewald spheres for LEED and RHEED due to the different wave lengths. The wave length of the low energy electrons is of the order of Å, but for the high energy electrons (i.e., 57 keV) it is of the order of $\frac{1}{20}$ Å, which means, the Ewald sphere for RHEED is 20 times larger than that for LEED. The different geometrical constructions of diffraction patterns with the Ewald sphere for LEED and RHEED are shown in Figure 2.2.

From such considerations diffraction conditions can be found under which the intensity of diffracted beams is most sensitive to the surface topography, the surface structure, and the arrangement of the atoms in the surface [28]. In an electron microscope such a beam can be used for imaging surface structures of the order of 10–100 Å [29], as will be shown. This capability can be of great interest since ordinary diffraction gives only averaged information from areas with a size depending on the diameter of the electron beam on the surface. This size is of the order of 100 μm.

In LEED we observe a vertical section through the rods of an area in the reciprocal space (Fig. 2.2a) that depends on the radius of the Ewald sphere and hence, on the energy. In RHEED, however, we see only a nearly tangential

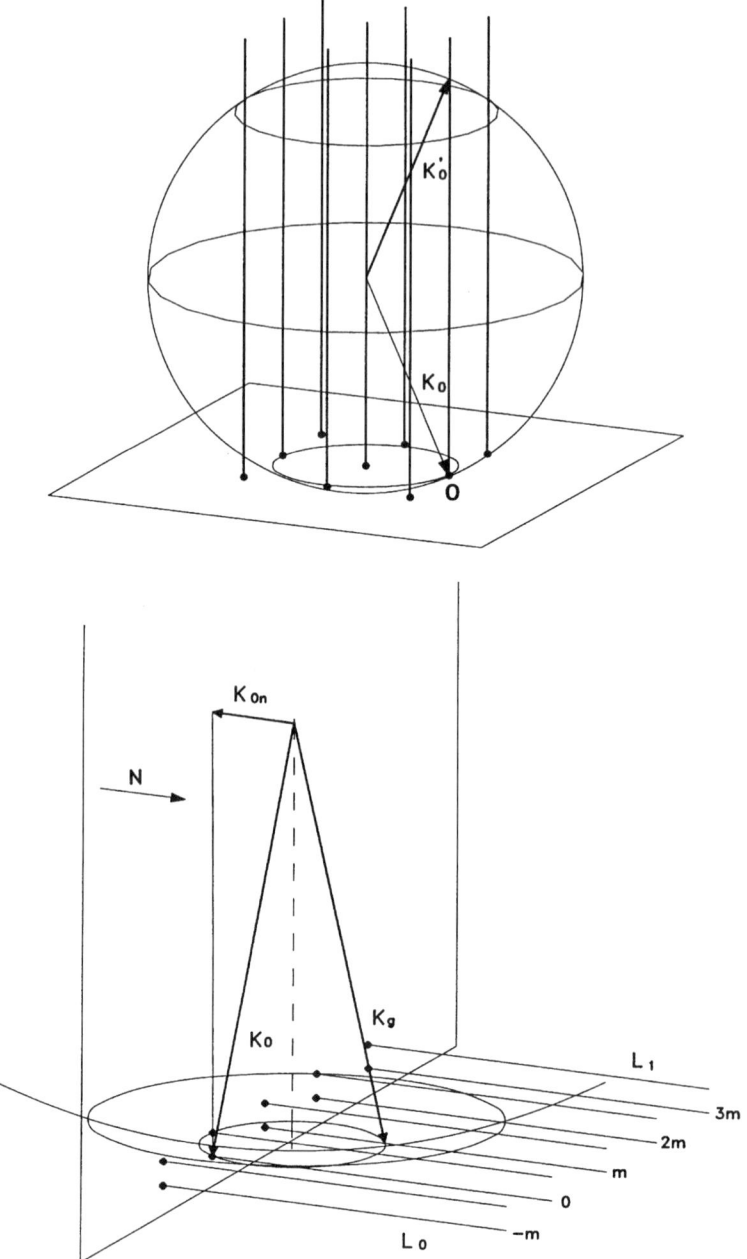

Figure 2.2. Reciprocal lattices with Ewald spheres for (a) LEED and (b) RHEED. In RHEED the rods for two Laue-zones, L_0 and L_1, are indicated. Diffracted beams are excited when the Ewald sphere intersects the rods. k_{0n} is the normal component of the wave vector k_0. In LEED k_0' is the mirror image of the incident beam k_0. In LEED patterns this is the so-called zero beam.

2.2 SOME BASIC CONSIDERATIONS OF DIFFRACTION

section through a reduced number of rods. In order to record sections through other rods, the azimuth of the incident beam has to be changed by rotating the crystal about an axis perpendicular to the surface. In a RHEED pattern, rods of different Laue zones can be seen as schematically indicated in Figure 2.b.

While LEED gives only information on the uppermost surface layers, one can obtain by RHEED also information from the bulk. The information depth in RHEED is approximately five or more layers. Surface roughness can be transmitted because of grazing incidence of the electron beam and so additional information on the bulk and the roughness itself can be obtained, which is not possible in LEED: The quality of the crystal can be seen from so-called Kikuchi patterns [26]. The methods complement each other. Since LEED requires UHV, one would expect that LEED cannot be applied to surface studies of an emersed electrode because a LEED pattern is not observed from a newly inserted specimen unless a cleaning procedure has been done. However, by well-defined electrochemistry on an electrode surface that had been cleaned by UHV treatment, the sample remains clean, except for electrolyte residues, allowing the observation of a LEED pattern without any further treatment. This result was unexpected and very surprising. Adsorbates can form ordered superstructures on the crystal surfaces. Such a superstructure leads to additional spots in the diffraction pattern. Examples of superstructures are shown in Figure 2.3. The diffraction pattern is the Fourier-image of the surface structure and hence, has to be inverted to an image in real space. The LEED and RHEED analyses by geometrical interpretation give information on the arrangement of the atoms in the surface and on the dimensions of the superstructure cell, but not on the

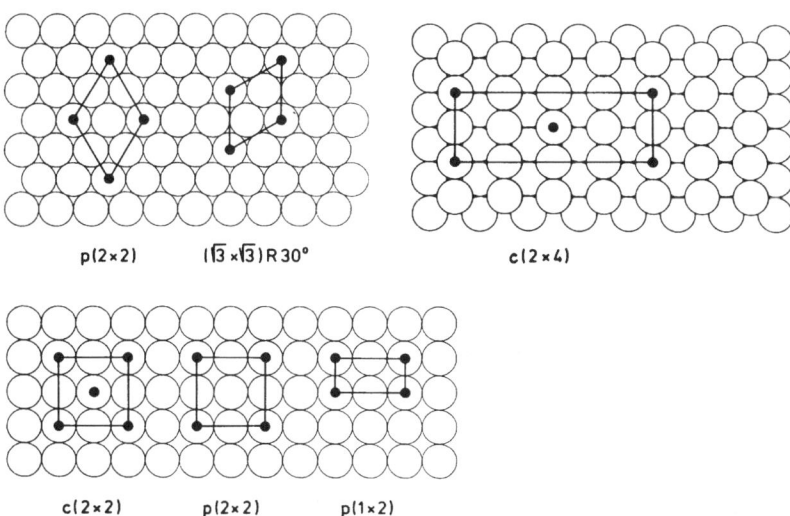

Figure 2.3. Typical arrangements of adatoms on the three low-index faces of an fcc metal, which lead to superstructures. The corresponding LEED notation is also shown.

exact position of the adatoms with respect to the substrate surface. Here, only an analysis of the intensity of the diffraction spots as function of the energy of the electrons allows the determination of the exact position of these atoms, as was demonstrated in a pioneering investigation by Forstmann et al. [30]. Additional information on the chemical nature of atoms is obtained by Auger electron spectroscopy (AES) [23].

2.3 EXPERIMENTAL

2.3.1 Diffraction

A schematic diagram of the setup for a combined LEED and RHEED investigation is shown in Figure 2.4. The specimen is mounted on a bakeable goniometer stage in front of the LEED system. The latter consists of an electron gun, and a spherical fluorescent screen with a set of grids at different potentials, which also allows retarding field Auger-electron spectroscopy (AES) to be performed. The electron beam is vertical to the specimen surface. The specimen and the first grid are grounded while the screen is at a high positive potential of several kV. The diffracted electrons are accelerated by this potential to produce a bright

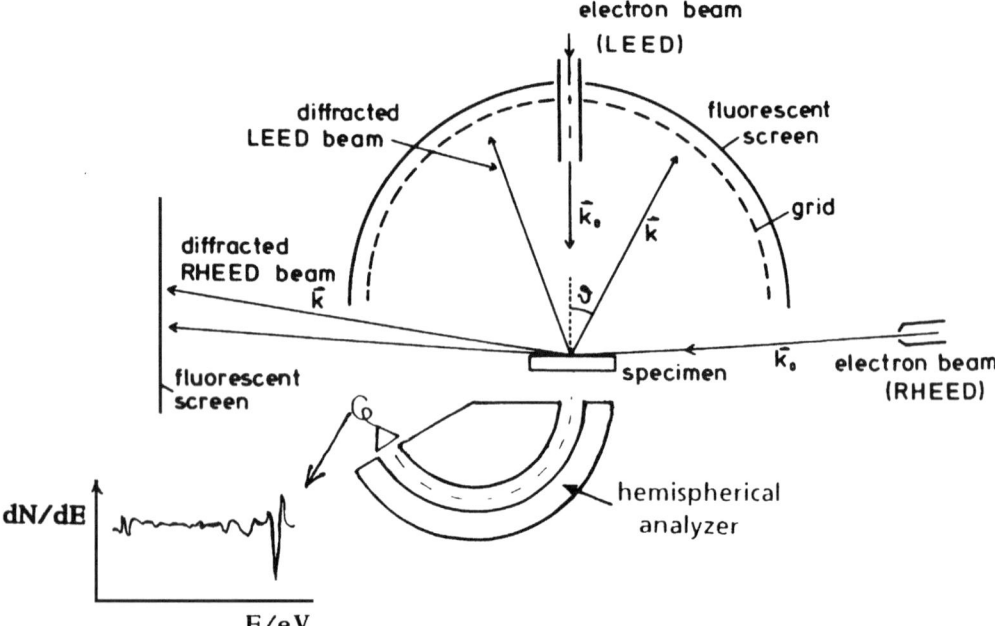

Figure 2.4. Schematic diagram of a combined LEED, RHEED, and AES setup for surface studies [22]. k_0 and k are the wave vectors of the incoming and the diffracted beam, respectively.

2.3 EXPERIMENTAL 65

spot on the fluorescent screen. The LEED pattern is normally viewed from the front (the center being blocked by the specimen); it can also be recorded from the back if a transparent spherical screen is employed (back-view LEED). The RHEED system consists of an electron gun with condenser lens and a transparent fluorescent screen. The electron beam of grazing incidence can be focused onto the specimen or onto the screen. The diameter of the focused beam on the specimen is of the order of 50 μm. Due to grazing incidence the illuminated specimen area is elongated in beam direction depending on the angle of incidence. The diffraction patterns can be recorded photographically from outside the vacuum chamber through a view port. Simultaneously with RHEED observation an Auger spectrum can be recorded from the same surface area via the LEED optics. For a higher surface sensitivity of the Auger spectroscopy a hemispherical analyzer in combination with RHEED can be used after the specimen has been rotated 180 degrees. The exposure time for RHEED patterns is of the order of seconds and that for LEED of the order of one minute. The diffraction patterns can also be recorded with video techniques [31].

2.3.2 Specimen Treatment

The samples are mounted in a bakeable UHV chamber, which contains various tools for surface treatment, such as an ion gun for argon ion bombardment; a gas inlet system for argon, oxygen, or other gases; a quartz microbalance; and a quadrupole mass spectrometer and pressure control. The chamber is connected by a gate valve with another UHV chamber for the electrochemical experiments under controlled conditions [32–36]. The specimen is transferred by a mechanically and magnetically coupled transfer system from one chamber to the other. For the electrochemical experiments the electrochemical chamber is flooded by pure argon gas up to atmospheric pressure. A sketch of the chamber and the electrochemical cell is shown in Figure 2.5. The electrolyte vessel is on top of a glass capillary through which the electrolyte is filled by controlled argon gas pressure. The glass capillary is clamped in a teflon sealing, which allows rotation and axial movement indicated by arrows in Figure 2.5, to bring the electrolytic cell through the gate valve into the electrochemical chamber. The working electrode is a single crystal disc, held by two isolated tungsten wires. By controlled operations the electrolyte is brought into the vessel until it forms a meniscus above the edge of the cell, and then the working electrode is brought into contact with the meniscus by movement of the transfer rod and the glass capillary.

After the electrochemical treatment the level of the electrolyte is lowered and with the axillary capillary, remaining droplets in case of wet emersion are removed. The glass tube with the electrolytic cell is then pulled down through the gate valve and the electrochemical chamber is evacuated. At a pressure of 10^{-7} torr, which may be reached within 10 minutes, the specimen is transferred to the main chamber. Now, LEED and RHEED observations of the surface can

66 ELECTRON DIFFRACTION AND ELECTRON MICROSCOPY

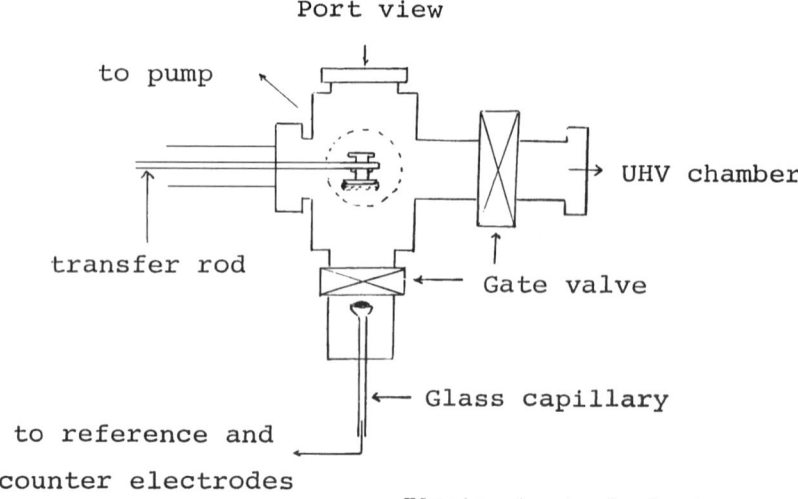

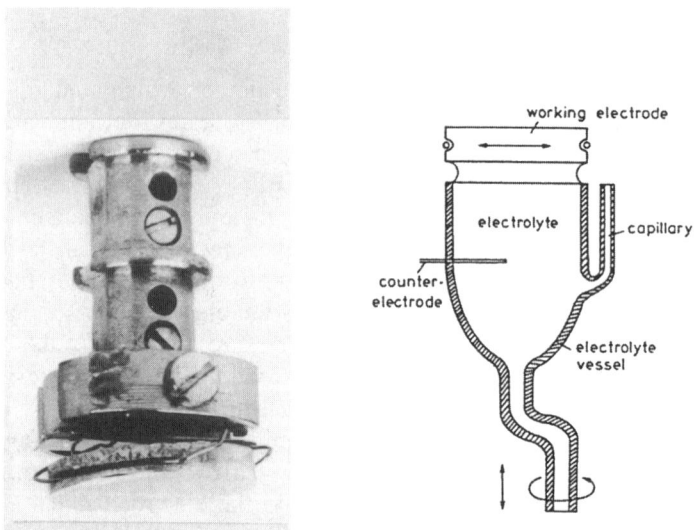

Figure 2.5. Electrochemical cell and specimen holder [22].

start at a pressure of 10^{-9} torr. In former experiments the whole vacuum chamber had to be flooded for inserting a new specimen. Now, specimen exchange can be carried out by an airlock system via the electrochemical chamber.

Before electrochemical experiments the single crystal surfaces are cleaned by several cycles of argon ion bombardment and subsequent annealing. Their cleanliness is checked by AES and the crystal quality and the ordering of the crystal surface is characterized by LEED and RHEED.

2.3 EXPERIMENTAL

2.3.3 Surface Microscopy

Electron diffraction gives an averaged information about a surface area, determined by the diameter of the electron beam, as already mentioned. For spatial resolution electron microscopy has to be applied using a diffracted beam. Under certain diffraction conditions, the intensity of a diffracted beam is most sensitive to the surface structure. Using such a beam the surface topography can be imaged. The principle of imaging in reflection is shown in Figure 2.6. The technique is called reflection electron microscopy (REM) [37]. The specimen has to be mounted on a goniometer stage and the direction of the incident beam has to be tilted by several degrees. The adjustment of crystal orientation and beam tilt should allow the Bragg-diffracted beam to follow the optical axis of the microscope. Such investigation should be done under UHV conditions, like the diffraction experiments, in order to avoid specimen contamination. Particularly

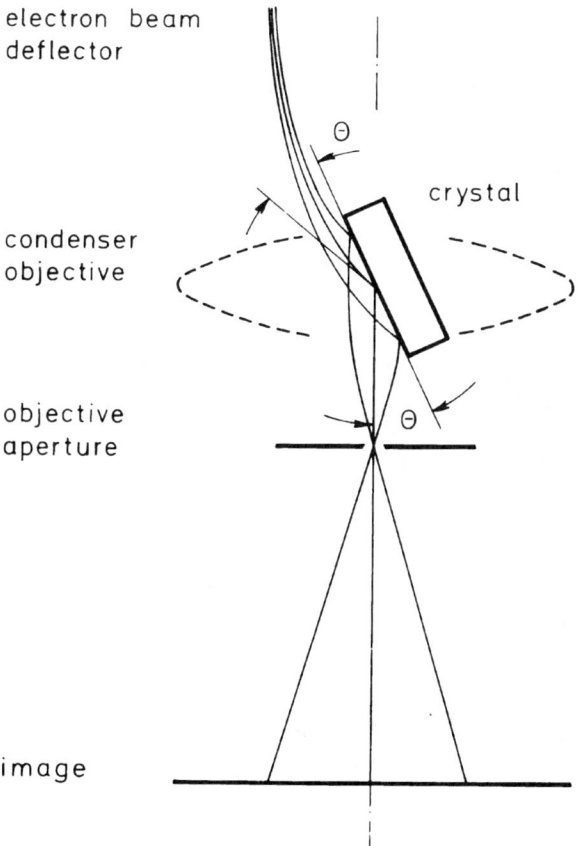

Figure 2.6. Schematic diagram for reflection imaging in an electron microscope (REM).

hydrocarbons on the surface from the residual gas will interact with the electron beam, obscuring the surface structure during observation. However, the number of truly-UHV electron microscopes worldwide is very small, and it may be sufficient when working with conventional high-vacuum equipment to ensure clean conditions at least in the vicinity of the specimen. This is achieved by an anticontamination shield around the specimen and a cryogoniometer stage at temperatures of liquid nitrogen, which acts as a local pump.

If REM is done at conditions under which the Bragg reflection for imaging shows an intensity enhancement, a bright surface image is obtained and surface imperfections that lead to small deviations from this condition (e.g., lattice-deformation, atomic steps, surface reconstruction, or adsorbates) can be detected with high contrast. A disadvantage of REM, however, is the reduced resolution due to foreshortening, because the surface is imaged at a very small angle (typically two or three degrees). A somewhat higher resolution is achieved in transmission mode (TEM), but this mode requires thin crystalline specimens [29].

Surface imaging can also be performed with low energy electrons in reflection mode, using a diffracted LEED beam, which stems from the surface structure under consideration. This technique, which is called LEEM (low energy electron microscopy), was introduced by Bauer and Telieps [38], but so far has not been applied to the study of emersed electrodes. Another surface imaging technique for UHV conditions, which is called "scanning RHEED," should also be briefly mentioned [39]. A small electron beam is scanned across the surface and the image is formed by the intensity of a Bragg reflection. The crystal has to be oriented, as in the case of REM, in such a way that the intensity of the Bragg reflection is most sensitive to the surface structure. The resolution is limited by the electron beam size.

At this point it may be appropriate to compare briefly the ex situ electron microscopy with in situ STM and AFM, which have been developed to most powerful tools for imaging electrode surfaces. The latter give structure information at very high magnification, but only on the uppermost layer of the surface, and the area of observation is rather restricted if atomic resolution is required. This makes it difficult to identify superstructures with large unit cell. Hence, REM in combination with diffraction, although under ex situ conditions, can give valuable additional information on the bulk conditions and on surfaces because of a wide-angle view, which is often helpful for a reliable STM image interpretation.

2.3.4 Specimen Preparation for Imaging

For diffraction studies the size of the single crystal electrodes is not critical. They are typically crystal discs of approximately 2 mm thickness and a diameter of 5 to 10 mm. Their cylindrical surface has small grooves, so the sample can be clamped between two tungsten wires for heating (see Fig. 2.5). Such specimens are cleaned by Ar-ion bombardment with subsequent annealing or by the so-called "flame-treatment" with a Bunsen burner and subsequent cooling in pyrolytic water [40, 41].

For electron microscopy, however, the specimens have to be much smaller, of the order of one mm. They were prepared by melting one end of a wire of 0.2 mm diameter by electron beam heating, or in the case of platinum or gold in a hydrogen-oxygen flame [42], the resulting droplet then being allowed to solidify to a sphere of approximately 0.5 mm diameter. Such a sphere often consists of only one crystal with fairly large (111) facets and smaller (100) facets. Although such a sphere acts like a polycrystalline electrode in an electrolyte, the individual facets are suitable for REM studies of single-crystal surfaces.

The evaporation of gold onto glass at elevated temperatures is yet another technique of single crystal electrode preparation [43]. The smooth gold film consists of small crystallites with (111) surfaces parallel to the glass plate, but randomly oriented. If the gold films are thin enough, they can be floated off the glass after electrochemical treatment and their surface structure can be studied in transmission mode in the electron microscope. It is the only exception when TEM is used for the investigation of electrode surfaces.

2.4 CASE STUDIES

2.4.1 Electron Diffraction

2.4.1.1 Cu Underpotential Deposition on Au(111) Deposition of a metal onto a foreign metal substrate (e.g., Cu on Au) often starts with the formation of a monolayer at potentials that are clearly positive of the respective Nernst potential for bulk metal deposition [44]. This so-called underpotential deposition (upd) simply reflects the fact that, in the case of Cu on Au, the interaction between Cu and Au exceeds that between Cu and Cu [45]. From the cyclic voltammograms of Cu upd on Au single crystal electrodes, which generally show monolayer formation to occur in several, energetically separated steps, electrochemists concluded already in the mid-1970s that ordered adsorption must take place in the upd process [46]. The multipeak structure in the cyclic current-potential curves of upd is particularly pronounced for Cu deposition onto Au(111) in sulfuric acid solutions. Such a curve is shown in Figure 2.7, where monolayer formation in two, energetically well-separated steps is clearly seen. In those days, ex situ electron diffraction was the only means of proving beyond doubt the existence of ordered adsorption (e.g., by observation of so-called superstructure reflexes in LEED or RHEED). The ex situ route for the study of electrode surfaces and strongly bound adlayers had been systematically exploited first by Hubbard in the early 1970s and was further explored several years later by a number of research groups including those of Winograd [4], Yeager [2], Kolb [12, 13], and Ross [32]. Underpotential deposition in general and Cu on Au(111) in particular were intensively investigated at the Fritz-Haber-Institut [13, 44, 48] and the latter system later became the

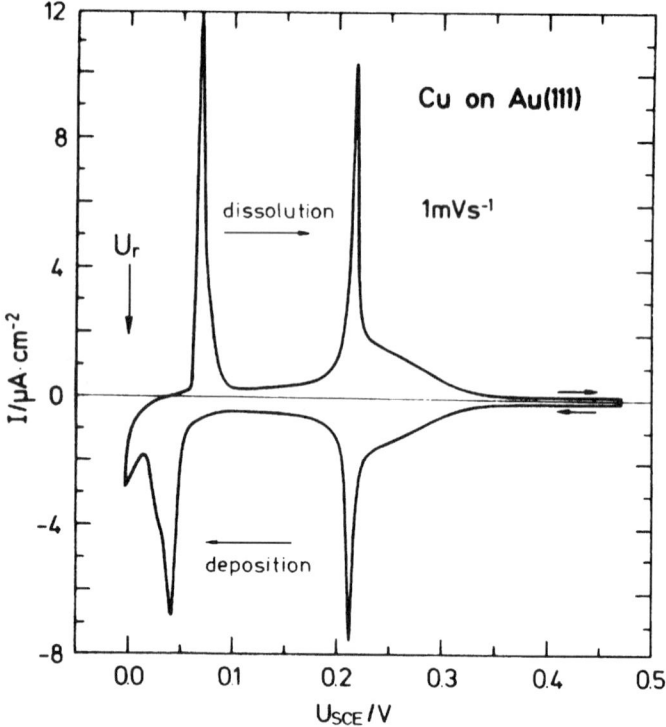

Figure 2.7. Cyclic current-potential curve for a Au(111) electrode in 0.05 M H_2SO_4 + 1 mM $CuSO_4$. Scan rate: 1mV/sec [47].

test object for almost every technique newly introduced in physical electrochemistry.

Before the electrochemical experiment, the bare Au(111) surface was prepared in the UHV chamber by several cycles of Ar-ion bombardment and subsequent annealing at about 500°C. Single crystallinity and cleanliness were tested by LEED or RHEED and AES. As will be discussed in more detail in Section 2.4.1.4, the clean Au(111) surface is reconstructed and shows a (1×23) superstructure in LEED [49–51]. It means the atoms in the uppermost layer are compressed in [110] direction by about 4%, so that every 24th surface atom is in registry with every 23rd bulk atom. Figure 2.8 shows LEED and RHEED patterns of clean reconstructed Au(111). The six superstructure spots around each fundamental spot in the LEED pattern (Figure 2.8), indicate the existence of three domains, each domain producing a pair of superstructure spots around the substrate spots. In the RHEED pattern the diffraction spots appear as long streaks, representing the intersection of the Ewald spherer with the rods of the 0th Laue zone, as shown in Figure 2.2b. If the rods would be very sharp due to a large perfect surface area, one would expect sharp diffraction spots even in RHEED. However, the long streaks in Figure 2.8 indicate

2.4 CASE STUDIES

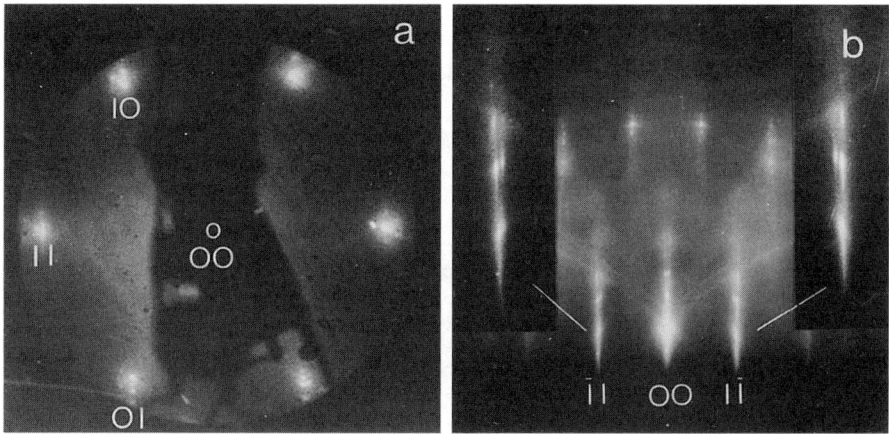

Figure 2.8. (a) LEED and (b) RHEED patterns of a clean, reconstructed Au(111) surface with (1 × 23) superstructure. The reflections are indexed by only two indices due to a two-dimensional lattice. In (b) the inserts show the superstructure streaks on an enlarged scale. RHEED pattern in [112] azimuth. The electron energy is 53 eV for LEED and 40 keV for RHEED [13].

that the surface is not perfectly ordered. Reflections due to the intersection of the Ewald sphere with the rods of the 1st Laue zone can also be clearly seen. The RHEED pattern is indexed in the very same way as the LEED pattern.

When the Au(111) electrode after potential cycling in a Cu^{++} containing sulphuric acid electrolyte was emersed in the potential region between the two prominent current peaks (see Figure 2.7), e.g., at 0.13 V vs. SCE, and transferred into the UHV chamber, LEED and RHEED patterns could be observed without any further surface treatment at a base pressure of about 10^{-9} torr. The corresponding LEED and RHEED patterns are shown in Figure 2.9. In LEED superstructure spots due to an ordered adlayer are clearly seen. Superstructure streaks are also visible in RHEED in the 0th and in the 1st Laue zone for the [112] azimuth. Superstructure streaks are not visible for the [110] azimuth in the 0th Laue zone, but they are in the two sub-Laue zones L10 and L20. The (1 × 23) superstructure has disappeared.

The superstructure due to the adlayer at medium Cu coverages has been identified as a $(\sqrt{3} \times \sqrt{3})R30°$ structure. It means the adlayer has also a hexagonal structure, but its unit cell is rotated by 30° with respect to the substrate unit cell, and it has unit cell vectors that are $\sqrt{3}$ times larger than those for the substrate as shown in Figure 2.10. Although the real adlayer structure could not be unequivocally determined on the basis of the diffraction patterns alone (the $\sqrt{3}$ structure can be produced by ordered Cu layers of either $\frac{1}{3}$ or $\frac{2}{3}$ ML coverage), a coverage estimate by AES led to the conclusion that the Cu adlayer has an honeycomb structure with an ideal coverage of $\frac{2}{3}$ ML [52]. Later this interpretation was questioned by in situ STM measurements, which clearly showed an ordered

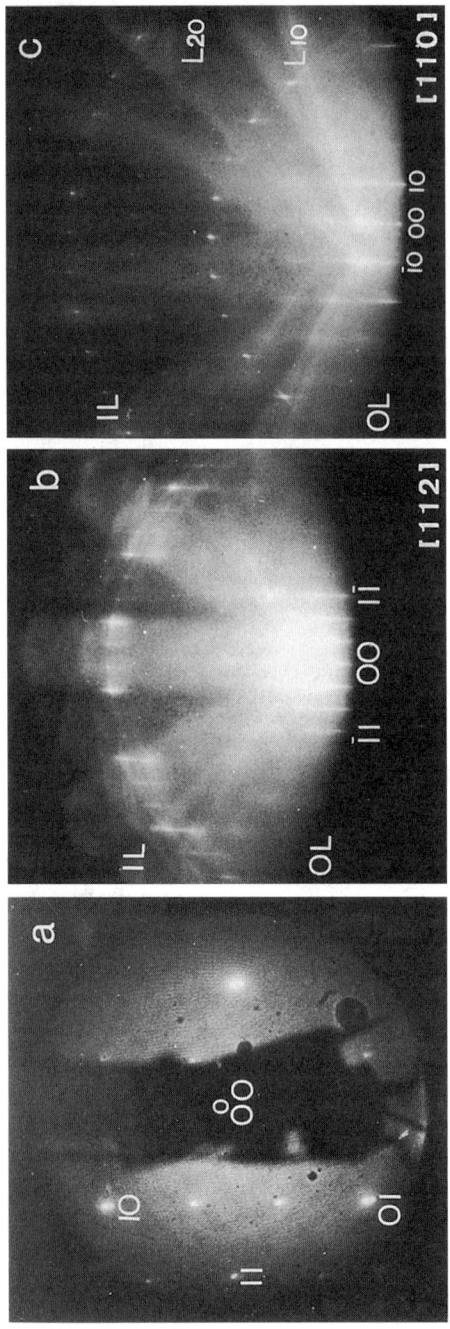

Figure 2.9. (a) LEED and (b, c) RHEED patterns of a Au(111) electrode covered with half a monolayer of Cu, showing a $(\sqrt{3}\times\sqrt{3})$R30° superstructure. In RHEED the superstructure is seen for the [11$\bar{2}$] azimuth in the 0th and the 1st Laue zone (b), while for the [1$\bar{1}$0] azimuth (c) it is visible only in the sub-Laue zones L10 and L20 [13].

2.4 CASE STUDIES

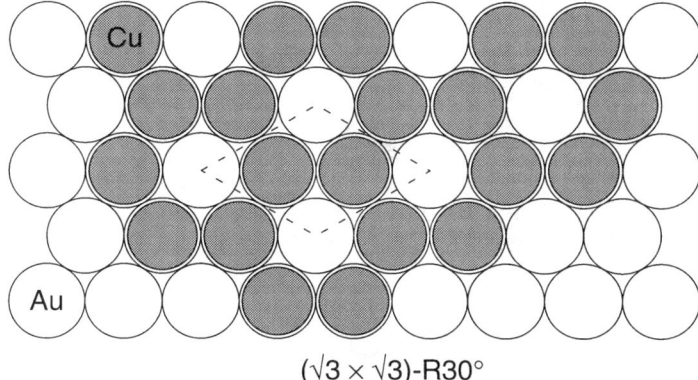

$(\sqrt{3} \times \sqrt{3})$-R30°

Figure 2.10. Proposed model for the ordered adsorption of Cu on Au(111), corresponding to the $(\sqrt{3} \times \sqrt{3})R30°$ superstructure [52].

adlayer with $\frac{1}{3}$ ML coverage [53]. Model calculations by Blum et al. [54, 55] and in situ X-ray measurements [56] finally gave a decisive answer: The Cu adatoms at medium coverages indeed form a honeycomb structure with $\frac{2}{3}$ ML limiting coverage. This rather unusual adatom structure is stabilized by coadsorbed sulfate ions nested inside the Cu hexagons. A careful coverage determination by Lipkowski et al. [57] based on charge measurements yielded $\frac{2}{3}$ ML for Cu and $\frac{1}{3}$ ML for the anion, thus confirming the hexagon structure of the Cu adlayer.

When the electrode was emersed with a full monolayer at 0.05 V vs. SCE, the $(\sqrt{3} \times \sqrt{3})R30°$ superstructure was still visible instead of the expected (1×1) structure, which was later confirmed by in situ STM [58]. This result may be due to the fact that a small fraction of the Cu monolayer is lost during emersion. Stripping curves after the diffraction experiments have indicated a loss of 10–20% of the full Cu monolayer.

It was found by ex situ AES that copper UPD is accompanied by substantial anion (HSO_4^- or SO_4^{--}) coadsorption, which ultimately is responsible for the $(\sqrt{3} \times \sqrt{3})R30°$ superstructure formation. Coadsorption of anions during copper UPD on polycrystalline Au has also been reported by Horanyi [59] in his radio tracer work and late by Varga et al. [60], and it has been nicely confirmed for single crystal surfaces by in situ SEXAFS experiments [61]. The influence of anion coadsorption in Cu UPD has been systematically studied using ex situ LEED and RHEED [62], demonstrating the structure-determining role of the supporting electrolyte. The superstructure patterns due to the ordered adlayer formed at medium Cu coverages in sulphate, perchlorate, or chloride solutions differ markedly from each other [48] and are altogether substantially different from the Cu monolayer growth mode during vacuum deposition [13]. While the $(\sqrt{3} \times \sqrt{3})R30°$ structure was observed exclusively in sulfate-containing electrolytes, an incommensurate (2.2×2.2) structure together with a rectangular $(\sqrt{3} \times 2)$ structure was obtained in perchlorate solution (Fig. 2.11). Because

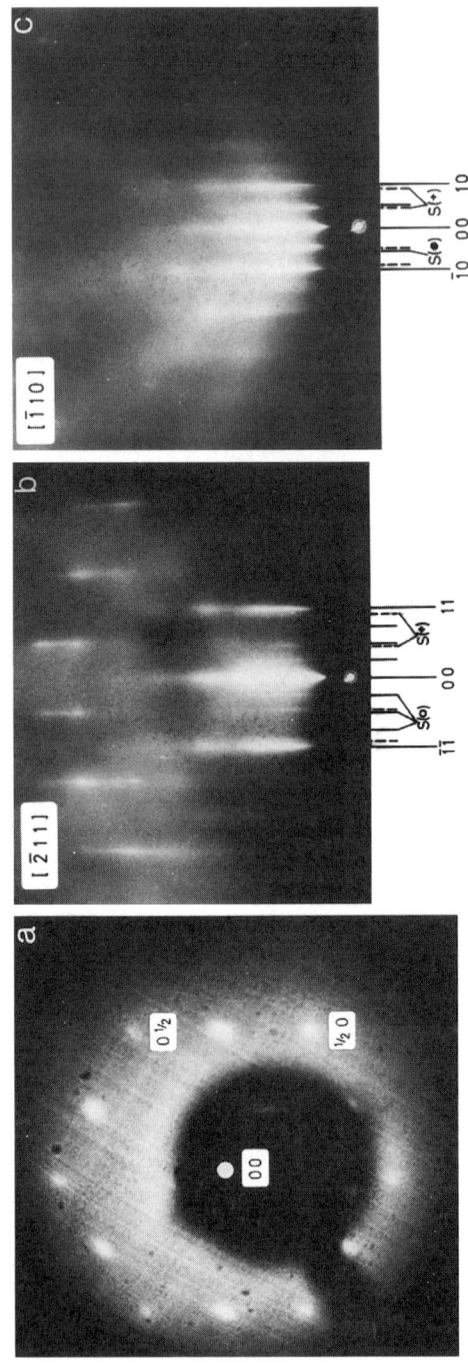

Figure 2.11. (a) LEED (22 eV) and (b, c) RHEED (40 keV) patterns of a Au(111) electrode covered with half a monolayer Cu, deposited from perchlorate solution. The superstructure spots are indicated [62].

2.4 CASE STUDIES

a similar adlayer structure was seen by in situ STM for chloride-containing electrolytes, it has been argued in the literature that the (2.2 × 2.2) structure reported for perchlorate solutions is due to Cl⁻ impurities [63]. A final decision on the correct structure of the Cu adlayer in perchloric acid is still awaited. The arrangement of Cu adatoms on Au(111) at medium coverages in sulfuric and perchloric acid, as derived from ex situ electron diffraction experiments, is shown in Figure 12.2, together with that for Cu evaporated on Au(111) in vacuum, which yields pseudomorphic island growth from the very beginning. In Figure 2.13 the cyclic voltammogram for Au(111) in 0.1 M $HClO_4$ + 1 mM $Cu(ClO_4)_2$ is shown, which demonstrates that the anions also have a profound influence on the deposition kinetics [64]. It reveals an extremely slow Cu deposition kinetics in perchloric acid solutions, which reacts sensitively to traces of chloride impurities by a drastic increase of the deposition rate.

2.4.1.2 Underpotential Deposition of Cu on Pt(111) and Pt(110) While the underpotential deposition of Cu on gold single crystal electrodes has been extensively investigated [see, e.g., 13, 46–48, 53–58], relatively few studies

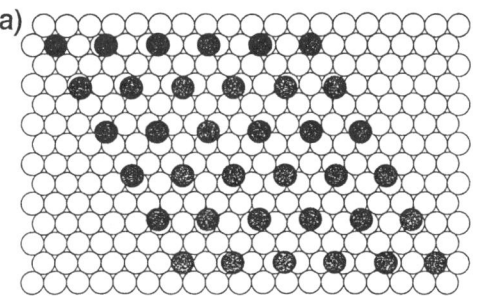
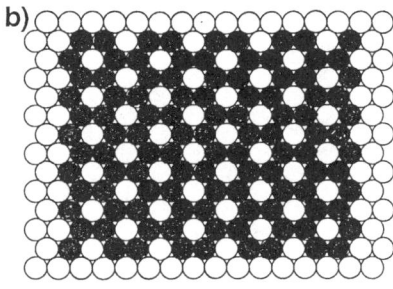
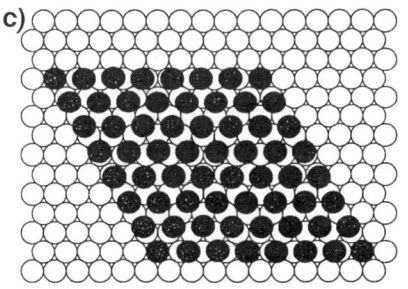
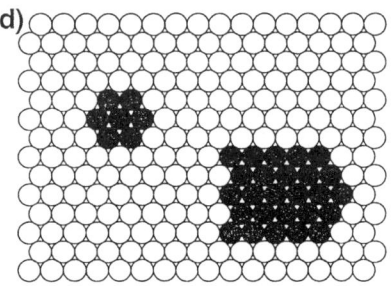

Figure 2.12. Structures proposed for Cu adlayers on Au(111) at medium coverages. (a) 2.2 × 2.2); deposition from perchlorate solution. (b) ($\sqrt{3} \times \sqrt{3}$)R30°; sulfate. (c) (1.29 × 1.29); chloride. (d) (1 × 1) island growth; vacuum deposited [48].

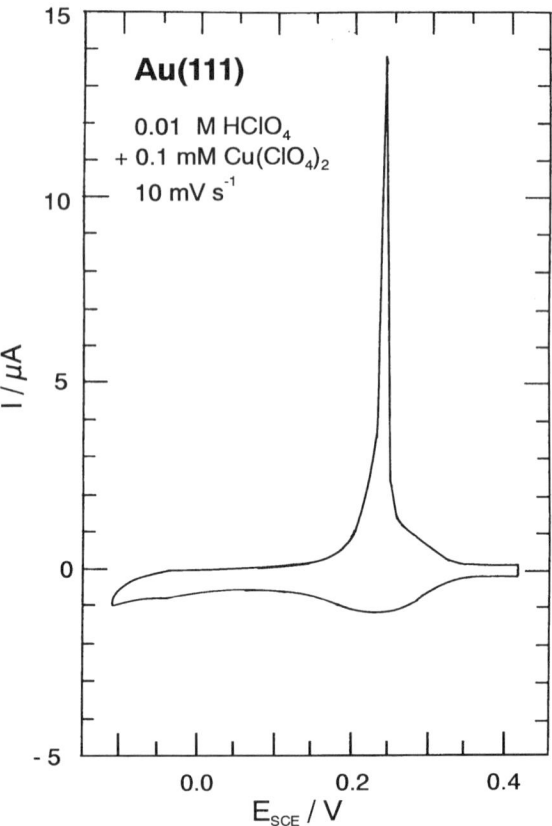

Figure 2.13. Cyclic current-potential curve for Au(111) in 0.01 M HClO₄ + 1 mM Cu(ClO₄)₂. Scan rate: 20 mV/sec [64].

were reported for Pt substrates [65, 68, 71–73], presumably due to difficulties encountered with preparation and handling of clean and structurally well-defined Pt surfaces in an electrochemical environment.

A pronounced effect of Cl$^-$ coadsorption on the Cu UPD onto Pt(111) has been observed [65], which is similar to that for Cu UPD on Au(111) [66, 67]. The cyclic current-potential curve for Cu UPD on Pt(111) in sulfuric acid solution is markedly changed with the addition of Cl$^-$, giving rise to two very sharp and well-separated voltammetric peaks (Fig. 2.14). The LEED pattern of the Pt(111) electrode, emersed at +0.4 V vs. Ag/AgCl (i.e., between the two dissolution peaks), showed a (4 × 4) superstructure (Fig. 2.15), which was assigned on the basis of XPS measurements for the Cu deposit to an ordered Cl adlayer on a full (1 × 1) Cu monolayer on Pt(111) [65]. Recent work by Tidswell et al. [68], employing a rotated ring-disk Pt(111) electrode, however, suggested that the Cu coverage is only about $\frac{1}{2}$ monolayer at that potential in the presence of chloride. Consequently the (4 × 4) structure with an interatomic spacing of

2.4 CASE STUDIES

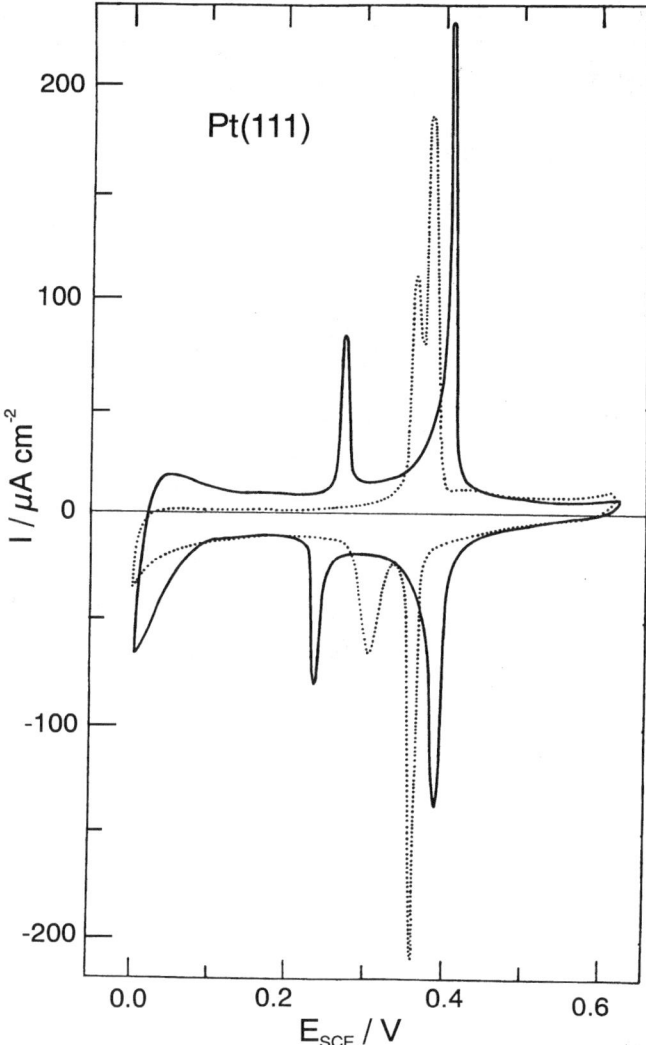

Figure 2.14. Cyclic current-potential curves for Pt(111) in 0.05 M H_2SO_4 + 1 mM $CuSO_4$, with (—) and without (----) addition of 0.1 mM HCl. Scan rate: 5 mV/s.

0.37 nm, which is in agreement with data from in situ anomalous X-ray diffraction [68] and in situ STM [66] and which was formerly attributed to the Cl-Cl spacing in the chloride adlayer, was now assigned to the Cu spacing in an Cu-Cl bilayer [68]. Unfortunately, the Cu-Cu distance underneath the Cl-adlayer cannot be determined by STM [66]. Recent work by Bludau et al. [69] using a full-dynamical LEED-intensity analysis of the (4 × 4) structure for the Pt(111) electrode emersed at a potential between the two voltammetric peaks suggested

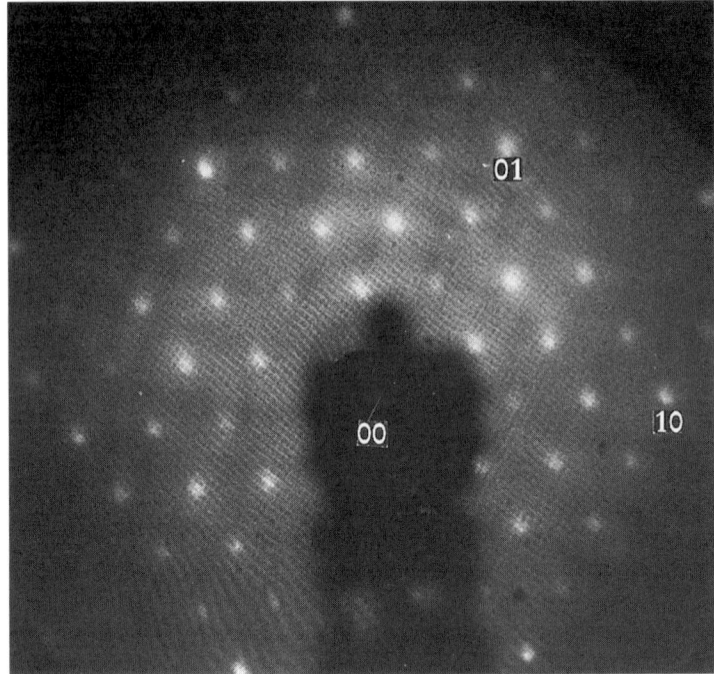

Figure 2.15. LEED pattern of a Pt(111) electrode emersed at +0.35 V vs. SCE from a Cl$^-$ containing solution (see Fig. 2.14). The (4 × 4) superstructure is clearly visible [65].

that the copper adatoms grow pseudomorphically on the Pt(111) electrode. The LEED I-V (intensity vs. beam energy) curves of the integer-order beams for the Pt(111) electrode emersed at +0.4 V are almost identical to that obtained from the electrode at +0.14 V where a full Cu monolayer on Pt(111) is generally accepted [65, 66, 68, 70]. It is therefore concluded that the arrangement of Cu adatoms at +0.4 V is indeed (1 × 1). The model of a Cl-Cu bilayer at +0.4 V with a Cu-Cu distance of 0.363 nm as proposed by Ross et al. [68] can therefore be considered unlikely on the basis of the LEED-intensity analysis [69].

Good agreement between the Br adlayer structures on Cu/Pt(111) as derived from STM [66], X-ray [68], and ex situ LEED investigations [65] was found. The Br-Br spacing of 0.388 nm for a (7 × 7) structure detected by ex situ LEED [65] agrees well with the value of 0.378 nm derived from in situ X-ray measurements [68].

Several studies of Cu UPD onto reconstructed and unreconstructed Pt(110) have been reported [71, 72]. For UHV-prepared reconstructed Pt(110)-(1 × 2) surfaces, the $\frac{1}{2}$-superstructure spots were observed up to half a monolayer of Cu [72]. Recently Zei et al. [73] reported the Pt(110)-(1 × 2) structure to be stable over the total potential range of Cu UPD (i.e., for deposition of Cu up to a full monolayer). From the variation of the relative intensity of the $\frac{1}{2}$-order beams with potential, differences in the Cu adlayer structures were inferred. Furthermore, the cyclic voltammograms for Cu UPD on Pt(110) (Fig. 2.16) shows two energetically well-separated current peaks, each associated with one full monolayer of Cu, the coulometric charge data being confirmed by AES measurements and diffraction data [73].

2.4.1.3 Structure of a H_2SO_4 Monolayer on Au(111) One crucial condition for a successful ex situ investigation of electrode surfaces is dry emersion. Under favorable conditions an electrode can be emersed with the electrochemical double layer intact and the bulk electrolyte absent [21]. This is often the case with $HClO_4$ as supporting electrolyte. Occasionally, small droplets of per-

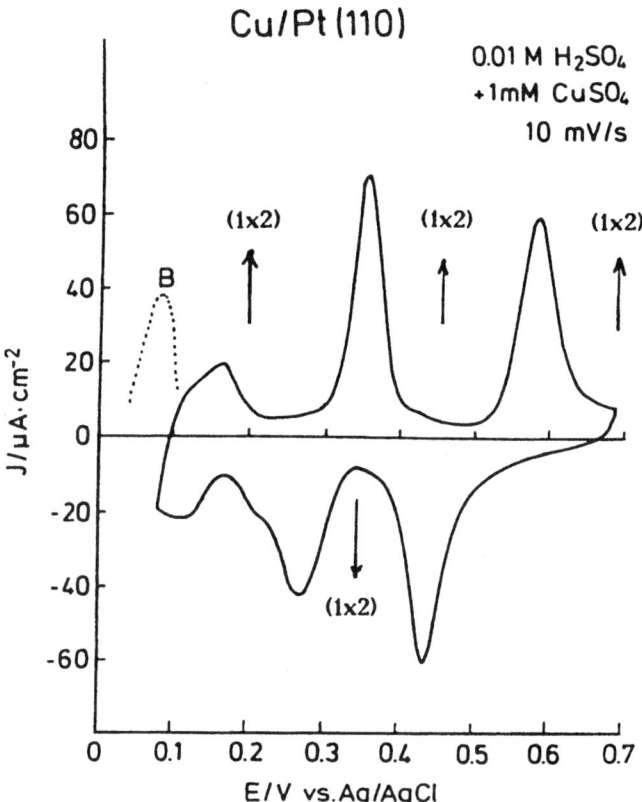

Figure 2.16. Cyclic current-potential curve for the Cu underpotential deposition onto Pt(110)-(1 × 2) in 0.01 M H_2SO_4 + 1 mM $CuSo_4$. Scan rate: 10 mV/s. B: bulk deposition. The emersion potentials and the corresponding superstructures are indicated in the figure [73].

chloric acid solution remain on the surface during emersion, which form small crystallites of hydronium perchlorate during evacuation of the vacuum chamber [22]. These crystallites do not seriously disturb the surface studies. A distinctly different behavior has been observed when using H_2SO_4 as electrolyte [74]. Residues of the bulk electrolyte after electrode emersion and evacuation form a clear superstructure on Au(111) (as well as on Au(100)), which has been analyzed by LEED and RHEED. Within a fairly wide range of -0.2 V $< E <$ $+0.6$ V vs. Ag/AgCl the superstructure was independent of the potential at which the Au(111) electrode was emersed. The superstructure obviously is a consequence of wet emersion from the sulfuric acid solution and is not related in any way to the emersed double layer. It is formed by a monolayer of H_2SO_4 molecules (the monohydride thereof), which solidifies only below 9°C [75]. The diffraction patterns of the superstructure in LEED and RHEED are shown in Figure 2.17. Their analysis showed that there is a superposition of three

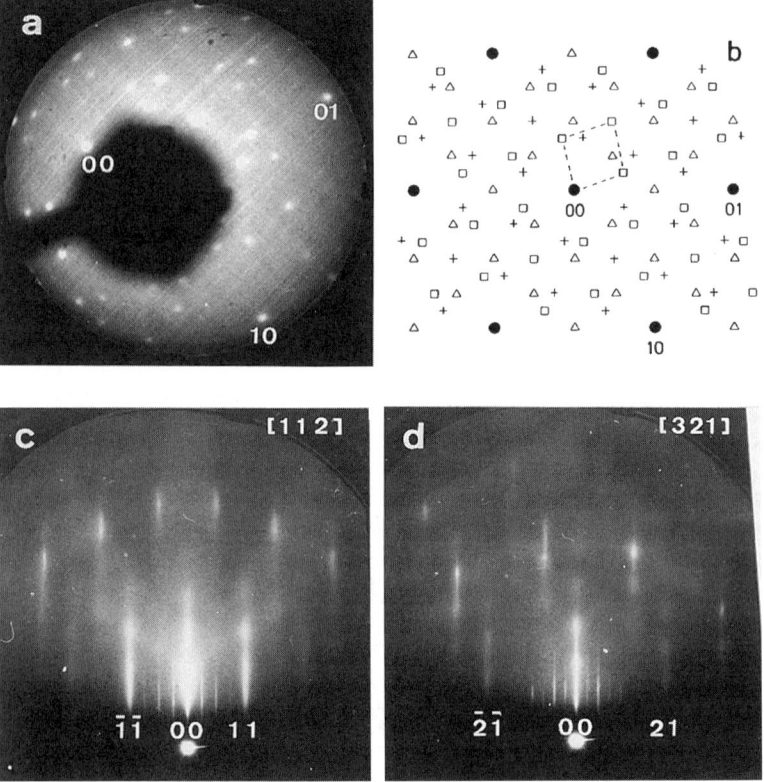

Figure 2.17. (a) LEED pattern (27 eV) of a Au(111) electrode after emersion from 5 mM H_2SO_4 at 0 V vs. SCE. The superstructure consists of the superposition of three different domains, which are shown schematically in (b). (c) RHEED pattern (40 keV) in [112] azimuth, and (d) in [321] azimuth [74].

2.4 CASE STUDIES

different domains rotated by 120° against each other. The corresponding LEED pattern is shown schematically. The REED patterns show the superstructure spots in two different azimuths, the [112] azimuth and the [321] azimuth. In [112] $\frac{1}{4}$-order superstructure streaks can be seen while in [321] four of the $\frac{1}{8}$-order superstructure streaks are clearly visible. The splitting of the (21) and ($\bar{2}\bar{1}$) streaks indicates that the gold surface is still reconstructed, which means that the (1 × 23) reconstruction is stable against such a treatment. Due to the presence of the adsorbate layer the (1 × 23) reconstruction can only be seen in RHEED and not in LEED. The proposed structure of one domain of the H_2SO_4 monolayer is shown in Figure 2.18.

2.4.1.4 Surface Reconstruction of Gold The low-index faces of gold are known to reconstruct when prepared under UHV conditions by Ar-ion bombardment and annealing in order to minimize their surface energy [76–80]. The same behavior is observed for gold single crystal electrodes [81]. Surface reconstruction means that the positions of the atoms in the top layer differ from those that one would expect for a parallel plane in the bulk. For example, the atoms of the (100) surface are rearranged in such a way that they form a hexagonal close packed layer, termed as (5 × 20) or (hex), which is a slightly buckled (111) layer. The structure of the complicated overlayer was investigated in detail by LEED (Van Hove et al. [80]) and REM (Wang et al. [82]). The (110) surface reconstructs to yield a (1 × 2) superstructure, which results from the so-called "missing-row" model, where every second row of gold atoms in [110] direction is missing [83–85]. This leads to a buckled surface with (111) micro facets. And even the close-packed (111) surface of gold reconstructs and forms a (1 × 23), or better ($\sqrt{3} \times 22$) superstructure as has been described in Section 2.4.1.1.

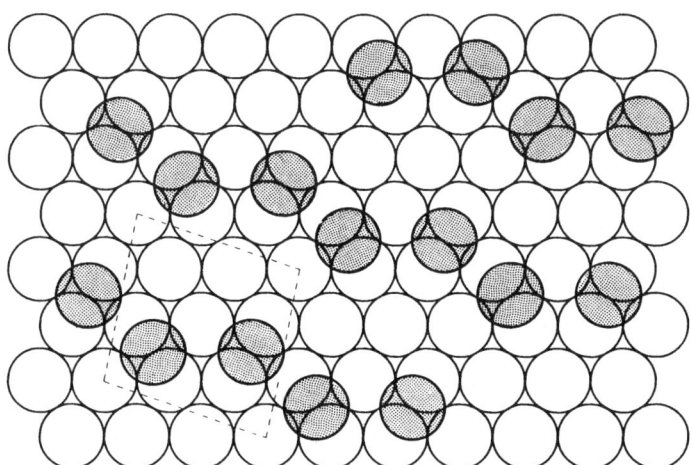

Figure 2.18. Model for the H_2SO_4 adlayer on Au(111), with two molecules per unit cell [74].

All three reconstructed gold surfaces could be imaged on an atomic scale by in situ STM with samples that were subjected to the so-called flame treatment [86–88].

LEED and RHEED have given first convincing proof that the reconstructed surfaces of gold such as (100)-(hex), (110)-(1 × 2) and (111)-(1 × 23) are stable in an aqueous electrolyte under certain potential conditions [89]. On the other hand, the reconstruction could be lifted at anodic potentials by specific adsorption of anions and it could be regained at more cathodic potentials [90]. As an example, the electrochemically induced lifting and reformation of the Au(100)-(hex) superstructure in 5 mM H_2SO_4 is shown in Figure 2.19 [89, 91]: (a) and (b) are the LEED and RHEED patterns of reconstructed Au(100) after preparation in UHV, while (c) and (d) are those obtained after cycling between U_{SCE} = −0.35 V and +0.25 V and emersion at +0.20 V. The surface is seen to be still reconstructed. However, reconstruction is lifted after cycling between −0.35 V and +0.45 V and emersion at −0.25 V as can be seen in (e) and (f). The superstructure due to the (hex) reconstruction reappears after applying a cathodic potential of −0.35 V for three minutes (figures (g) and (h)). The structure transitions between Au(100)-(hex) and (1 × 1) have also been investigated by double-layer capacity and by electroreflectance measurements [41, 90] and the results were confirmed by in situ STM investigations [87, 88, 92].

Similar investigations have been performed with Pt(100)-(5 × 20) by Zei et al. [93]. It was found by LEED and mainly by RHEED, that the reconstruction is stable against the exposure to Ar, to water, and to H_2SO_4 solutions. The ex situ observations were compared with (rather preliminary) in situ STM experiments. Most recently, a correlation of the voltammetric changes with the electrode surface structure in 0.01 M H_2SO_4 solution has been studied by AES, LEED, and RHEED. As with gold, the reconstruction is lifted by specific adsorption of anions, however, unlike Au(100)-(hex), the changes in the cyclic voltammogram and in the surface structure of Pt(100) are irreversible [94].

2.4.2 Surface Imaging by Diffracted Electrons

2.4.2.1 Surface Reconstruction The reconstruction of gold single crystal surfaces has been imaged directly with an electron microscope in transmission

---

Figure 2.19. LEED and RHEED patterns of a Au(100) surface. (a, b) after preparation in UHV with the (hex) superstructure. RHEED in [110] azimuth, 40 keV; LEED: 41 eV. (c, d) after potential cycling in 5 mM H_2SO_4 between −0.35 V and +0.25 V; emersed at +0.2 V. The surface is still reconstructed as is clearly seen in the diffraction patterns. (e, f) after potential cycling between −0.35 V and +0.45 V and emersion at −0.25 V. The reconstruction is lifted, but a (2 × 2) superstructure due to an ordered H_2SO_4 adlayer can also be seen [22]. (g, h) after applying a potential of −0.35 V for 3 minutes. The (hex) reconstruction has reappeared [22, 89].

2.4 CASE STUDIES

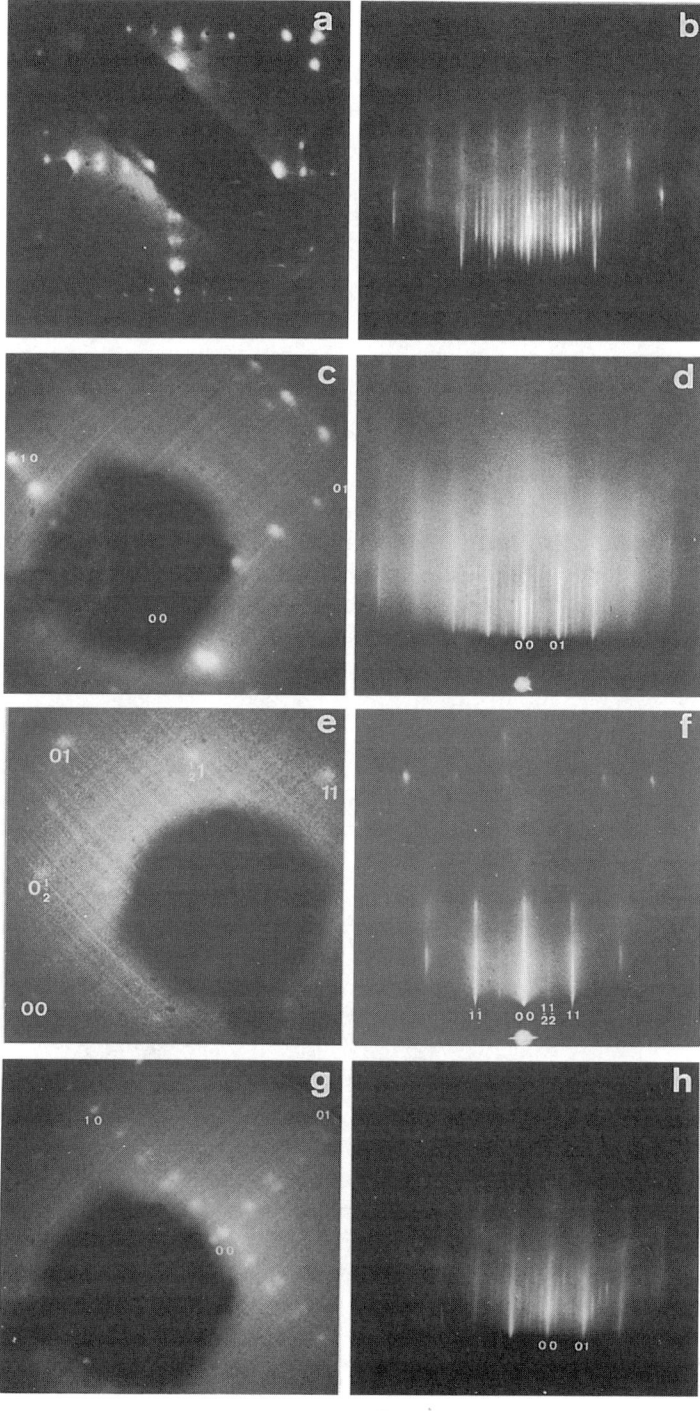

(TEM) and in reflection (REM). One example of each will be shown. Figure 2.20 is the TEM image of a thin gold film with (111) orientation that had been evaporated onto glass and subsequently floated off the substrate. The so-called (1 × 23) superstructure with three domains, rotated by 120° against each other, are clearly seen on one single grain. Because of the poor vacuum conditions in the conventional electron microscope, the original superlattice spacing of 6.6 nm is continuously growing due to an electron beam induced interaction of the surface with residual gas molecules. After half a minute of electron beam bombardment the superstructure has disappeared. Similar observations were reported by Nihoul et al. [95] for a conventional electron microscope and by Takayanagi and coworkers [96, 97] for an UHV microscope.

Figure 2.21 is the REM image of a partly (hex)-reconstructed (100) facet of a single-crystal Au sphere. The dark areas are the reconstructed domains where the reconstruction rows with a periodicity of 1.45 nm are clearly seen in (b).

2.4.2.2 Surface Faceting by Fast Potential Cycling It has been shown by Arvia et al. that the surface topography of Pt electrodes can be changed by fast oxidation-reduction cycles in 1 M H_2SO_4 [98–100]. After several hours of

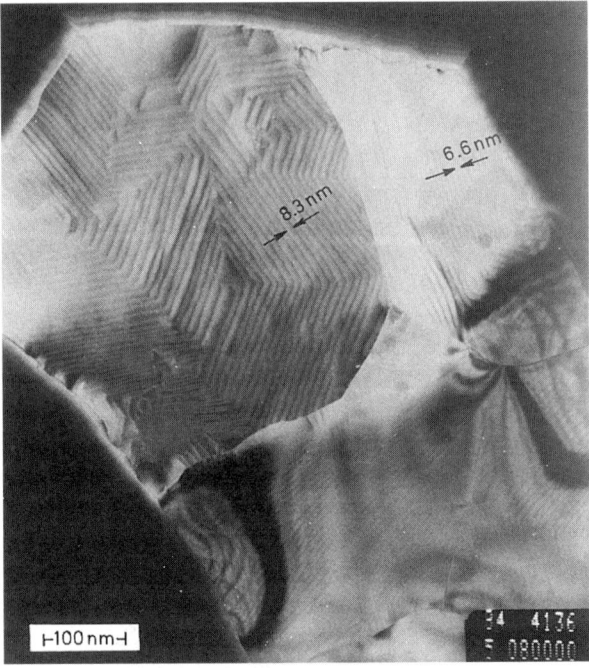

Figure 2.20. Transmission electron microscopy (TEM) image of a 400 Å thick Au(111) film, floated off the glass onto which it had been grown epitaxially. The (1 × 23) superstructure is clearly seen by the corresponding reconstruction rows.

2.4 CASE STUDIES

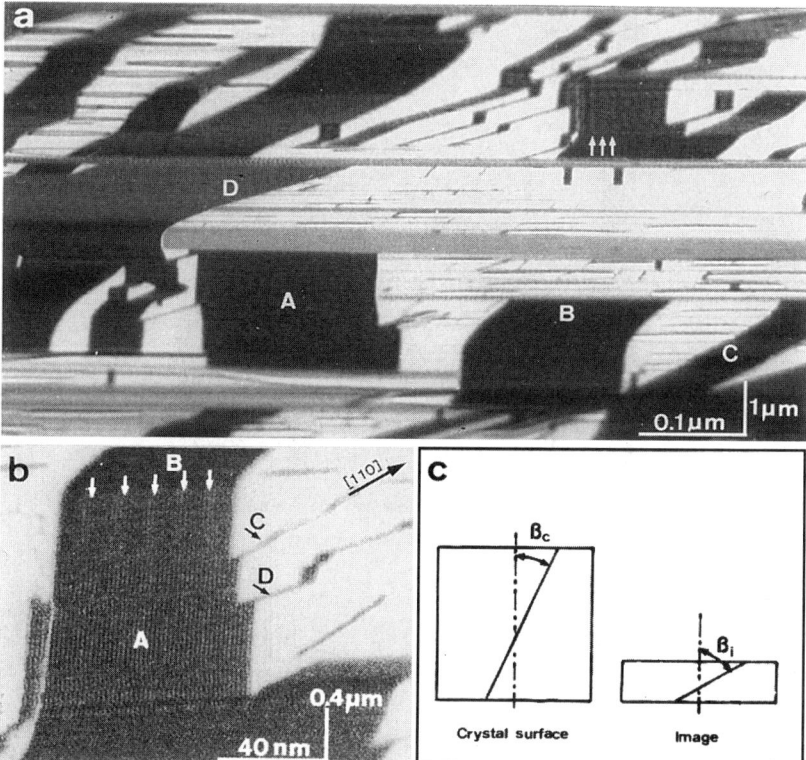

Figure 2.21. Reflection electron microscopy (REM) image of the "(hex)" reconstruction of a Au(100) surface. Different domains of reconstruction, indicated by A-D, can be seen as dark areas [82]. Domain A at higher magnification with lattice modulations is shown in (b). (c) indicates the principle for angle enlargement in REM.

potential cycling in the kHz range an initially polycrystalline electrode develops a preferential crystallographic orientation, which is reflected in cyclic voltammograms by certain characteristic features. This effect has been investigated ex situ with single crystal Pt electrodes by electron diffraction and reflection electron microscopy [101]. The electrodes were single crystal spheres of 0.3–0.5 mm diameter, prepared by melting a Pt wire as mentioned. Such spheres often consist of only one crystal with fairly large (111) and smaller (100) facets. Other facets are very small so that in an electrochemical cell the sphere acts as a polycrystalline electorde.

The scanning electron microscope (SEM) image of a freshly prepared Pt single-crystal sphere is shown in Figure 2.22. Different (111) facets can be seen and the position of a smaller (100) facet can be recognized. RHEED patterns from such facets are shown in Figure 2.23. Diffraction pattern (a) is from a (111) facet with the intensity-enhanced 555 reflection near the [110] azimuth; (b) is the corresponding diffraction pattern from the smaller (100) facet obtained with

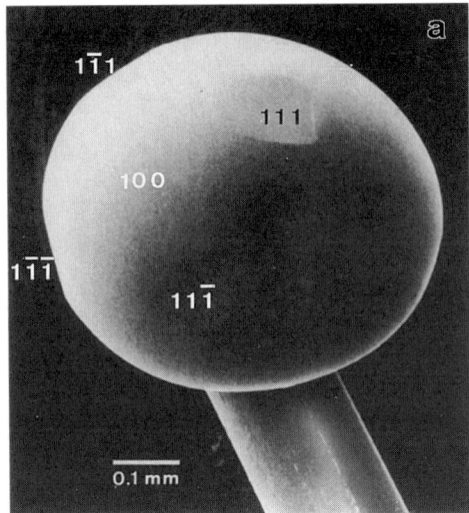

Figure 2.22. Scanning electron microscopy image of a Pt single crystal sphere with large (111) facets [101].

the intensity-enhanced 800 reflection [101]. Using intensity enhanced reflections for imaging in the REM mode, the corresponding surfaces can be shown with monoatomic high steps as demonstrated in Figure 2.24. Due to the small angle of observation, the images are strongly foreshortened as indicated by the two different magnification bars. The direction of the incident electron beam is

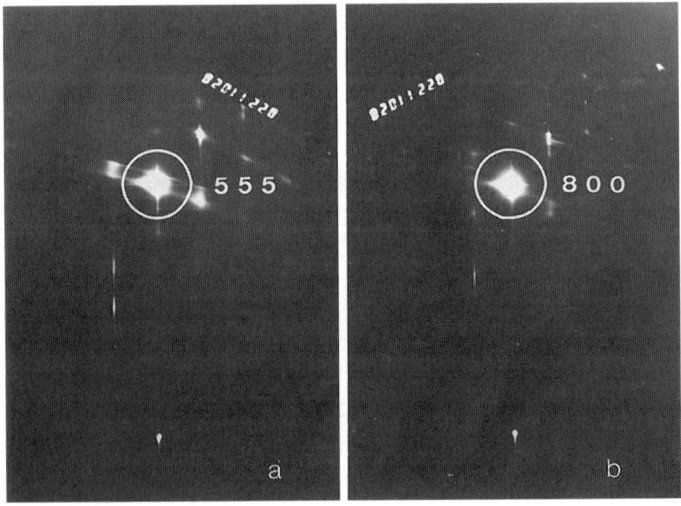

Figure 2.23. RHEED patterns at 100 keV near the [110] azimuth from (a) the (111) facet and (b) the (100) facet of Pt, showing intensity enhanced Bragg reflections (encircled) [101].

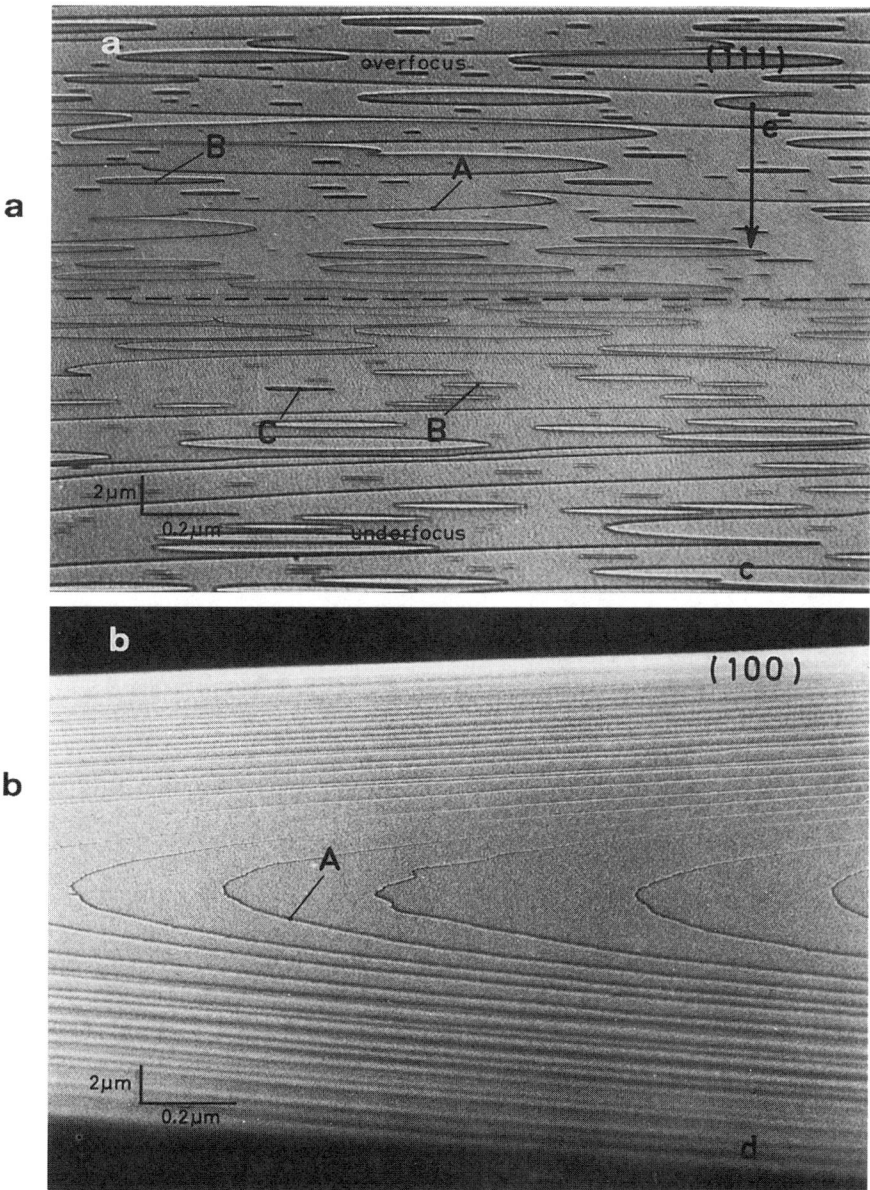

Figure 2.24. Reflection electron microscopy images from (a) the (111) facet and (b) the (100) facet of a Pt single crystal sphere, obtained with the diffraction conditions shown in Figure 2.23. The REM image of the (111) surface shows monoatomic high steps (A), atomic height extrusions (B), and protrusions (C). The foreshortening is indicated by the different magnification bars, the direction of the incident beam by an arrow. The (100) surface shows monoatomic steps (A) only [101].

indicated by the arrow. Along the dotted line in Figure 2.24(a) the image is in focus. Above this line the image is in overfocus and below in underfocus. On the (111) surface circular islands of monoatomic height can be seen as ovals due to foreshortening. They can be positive (protrusions, C) or negative (extrusions, B). On the (100) surface on the other hand, only large atomically flat terraces and monoatomic high steps were found. After 14 hurs of fast potential cycling in 1 M H_2SO_4 with 3.5 kHz between 0.23 and 1.23 V, the single crystal electrode is strongly modified as shown in Figure 2.25 by SEM. The (100) poles, for example, have clearly been developed. This cycling procedure is called "(100)" treatment because it leads to a current-potential curve characteristic for a (100) surface. Since it is very difficult to image by REM the macroscopically rough surface after a 14-hour treatment, the initial stages of surface modification were investigated, especially the attack of the fast potential cycling on the monoatomic high steps [101]. In case of "(100)" treatment, the atomic steps on the (111) facet disappeared after 5 sec, while the steps on the (100) facet were still visible after 15 sec. A corresponding observation was made for the so-called "(111)" treatment, which involved cycling the potential

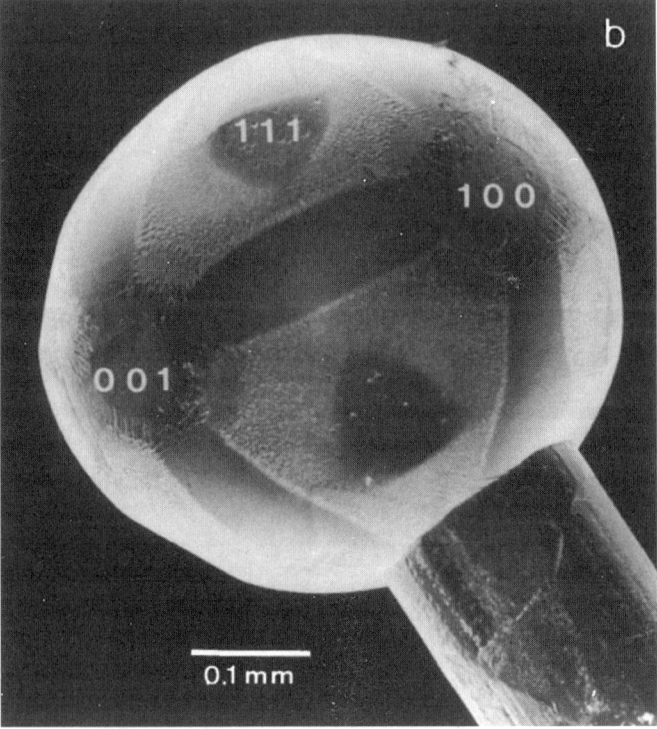

Figure 2.25. Scanning electron microscopy image of a Pt single crystal sphere after fast potential cycling in 1 M H_2SO_4 at 3.5 kHz between 0.23 V and 1.23 V vs. RHE ("(100)" treatment; see text) [101].

2.4 CASE STUDIES

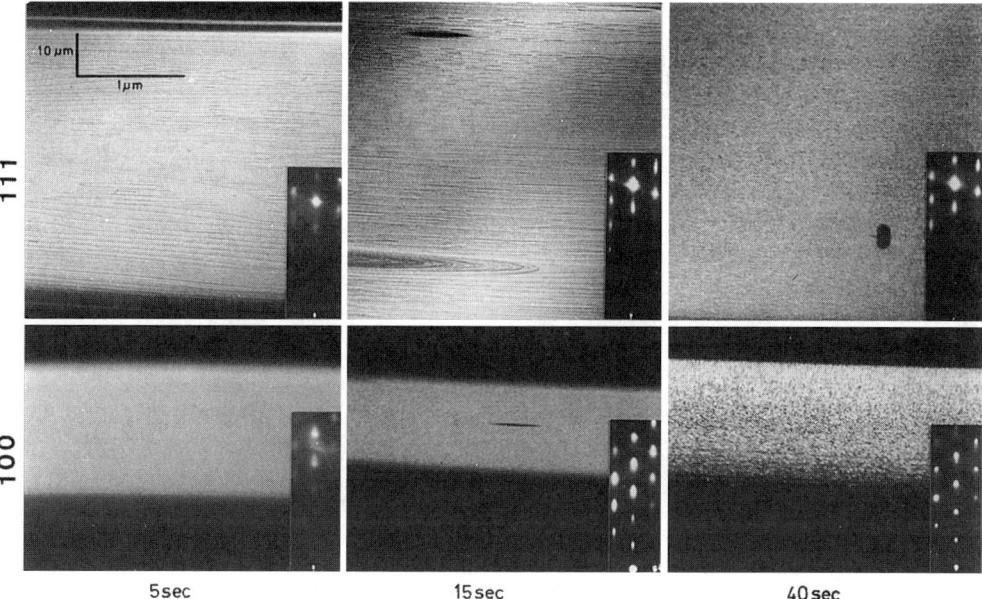

Figure 2.26. REM images and RHEED patterns from (111) and (100) facets of a Pt single crystal sphere after various perods of "(111)" treatment [101].

between 0.68 and 1.38 V. Here, the steps on the (111) surface could still be seen after 15 sec of cycling in the kHz range, while the steps on the (100) surface were no longer visible after 5 sec. An example of the "(111)" treatment (cycling between 0.68 and 1.38 V) is shown in Figure 2.26. The different sizes of the imaged facets are due to the fact that after each REM observation a new Pt single-crystal sphere had to be used in order to avoid artifacts by contamination. However, the images of the (111) and (100) facets obtained after a certain treatment time stem from the very same sphere.

With increasing treatment time the surface becomes more and more roughened. During the "(100)" treatment positive islands are developed on the (100) surface [101]. They are obviously formed from excess Pt atoms during lifting of the (hex)-reconstruction as observed in STM investigations by Hössler et al. [102]. The average size of the islands increases with time, reflecting the mobility of the Pt atoms on the surface. An example of a Pt(100) surface with a granular structure formed by equally sized Pt islands of approximately 6.0 nm diameter is shown in Figure 2.27. The light diffraction pattern of the granularity is shown in the inset [101]. The conclusion that this roughness is a consequence of lifting the (hex)-reconstructed (100) surface is also supported by the observation that no such structure developed on the (111) surface, which does not reconstruct. After extensive "(100)" treatment (14 hours) new facets had been developed near the (100) poles, which were close to (511) or (711) orientation. They were identified by an analysis of the diffraction patterns.

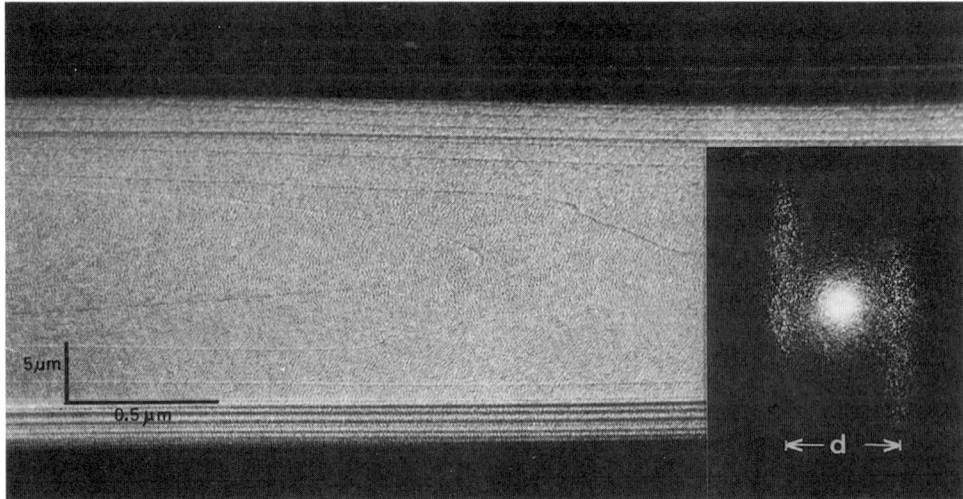

Figure 2.27. REM image of Pt(100) after 10 sec "(100)" treatment. Regularly sized islands are formed with a diameter of about 10 nm. The light diffraction pattern from the electron micrograph in the insert shows clearly this situation [101].

During the initial stage of electrochemical faceting (e.g., within the first second of "(100)" treatment), a change of the surface structure of the (111) facet was observed mainly in the center of flat terraces, while the areas near the atomic steps remained intact. The change of the surface structure has been identified as due to a statistical collection of adatoms [101]. The atomic steps act as a sink for the adatoms leading to an unperturbed surface area (bright region) along the steps as shown in Figure 2.28. From the width of this unperturbed area and its temperature dependence the diffusion length of the adatoms and an activation energy for their diffusion were derived [103]. The activation energy for surface diffusion of Pt on Pt(111) in sulfuric acid solution was determined to be 0.20 ± 0.05 eV. This mobility of the adatoms appears to be the reason for the electrochemically induced faceting of surfaces by fast potential cycling. A similar observation of faceting on gold by fast potential cycling was reported by Twomey [104].

2.4.2.3 The Structure of Flame-Treated Pt Surfaces After Rapid Quenching

Figure 2.29 shows the REM image of a clean Pt(111) surface. The small spherical single crystal was cleaned in UHV by Ar-ion bombardment, subsequently annealed up to about 1000°C and cooled down very slowly to room temperature. The surface is seen to be very smooth, and no significant defect structure can be observed in contrast to the surface shown in Figure 2.24(a). In the latter case, the crystal was annealed in the flame of a Bunsen burner (temperature of the specimen was measured to be 1400°C) and then removed from the flame and cooled in air. In this case, the initial cooling rate has been

2.4 CASE STUDIES

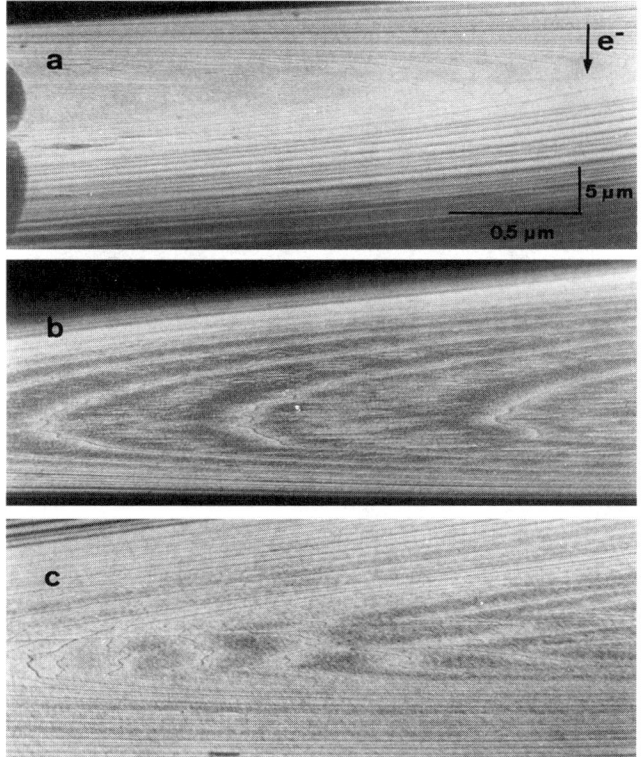

Figure 2.28. REM images from the (111) facets of Pt single crystal spheres after 1 s of "(100)" treatment at three different treatment temperatures: (a) 0°C, (b) 20°C, (c) 50°C. The bright areas near the steps are still unperturbed, while the dark areas indicate regions of roughness. The direction of the incident beam is indicated by the arrow. The different magnification bars show the effect of foreshortening [103].

Figure 2.29. REM image of Pt(111), which was cleaned by Ar ion-bombardment, annealed at 1000°C for 15 min and cooled slowly within 20 min. This surface is very smooth and looks similar to the (100) surface in Figure 2.24(b).

Figure 2.30. REM image of Pt(111) for a specimen, which was annealed in the flame of a Bunsenburner for 10 min and quenched by injecting a stream of water. Straight lines indicate atomic steps formed by moving screw dislocations [105].

estimated to about 2×10^{3}°C/sec. Many loops with monoatomic high steps are observed on such a Pt(111) surface. The loops were analyzed to be intrusive. Therefore they must have been formed by coagulation of advacancies during the cooling procedure. The advacancy concentration is a function of sample temperature and increases with the latter [105].

In another experiment the flame-annealed Pt sample was subjected to rapid quenching, a treatment often advocated by electrochemists. This rapid cooling was performed by water injection on the Pt specimen in the flame. The initial cooling rate of the specimen was estimated to be about 4×10^{4}°C/sec. Figure 2.30 shows a REM image of the water-quenched Pt(111) surface. An immense number of small monoatomic deep loops has developed. The size of these loops is of the order of the resolution of REM imaging. In addition to these loops straight lines are observed, which correspond to atomic steps formed by moved screw dislocations due to the thermal strain induced by water quenching. A systematic study [105] of the influence of the cooling rate on the quality of Pt single crystals has confirmed the notion that rapid cooling leads to rough surfaces, and ultimately it will even destroy the single crystallinity of the bulk.

2.4.2.4 Electrolytic Deposition of Silver Another form of electrochemically induced faceting is achieved by electrolytic metal deposition [106]. This is shown for silver deposited onto a silver single-crystal sphere from a $AgNO_3$ solution (Figure 2.31). The single crystal sphere was prepared in UHV by melting one end of a silver wire by electron bombardment. After Ag deposition large (112) facets are clearly seen in the SEM image, besides the (100), (110) and (111) facets. The REM images of the (111) and the (112) facets are shown in Figure 2.32. Such "multi-surface" single crystals are of great interest for the investigation of catalytic processes with respect to structure-reactivity relations, and of process-induced surface structure modifications.

2.4 CASE STUDIES

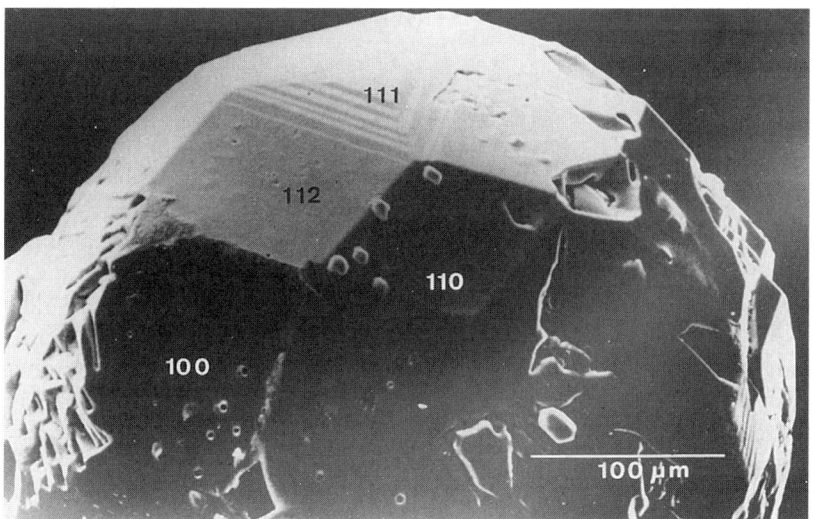

Figure 2.31. Scanning electron microscopy image of an electrochemically treated silver single crystal sphere with different facets, as indicated in the figure.

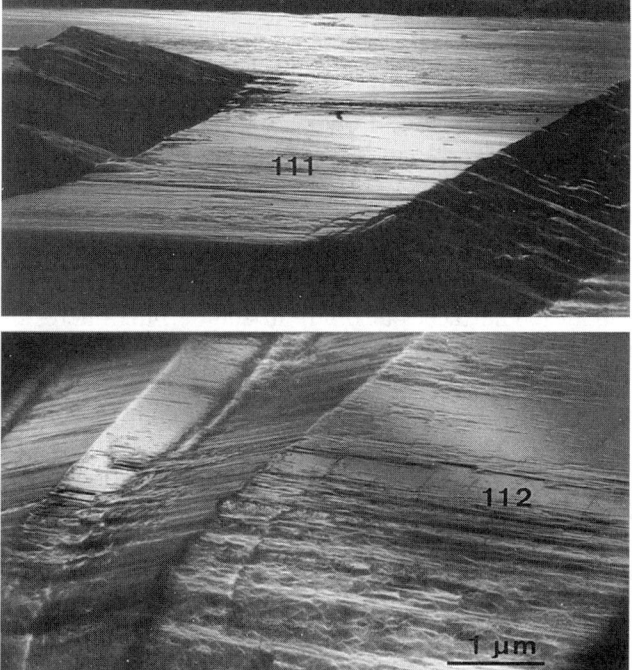

Figure 2.32. REM images of two facets of the Ag single crystal sphere shown in Figure 2.31; (a) (111) and (b) (112).

2.5 CONCLUSIONS AND OUTLOOK

The initial intention of ex situ studies was the structural characterization of bare and adsorbate-covered single crystal surfaces, with special emphasis on ordered adsorption. The latter was investigated by electron diffraction (LEED and RHEED), which yielded unique information on adlayer structures and particularly on the growth mode of metal monolayers. However, with the recent introduction of powerful structure-sensitive in situ techniques, such as surface X-ray diffraction and scanning tunneling microscopy, the ex situ route lost its leading role in electrochemical structure studies. In addition, the preparation of clean and structurally well-defined single crystal electrodes by flame annealing was advanced over the years to such a point that the quality of flame-annealed surfaces became comparable to those of UHV-prepared ones. Furthermore, the flame treatment has become so reliable that a direct structural test of the outcome seems no longer mandatory. Particularly for Au(hkl) cyclic voltammograms can now be used as quick and rather safe quality checks since there are several structure-sensitive features that have been identified as such by a comparison with STM data [107]

Besides the fact that UHV studies of emersed electrodes are interesting in their own right, electron diffraction remains a valuable addition to the in situ structure techniques, especially since LEED investigations may now include an intensity analysis (so-called I-V curves). Past experiences have demonstrated that even the seemingly most straightforward and direct imaging technique like the STM can benefit from cross-checks with other methods. The most promising future application of the ex situ route, however, may be the combination of structural and chemical information. In this respect photoelectron spectroscopy, particularly ESCA, can yield extremely useful information on the chemical nature in general and on the valence state of a certain species in particular (chemical shift), which otherwise would be almost impossible to obtain in situ [108]. Auger electron spectroscopy (AES) gives similarly important chemical information about the surface, but offers in addition the interesting possibility of imaging the surface with lateral resolution in the submicron region (scanning AES).

As the imaging of surfaces in real space has become an important issue in recent years, the number of scanning techniques has increased tremendously. Leaders in this field are nowadays again the scanning probe microscopy methods, but electron microscopy will continue to play an important role in surface science, particularly in areas relevant to catalysis. Since the availability of high-resolution microscopes (REM or TEM) is fairly restricted, particularly for the electrochemistry community, their use in electrode studies may be limited to surface science centers like the Fritz-Haber-Institut where different areas of surface science meet. Progress in science has often been boosted by employing a combination of different techniques, and in this regard it is our belief that ex situ structural information will continue to contribute in a useful way to our electrochemical knowledge.

REFERENCES

1. G. A. Somorjai, *Introduction to Surface Chemistry and Catalysis* (Wiley, New York, 1994).
2. W. E. O'Grady, M. Y. C. Woo, P. L. Hagans, and E. Yeager, *J. Vac. Sci. Technol.*, **14** (1977) 365.
3. A. T. Hubbard, J. L. Stickney, S. D. Rosasco, M. P. Soriaga, and D. Song, *J. Electroanal. Chem.*, **150** (1983) 165.
4. J. S. Hammond and N. Winograd, *J. Electroanal. Chem.*, **80** (1977) 123.
5. J. S. Hammond and N. Winograd, *J. Electroanal. Chem.*, **78** (1977) 55.
6. G. C. Allen, P. M. Tucker, A. Capon, and R. Parsons, *J. Electroanal. Chem.*, **50** (1974) 335.
7. A. T. Hubbard, *Crit. Rev. Anal. Chem.*, **3** (1973) 201.
8. E. Yeager, A. Homa, B. D. Cahan, and D. Scherson, *J. Vac. Sci. Technol.*, **20** (1982) 628.
9. P. N. Ross, *J. Electrochem. Soc.*, **126** (1979) 67.
10. P. N. Ross, *Surf. Sci.*, **102** (1981) 463.
11. F. T. Wagner and P. N. Ross, *J. Electroanal. Chem.*, **150** (1983) 141.
12. H. O. Beckmann, H. Gerischer, D. M. Kolb, and G. Lehmpfuhl, *Symp. Faraday Soc.*, **12** (1977) 51.
13. Y. Nakai, M. S. Zei, D. M. Kolb, and G. Lehmpfuhl, *Ber. Bunsenges. Phys. Chem.*, **88** (1984) 340.
14. W. N. Hansen, C. L. Wang, and T. W. Humpherys, *J. Electroanal. Chem.*, **93** (1978) 87.
15. W. N. Hansen and D. M. Kolb, *J. Electroanal. Chem.*, **100** (1979) 493.
16. D. M. Kolb, D. L. Rath, R. Wille, and W. N. Hansen, *Ber. Bunsenges. Phys. Chem.*, **87** (1983) 1108.
17. T. E. Furtak, *Surf. Sci.*, **299/300** (1994) 945.
18. W. E. O'Grady, M. Y. C. Woo, P. L. Hagans, and E. Yeager, *J. Vac. Sci. Technol.*, **14** (1977) 365.
19. A. T. Hubbard, R. M. Ishikawa, and J. Katekaru, *J. Electroanal. Chem.*, **86** (1978) 271.
20. P. N. Ross, Jr., *J. Electroanal. Chem.*, **76** (1977) 139.
21. D. M. Kolb, *Z. Phys. Chem.*, N. F. **154** (1987) 179.
22. D. M. Kolb, G. Lehmpfuhl, and M. S. Zei, in *Spectroscopic and Diffraction Techniques in Interfacial Electrochemistry*, C. Gutiérrez and C. A. Melendres, eds. (Kluwer, Dordrecht, 1990) p. 361.
23. G. Ertl and J. Küppers, *Log Energy Electrons and Surface Cehmistry* (VCH, Weinheim, 1985).
24. E. Bauer, *Elektronenbeugung*, Verlag Moderne Industrie, W. Dummer und Müller, München 2, 1958.
25. J. M. Cowley, *Diffraction Physics*, North Holland Publishing Company, 1981.
26. L. Reimer, *Transmission Electron Microscopy*, Springer Series in Optical Sci-

ences, Vol. 36, 2nd ed., p. 266 ff, Springer Verlag, Berlin, Heidelberg, New York, Tokyo, 1989.
27. M. v. Laue, *Materiewellen und ihre Interferenzen*, p. 161 ff, Akademische Verlagsgesellschaft, Geest & Portig, Leipzig 1948.
28. G. Lehmpfuhl and W. C. T. Dowell, *Acta Cryst.*, **A42** (1986) 569.
29. K. Kambe and G. Lehmpfuhl, *Optik*, **42** (1975) 187.
30. F. Forstmann, W. Berndt, and P. Büttner, *Phys. Rev. Lett.*, **30** (1973) 17.
31. P. Heilmann, E. Lang, K. Heinz, and K. Müller, *Appl. Phys.*, **9** (1976) 247.
32. P. Ross, Jr. and F. Wagner in *Advances in Electrochemistry and Electrochemical Engineering*, H. Gerischer and C. W. Tobias, eds. (Wiley, New York, 1985), p. 69, Vol. 13.
33. A. S. Homa, E. Yeager, and B. D. Cahan, *J. Electroanal. Chem.*, **150** (1983) 181.
34. B. C. Schardt, J. L. Stickney, D. A. Stern, A. Wieckowski, D. C. Zapien, and A. T. Hubbard, *Surf. Sci.*, **175** (1986) 520.
35. A. T. Hubbard, *Surf. Sci.*, **175** (1986) 520.
36. D. Aberdam, R. Durand, R. Faure, and F. El-Omar, *Surf. Sci.*, **162** (1985) 782.
37. T. Hsu and J. M. Cowley, *Ultramicroscopy*, **11** (1983) 239.
38. W. Telieps and E. Bauer, *Ultramicroscopy*, **17** (1985) 57; W. Telieps, M. Mundschau, and E. Bauer, *Surf. Sci*, **159** (1985).
39. M. Ichikawa, T. Doi, and K. Hayakawa, *Surf. Sci.* **159** (1985).
40. J. Clavilier, R. Fauré, G. Guinet, and R. Durand, *J. Electroanal. Chem.*, **107** (1983) 181.
41. D. M. Kolb and J. Schneider, *Electrochim. Acta* **31** (1986) 929.
42. Y. Uchida, G. Lehmpfuhl, and J. Jäger, *Ultramicroscopy*, **15** (1984) 119.
43. M. S. Zei, Y. Nakai, G. Lehmpfuhl, and D. M. Kolb, *J. Electroanal. Chem.*, **150** (1983) 201.
44. D. M. Kolb, in *Advances in Electrochemistry and Electrochemical Engineering*, H. Gerischer and Ch. W. Tobias, eds. (Wiley-Interscience, New York, 1978), Vol. 11, p. 125.
45. D. M. Kolb, M. Przasnyski, and H. Gerischer, *J. Electroanal. Chem.*, **54** (1974) 25.
46. J. W. Schultze and D. Dickertmann, *Surf. Sci.* **54** (1976) 489.
47. D. M. Kolb, K. Al Jaaf-Golze, and M. S. Zei, *Dechema-Monographien*, VCH, Weinheim, Vol. 102, p. 53.
48. D. M. Kolb, in *Schering Lecture's Publications* (Schering, Berlin, 1991), Vol. 2, p. 5.
49. H. Melle and E. Menzel, *Z. Naturforsch. 33a* (1978) 282.
50. K. Takayanagi, *Ultramicroscopy* **8** (1982) 145.
51. V. Horten, A. Lahee, J. Toennies, and Ch. Wöll, *Phys. Rev. Lett.*, **54** (1985) 2619.
52. D. M. Kolb, *Ber. Bunsenges Phys. Chem.*, **92** (1988) 1175.
53. O. M. Magnussen, J. Hotlos, R. J. Nichols, D. M. Kolb, and R. J. Behm, *Phys. Rev. Lett.*, **64** (1990) 2929.
54. L. Blum and D. A. Huckaby, in *Proc. Symp. on Microscopic Models of Electrode*

REFERENCES

Electrolyt Interfaces, J. W. Halley and L. Blum, eds., The Electrochem. Society, Inc. Pennington, N.Y. 1992.

55. D. A. Huckaby and L. Blum, *J. Electroanal. Chem.*, **315** (1991) 255.
56. M. F. Toney, J. N. Howard, J. Richer, G. L. Borges, J. G. Gordon, O. R. Melroy, D. Yee, and L. B. Sorensen, *Phys. Rev. Lett.*, **75** (1995) 4472.
57. Z. Shi and J. Lipkowski, *J. Electroanal. Chem.*, **365** (1994) 303.
58. N. Batina, T. Will, and D. M. Kolb, *Disc. Faraday Soc.*, **94** (1992) 93.
59. G. Horányi, *J. Electroanal. Chem.*, **45** (1973) 63.
60. K. Varga, P. Zelenay, and A. Wieckowski, *J. Electroanal. Chem.*, **330** (1992) 453.
61. O. R. Melroy, M. G. Samant, G. L. Borges, J. G. Gordon, L. Blum, J. H. White, M. J. Albarelli, M. McMillan, and H. D. Abruna, *Langmuir*, **4** (1988) 728.
62. M. S. Zei, G. Qiao, G. Lehmpfuhl, and D. M. Kolb, *Ber. Bunsenges Phys. Chem.*, **91** (1987) 349.
63. J. Hotlos, O. M. Magnussen, and R. J. Behm, *Surf. Sci.*, **335** (1995) 129.
64. T. Twomey and D. M. Kolb, *Proc. 38th ISE Meeting*, Maastricht, 1987, p. 334.
65. R. Michaelis, M. S. Zei, R. S. Zhai, and D. M. Kolb, *J. Electroanal. Chem.*, **339** (1992) 299.
66. H. Matsumoto, J. Unukai, and M. Ito, *J. electroanal. Chem.*, **379** (1994) 223.
67. Z. Shi, S. Wu, and J. Lipkowski, *Electrochim. Acta* **40** (1995) 9.
68. I. M. Tidswell, C. A. Luca, N. M. Markovic, and P. N. Ross, *Phys. Rev. B*, **51** (1995) 10205.
69. H. Bludau, K. Wu, M. S. Zei, M. Eiswirth, H. Over, and G. Ertl, *Surf. Sci.*, 402–404 (1998 786.
70. H. S. Yee and H. D. Abruna, *Langmuir* **9** (1993) 2460.
71. D. Aberdam, R. Durand, R. Fauré, and F. El-Omar, *Surf. Sci.*, **162** (1985) 782.
72. R. Michaelis and D. M. Kolb, *J. Electroanal. Chem.*, **328** (1992) 341.
73. M. S. Zei and G. Ertl, *Z. Phys. Chem.*, **202** (1997) 5.
74. M. S. Zei, D. Scherson, G. Lehmpfuhl, and D. M. Kolb, *J. Electroanal. Chem.*, **229** (1987) 99.
75. A. F. Hollemann and N. Wiberg, *Lehrbuch der Anorganischen Chemie*, Walter de Guyter, Berlin, 1995, p. 585.
76. G. A. Somorjai, in *Chemistry in Two Dimensions: Surfaces*, Cornell University Press, Ithaca, New York (1981).
77. K. Müller, Ber. Bunsenges. *Phys. Chem.*, **90** (1986) 184.
78. G. A. Somorjai and M. A. Van Hove, *Prog. Surf. Sci.*, **30** (1989) 201.
79. D. G. Fedak and N. A. Gjostein, *Surf. Sci.*, **8** (1967) 77.
80. M. A. Van Hove, R. J. Koestner, P. C. Stair, J. P. Biberian, L. L. Kesmodel, I. Bartos, and G. A. Somorjai, *Surf. Sci.*, **103** (1981) 189, 218.
81. D. M. Kolb, *Prog. Surf. Sci.*, **51** (1996) 109.
82. N. Wang, Y. Uchida, and G. Lehmpfuhl, *Surf. Sci.*, **284** (1993) L419.
83. W. Moritz and D. Wolf, *Surf. Sci.* **163** (1985) L655.
84. J. Möller, H. Niehus, and W. Heiland, *Surf. Sci.*, **166** (1986) L111.
85. G. Binnig, H. Rohrer, Ch. Gerber, and E. Weiberl, *Surf. Sci.*, **131** (1983) L379.

86. X. Gao, A. Hamelin, and M. J. Weaver, *J. Chem. Phys.*, **95** (191) 6993.
87. O. M. Magnussen, J. Hotlos, R. J. Behm, N. Batina, and D. M. Kolb, *Surf. Sci.*, **296** (1993) 310.
88. X. Gao, G. J. Edens, A. Hamelin, and M. J. Weaver, *Surf. Sci.*, **296** (1993) 333.
89. M. S. Zei, G. Lehmpfuhl, and D. M. Kolb, *Surf. Sci.*, **221** (1989) 23.
90. D. M. Kolb, in *Structure of Electrified Interfaces*, J. Lipkowski and P. N. Ross, eds. (VCH, New York, 1993), p. 65.
91. D. M. Kolb, G. Lehmpfuhl, and M. S. Zei, *J. Electroanal. Chem.*, **179** (1984) 289.
92. X. Gao, A. Hamelin, and M. J. Weaver, *Phys. Rev. Lett.*, **67** (1991) 618.
93. M. S. Zei, N. Batina, and D. M. Kolb, *Surf. Sci.*, **306** (1994) L519.
94. K. Wu and M. S. Zei, *Surf. Sci.*, **415** (1998) 242.
95. G. Nihoul, K. Abdelmoula, and J. J. Metois, *Ultramicroscopy* **12** (1984) 353.
96. K. Yagi, K. Takyanagi, and G. Honjo, in *Crystal Growth, Properties and Applications*, H. C. Freyhardt, ed. (Springer, Berlin, Heidelberg, 1982, Vol. 7, p. 47.
97. Y. Tanishiro, H. Kanamori, K. Takayanagi, K. Yagi, and G. Honjo, *Surf. Sci.* **111** (1981) 395.
98. A. M. Cerviño, W. A. Triaca, and A. J. Arvia, *J. Electrochem. Soc.*, **132** (1985) 266.
99. J. C. Canullo, W. E. Triaca, and A. J. Arvia, *J. Electroanal. Chem.*, **175** (1984) 337.
100. R. M. Cerviño, A. J. Arvia, and W. Vielstich, *Surf. Sci.*, **154** (1985) 623.
101. J. Canullo, Y. Uchida, G. Lehmpfuhl, T. Twomey, and D. M. Kolb, *Surf. Sci.*, **188** (1987) 350.
102. W. Hössler, E. Ritter, and R. J. Behm, *Ber. Bunsenges Phys. Chem.*, **90** (1986) 205.
103. T. A. Twomey, Y. Uchida, G. Lehmpfuhl, and D. M. Kolb, *Z. Phys. Chem. N. F.*, **160** (1988) 1.
104. T. A. Twomey, *J. Electroanal. Chem.*, **270** (1989) 465.
105. Y. Uchida and G. Lehmpfuhl, *Surf. Sci.*, **243** (1991) 193.
106. U. Bethge and M. Klaua, *Ann. Physik 7. Folge*, **17** (1966) 177.
107. M. H. Hölzle, Ph.D. Thesis, University of Ulm (1995).
108. U. W. Hamm, D. Kramer, R. S. Zhai, and D. M. Kolb, *Electrochim. Acta*, **43** (1998) 2969.

3

IMAGING METAL ELECTROCRYSTALLIZATION AT HIGH RESOLUTION

RICHARD J. NICHOLS

Nature presents a multitude of crystallographic shapes and forms, which provides the basis for much artistic creativity and scientific discovery. Humans, in turn, have developed and utilized crystallization processes for their own technological benefit. By controlling the conditions, people have been able to produce materials with the desired compositional, structural, mechanical, and morphological characteristics.

Metal electrocrystallization is the science and technology concerned with the formation of solid metallic deposits at the cathode of an electrochemical cell. In practical terms, it forms the basis of commercially important metal plating, electrowinning, electroforming, and metal finishing industries [1]. Electrocrystallization involves a chain of distinct events, starting from the metal ion in solution and leading to the solid deposit [2]. The chain of events for metal electrocrystallization are illustrated in Figure 3.1. The metal ion, in its solvated or complexed form, diffuses to the electrode surface where it loses part or all of its solvation sphere or complexing agents. This process is accompanied by electron transfer to form a partially or completely discharged metal adatom, which is followed by surface diffusion and aggregation of sufficient adatoms to form a critical cluster of metal atoms. For a nucleus to evolve into a stable entity it must exceed this critical number of adatoms. The new phase then grows by incorporation of adatoms at lattices sites of the deposit. The nature

---

Imaging of Surfaces and Interfaces (Frontiers of Electrochemistry, Volume 5).
Edited by Jacek Lipkowski and Philip N. Ross.
ISBN 0-471-24672-7. © 1999 Wiley-VCH, Inc.

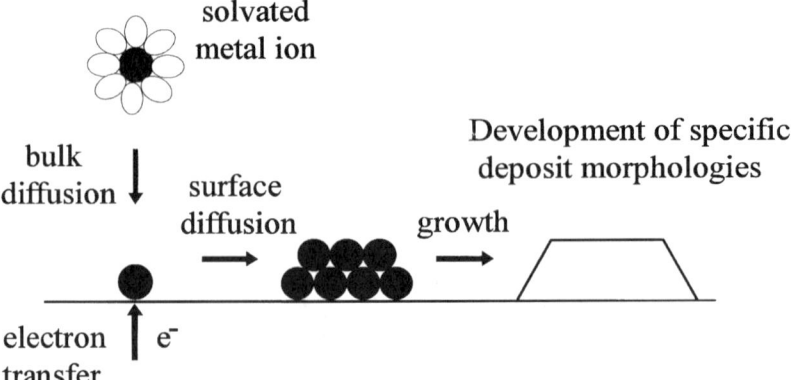

Figure 3.1 A schematic representation of the chain of processes leading to the formation of a bulk metal electrodeposit.

of this growth controls the crystallographic and morphological characteristics of the resulting deposit.

Figure 3.1 and the accompanying discussion clearly illustrate that metal electrocrystallization processes are characterized by highly localized processes occurring on the atomic scale. Although metal electrocrystallization has been studied extensively for many years, only relatively recently have methods become available to examine metal deposition processes at the atomic scale. In particular, local probe methods, such as scanning tunnelling microscopy (STM) and atomic force microscopy (AFM), have opened a new horizon in the investigation of mechanisms of metal electrodeposition, by allowing true contact to be made with microscopic theories for the deposition and growth. Just as important, local probe methods facilitate a view of the growth process over a staggeringly high dynamic range, from atomic resolution to fields of view over a hundred micrometers. As such they enable relationships to be established between microscopic aspects of the growth and macroscopic features of the resulting deposit.

In situ scanning probe microscopy (SPM) methods have been used to examine a wide variety of metal deposition systems. There are more than 100 publications of the application of in situ SPM; to foreign metal adlayers formed at underpotential, as well as bulk metal electrocrystallization. A listing of such publications is shown in Table 3.1. The discussion in this chapter focuses on just a few of these systems, with a view to illustrating the use of local probe methods in studying metal deposition processes. The systems discussed in detail here are copper deposition on gold electrodes and lead deposition on silver electrodes (see table). These systems are significant in that both the formation of metal adlayers at underpotential, as well as the overpotential metal deposition, have been studied extensively using SPM, as well as other conventional and modern electrochemical methods.

TABLE 3.1 A Listing of Single Metal Deposition Systems Studied by In Situ Scanning Probe Microscopy

Deposit	Substrate	STM or AFM	UPD or OPD or ECALE Studied	References
Ag	Ag	STM	opd	3
Ag	Au	STM	opd	4
Ag	Au(100)	STM	upd, opd	5, 81, 173
Ag	Au(111)	STM	upd	6–9, 195
Ag	Au(111)	AFM	upd	10–11
Ag	Au(111)	STM	opd	12–13, 174
Ag	HOPG	STM	opd	14–19, 168, 205
Ag	Pt	AFM	opd	20
Ag	Pt(111)	STM	upd	21–22
Ag	Pt(110)	STM	upd, opd	163
Au	Au(111)	STM	opd	23
Au	HOPG	STM	opd	189
Bi	Au(111)	AFM	upd	24–25
Cd	Cu(111)	AFM	upd, opd	26
Cd	Pt(110)	STM	upd	158
CdS	Ag(111)	STM	ECALE	172
CdS	Au(100)	STM	ECALE	175
CdTe	Au(110)/Au(111)	STM	ECALE	178, 200
Cu	Ag(111)	STM	opd	27–28
Cu	Ag(100)	STM	opd	164
Cu	Au	STM	opd	4, 29–30, 103, 155–156
Cu	Au	AFM	opd	198
Cu	Au(100)	STM	upd	31, 32, 33
Cu	Au(110)	STM	upd	32, 38
Cu	Au(111)	STM	upd	28, 32, 33, 39–46, 63
Cu	Au(100)	STM	opd	34, 35, 36
Cu	Au(100)	AFM	opd	37
Cu	Au(111)	STM	opd	28, 35, 36, 41, 45, 46, 49–53, 161, 162
Cu	Au(111)	AFM	upd	47, 48
Cu	Au(111)	AFM	opd	47, 53, 166–167
Cu	Cu	STM	opd	54–55
Cu	HOPG	STM	opd	56, 57
Cu	p-GaAs(100)	AFM	opd	58, 184, 203
Cu	polypyrrole	AFM	opd	59
Cu	Pt	STM	opd	60–62
Cu	Pd	STM	opd	62
Cu	Pt(100)	STM	opd	63
Cu	Pt(100)	AFM	opd	64–65

TABLE 3.1 (*Continued*)

Deposit	Substrate	STM or AFM	UPD or OPD or ECALE Studied	References
Cu	Pt(111)	AFM	opd	64
Cu	Pt(110)	STM	upd	66
Cu	Pt(111)	STM	upd	44, 63, 68–70, 181, 194
Cu	Pt(110)	STM	opd	67
Cu	Pt(111)	STM	opd	67, 71
GaAs	Au(100)/Au(111)	STM	ECALE	204
Hg	Au(111)	AFM	upd, opd	72–73
Hg	Au(111)	STM	upd	180
Li	HOPG	STM	opd	74
Ni	Au(111)	STM	opd	75, 190–191
Ni	Cu(100)	STM	opd	187
Ni	HOPG	STM	opd	76
Ni	beta-brass	STM	opd	192
Ni	n-GaAs	STM	opd	182
Pb	Ag(100)	STM	upd	77–83, 197, 199
Pb	Ag(100)	STM	opd	78–83
Pb	Ag(111)	STM	upd	78–88, 197
Pb	Ag(111)	STM	opd	78–83, 88–89
Pb	Au(111)	STM	upd	90–94
Pb	Au(111)	STM	opd	94
Pb	Au(111)	AFM	upd	95
Pb	Au(100)	STM	upd	199
Pb	HOPG	STM	opd	96–98
Pb	Pt(100)	STM	upd	99
Pb	n-Si(111)	STM	opd	196
Pd	Au(111)	STM	opd	100
Pd	Au(111)	STM	opd	193
Pd	HOPG	STM	opd	101, 201
Pt	Au(111)	STM	opd	202
Pt	HOPG	STM	opd	96, 102, 103
Se	Au(100)	STM	—	179
Se	Au(111)/Au(110)	STM	—	188
Sn	n-GaAs	STM	opd	104
Te	Au(111)/Au(100)/Au(110)	STM	upd	177
Tl	Ag(111)	STM	upd	85, 87
Tl	Pt(111)	STM	upd	194
Tl	Au(111)	STM	upd	105
Zn	HOPG	STM	opd	106, 107

upd = underpotential deposition.
opd = overpotential deposition.
ECALE = electrochemical atomic layer epitaxy.

3.1 METHODOLOGY

Both in situ STM and AFM have been used to study metal electrodeposition, although the number of STM studies of metal electrodeposition greatly outnumbers the number of AFM studies. As such this chapter focuses primarily on in situ STM studies. There are of course many common aspects between imaging the structure of electrode surfaces and imaging dynamic processes such as bulk metal electrocrystallization. Several good articles deal with STM imaging of electrode surfaces [45,108,109], and hence this section will primarily focus on the considerations for STM imaging of metal electrocrystallization.

In situ STM imaging of electrocrystallization processes has been performed by a number of groups worldwide, using both commercial and homemade instrumentation. The STM tip is immersed in the electrolyte for in situ imaging, and it effectively forms a fourth electrode along with the working, reference, and counter electrodes of the electrochemical cell. Typically a bi-potentiostat has been used, so that the tip and working electrode potentials can be controlled independently with respect to the reference electrode. For metal deposition studies, metal wire reference electrodes are most convenient due to their small size and stability. They offer additional convenience because all potentials are then measured with respect to the Nernst potential for M/M^{n+} for the electrolyte under study. For instance, high purity copper wire reference electrodes have been used for Cu electrocrystallization studies [49], while Ag wires and Pb covered Pt wires have been used for Ag [13] and Pb [78] deposition studies, respectively.

A variety of STM tips have been used including tungsten, Pt-Ir and Ir. These tips have to be isolated with a nonconductive coating to leave only a small apex of the tip open to the electrochemical environment. Many tip coating materials and methods are now available, including apiezon wax, polymers, nail varnish, glass, or electropaint. With use of the latter method, the effective radius of the exposed tip apex has been reduced down to typically significantly less than 0.1 μm [110].

The methods used to investigate the in situ growth of bulk electrodeposits can be divided into two categories. In the first method the area of interest is typically imaged prior to metal deposition. The tip is then retracted from the surface, and metal deposition is initiated by applying an appropriate electrode potential to the sample. After a certain deposition time the tip is reengaged and the deposit imaged. This method has been used by a number of groups to investigate, for instance, the deposition of Cu on Pt [67], Au on Au(111) [23], and Ag [17] and Pt [102] electrodeposition on graphite. The second method involves a continuous real-time imaging of the electrodeposit during its development, following a potential excursion to negative overpotentials. This method has been used to investigate a wide variety of bulk metal electrodeposition processes, including Cu^{2+}/Au [34–36,49–53], Cu^{2+}/Ag [27,28], Cu^{2+}/Cu [54], Cu^{2+}/Pt [60], Pb^{2+}/Ag [77], and $Ag^{2+}/HOPG$ [18,19].

The advantage with the method involving retraction of the tip during deposition is that the influence of the tip on the deposition process can be eliminated if the tip is retracted far enough from the substrate surface. Disadvantages include the method's inability to follow the deposition in real time, the difficulty in "catching" the initial stages of deposition, and sometimes difficulties in precisely relocating the same scanning area after the tip has been retracted and then re-engaged in tunnelling mode.

As will be shown in the following sections, impressive images of the progressive growth of electrodeposits have been obtained using the continuous scanning method. Of course, it is particularly attractive to be able to directly follow the growth of the deposit at high resolution, right from the initial stages. This feature has made it the preferred method of several groups. However, the method does have its drawbacks. The tip's close proximity to the surface means that it has an unavoidable effect on the bulk deposition process itself. The physical dimensions of the tip and the typically relatively small scanning area mean that the tip can limit the diffusive flux of metal ions in solution to the growth centers under observation. In this respect, it should be noted that there is effectively a thin layer of electrolyte between the tip and the area of the surface which is being scanned [36,51]. Just as in conventional thin layer cells, a limited amount of metal ion is available for deposition, and depletion of this metal ion in solution can occur during sustained deposition. Such depletion effects are particularly marked during rapid deposition at relatively high overpotentials. Under such conditions markedly reduced growth rates have been observed under the tip. These factors mean that microscopic growth rates observed by SPM for bulk deposition on areas covered by the tip can be far less than those on "free" areas away from the tip [103,165]. It has also been observed that the deposit under imaging can reach certain dimensions and then cease to grow significantly further, presumably due to depletion of metal ions from the thin layer.

These drawbacks initially seem to paint a rather discouraging picture for successfully imaging bulk metal electrodeposition in situ and in real time with STM. Clearly, in such in situ STM experiments the deposition conditions are significantly different from those that occur during "conventional" metal plating. As such, in situ SPM can be deemed unsuitable for the quantitative measurement of reliable microscopic growth rates for electrodeposits. However, if experimental limitations are understood and respected, then carefully, well-defined experiments in which given parameters are individually and systematically varied will yield significant qualitative information.

Using in situ STM, it has been possible to follow, for instance, Cu electrodeposition in real time, right from the initial stages up to the deposition of hundreds of monolayer thick deposits [36,51]. Certain experimental precautions can aid such imaging. For instance, tips with a sharp macroscopic radius of curvature of the tip apex will reduce depletion effects. With such tips, evidence indicates that the thin electrolyte layer between the tip and the surface can be replenished, during slow or moderate rates of deposit growth, by diffusion of metal ion from the bulk electrolyte in the proximity of the tip [36,51]. In this respect, factors such as micro-scale convection arising from tip scanning may help with the replenishment of the thin layer. Another wise precaution is to limit the potential difference between the tip and the substrate surface (tunnelling voltage) to minimize the electrostatic influence of the STM tip [51]. This electrostatic influence of the STM tip is most apparent at high tunnelling voltages and is presumably caused by overlap of two interfering double layers, those of the tip and substrate respectively [28]. Practically, this perturbation can be reduced by holding the tip potential (E_{tip}) near to the Nernst potential for M/M^{n+}. For instance, when $E_{tip} = 0$ mV and deposition is studied at -100 mV, the tunnelling voltage is only 100 mV.

For the observation and understanding of the microscopic aspects of metal deposition processes, well-defined single crystalline electrodes are indispensable. Conventional single crystalline electrodes, prepared by well-established methods, have been used to great effect for examining metal electrodeposition with STM. Notable examples include the use of massive gold [32,33,35], silver [77–89], platinum [21,22,44,63–71], and copper single crystals [26]. Haiss and Sass introduced the use of well-defined metal films on glass electrodes for electrochemical STM studies [111]. Such electrodes yield films with (111) orientation after an appropriate heat treatment and are much cheaper than conventional single crystals. Suitable gold-on-glass films may be prepared by vacuum evaporation at 10^{-6} torr of 2 nm of chromium followed by 200 nm of gold. The chromium layer promotes adhesion of the gold layer to the glass substrate. A commercially available flame polished glass (e.g., "Tempax"-Glass, Berliner Glas K.G.) has been used as the glass substrate. Prior to an experiment an Au film sample is flame annealed by heating the sample just to yellow heat for 30–120 sec in a Bunsen burner flame, or better still a pocket-sized "microburner" (such as the ones used for hobby applications). Care must be taken that the film does not exceed yellow heat, which can be ensured by pushing the sample (which is held at the edge by ceramic-ended tweezers) in and out of the flame. The sample is then left to cool in air. STM images of well-prepared Au films are presented in Figure 3.2 [111,112]. These images show the step and terrace surface structure of gold microcrystallites formed after flame annealing (Fig. 3.2a) and at higher resolution the reconstructed surface indicative of clean, well-prepared Au(111) films (Fig. 3.2b). Ag films on glass have been produced by a similar method [27,28].

3.2 METAL ADLAYERS

The electrochemical formation of the first layer of a metallic deposit on a foreign metal substrate occurs, in many circumstances, at potentials that are positive of the equilibrium potential for deposition of the bulk phase [113,114]. Such metal adlayers, formed at underpotential, may be considered as precursor layers on top of which the bulk metal phase nucleates. As such they can

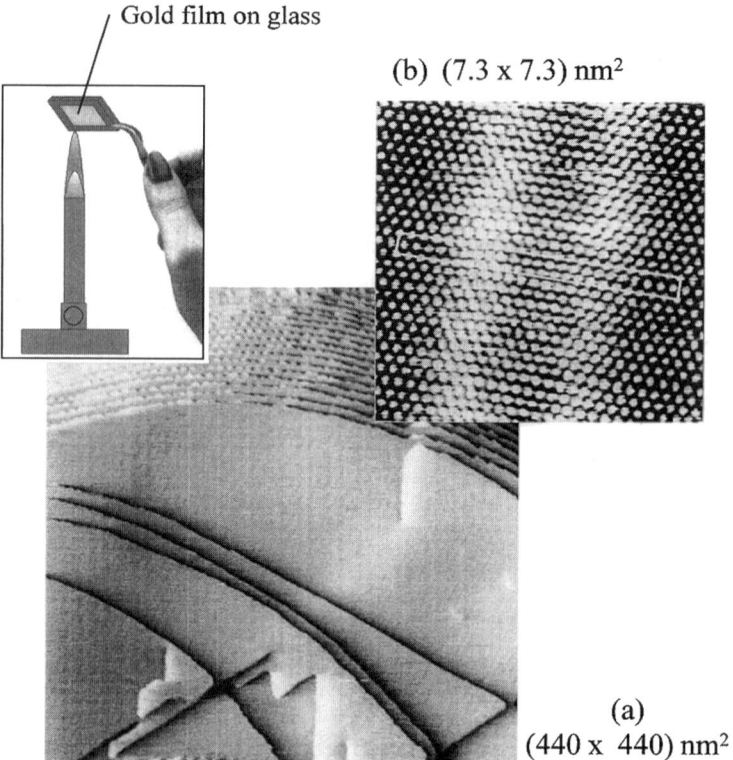

Figure 3.2 A simple flame annealing method can be used to produce well-defined gold films (see insert). (a) STM image of a gold film electrode after flame annealing, taken in air. (b) High-resolution STM image of a gold film showing the reconstructed Au(111) surface after flame annealing. The image was taken in glycerine. Adapted from [112], with permission.

play a major role in the development of the bulk electrodeposit. The electrochemical formation of such foreign metal adlayers at potentials positive of the Nernst potential for deposition of the bulk metal phase is termed underpotential deposition (UPD) [113,114]. UPD can involve the deposition of a foreign metal adatom in sub-monolayer amounts up to one or even several monolayers. The formation of UPD adlayers is related to the energetics of the foreign metal (M)–substrate (S) bonding. UPD typically occurs where the M-S bonding is significantly stronger than that in the bulk metal deposit (M-M).

In this section two UPD systems will be discussed, namely $Cu^{2+}/Au(111)$ [32,33,39–48] and $Pb^{2+}/Ag(111)$ [78–88]. These two systems are particularly interesting and relevant here because they have been extensively studied, and they show markedly differing behavior with respect to metal adlayer formation, which in turn reflects on the bulk metal electrocrystallization on these surfaces.

The underpotential deposition of Cu^{2+} on Au(111) has been extensively

3.2 METAL ADLAYERS   107

studied using a diverse variety of techniques, including the ex situ methods of LEED, RHEED, and AES [115–117], electrochemical measurements and thermodynamic analysis [118,119], in situ X-ray diffraction [120,121], surface extended X-ray absorption fine structure (SEXAFS) [122–124], X-ray absorption near edge structure (XANES) [125], radiotracer measurements [126], as well as in situ STM [32,33,40] and AFM [47,48]. A cyclic voltammogram of Au(111) in 0.1 H_2SO_4 + 1 mM $CuSO_4$ is shown in Figure 3.3 [112]. The two sets of voltammogram peaks indicate the formation of two adlayers, at medium

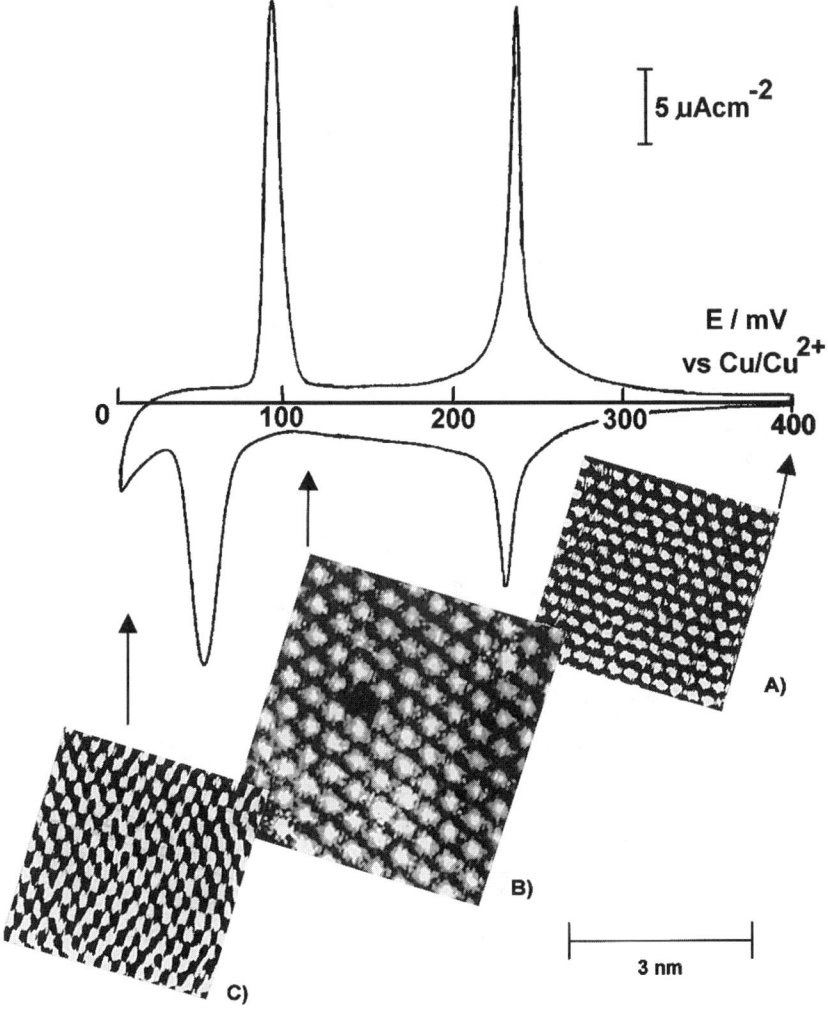

Figure 3.3 A cyclic voltammogram of Au(111) in 0.1 M H_2SO_4 + 1 mM $CuSO_4$ shown in conjunction with high-resolution STM images recorded at the marked electrode potentials. Used with permission of [112].

and high coverage, respectively. These adlayers were characterized some time before in situ STM investigation, using the ex situ diffraction methods of low energy electron diffraction (LEED) and reflection high energy electron diffraction (RHEED) as well as Auger electron spectroscopy (AES) [115–117]. The LEED pattern indicated that the peak at low underpotential corresponds to the formation of a (1×1) adlayer, while a ($\sqrt{3} \times \sqrt{3}$)R30° superstructure pattern was observed for the medium coverage structure, which is formed at potentials between these two peaks [115-117]. This latter LEED pattern could correspond to one of two Cu superstructures, and open ($\sqrt{3} \times \sqrt{3}$)R30° structure with a coverage of 1/3 of a monolayer, or a ($\sqrt{3} \times \sqrt{3}$)R30° honeycomb structure with a coverage of 2/3 monolayers. AES pointed to a honeycomb structure by indicating a copper coverage closer to 2/3 monolayer [116,117].

STM images at potentials corresponding to the bare gold substrate (Fig. 3.3a) and potentials where the medium (Fig. 3.3b) and high coverage (Fig. 3.3c) structures are formed are shown along with the voltammogram in Figure 3.3. The structure observed in image (Fig. 3.3a) clearly corresponds to that of an unreconstructed Au(111) substrate. The structure imaged in Figure 3.3b corresponds to an open ($\sqrt{3} \times \sqrt{3}$)R30° structure, with a coverage of $\frac{1}{3}$ monolayer. This result was originally misinterpreted as direct imaging of copper adatoms in an open ($\sqrt{3} \times \sqrt{3}$)R30° adlayer with a $\frac{1}{3}$ monolayer coverage. However, layer X-ray spectroscopic evidence [120,121], careful chronocoulometric studies [118,119], and model calculations [127] refuted this conclusion in favour of the higher coverage ($\sqrt{3} \times \sqrt{3}$)R30° honeycomb structure, originally suggested by ex situ LEED and AES studies. The model developed by Blum and Huckelby [127] for the medium coverage structure is shown in Figure 3.4.

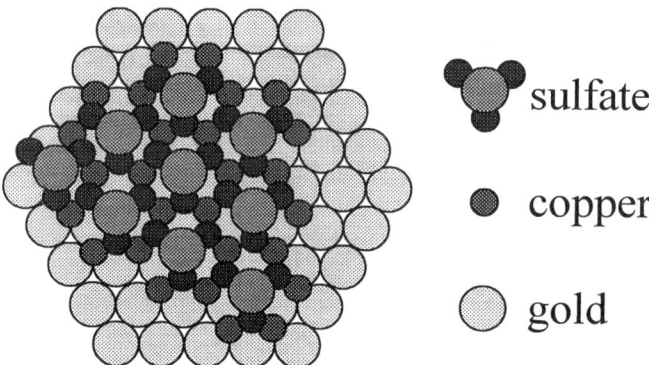

Figure 3.4 The model developed for the medium coverage structure of underpotentially deposited Cu on Au(111). The co-adsorbed sulphate forms a ($\sqrt{3} \times \sqrt{3}$)R30° structure, with a $\frac{1}{3}$ monolayer coverage, effectively "locking" the Cu adatoms into a honeycomb structure, which also has a ($\sqrt{3} \times \sqrt{3}$)R30° unit cell but with a coverage of $\frac{2}{3}$ monolayer. Reprinted with permission of [112].

3.2 METAL ADLAYERS

This $\frac{2}{3}$ coverage honeycomb structure of Cu adatoms is anchored into place by strongly adsorbed sulphate anions, occupying the honeycomb centers, with a coverage of $\frac{1}{3}$ of a monolayer. One must now conclude that, in this case, STM images the co-adsorbed anions that have the open $(\sqrt{3} \times \sqrt{3})R30°$ structure [120]. This clearly demonstrates that high-resolution STM images must be interpreted with care, particularly in the electrochemical situation where one or more co-adsorbed species may be present.

Figure 3.3c shows the in situ STM image of the (1×1) Cu adlayer. Structural parameters for this adlayer have been measured by EXAFS, with the Cu-Cu nearest neighbor spacing being determined as 0.292 +/− 0.003 nm [123]. In addition, surface EXAFS has been used to determine the Cu-Au bond length as 0.258 +/− 0.003 nm, which suggest that the Cu adatoms occupy a threefold hollow site [123]. Therefore, the clear evidence is that the metal adlayer formed at potentials just positive of the Nernst potential is a (1×1) Cu monolayer commensurate with the underlying Au(111) substrate. The Cu-Cu distance in Cu(111) at 0.256 nm is significantly shorter than that of Au-Au in Au(111), of 0.289 nm. This disparity between the Cu-Cu distance in the underpotentially deposited monolayer and that of a bulk Cu(111) surface has marked influences on the growth of the bulk electrodeposit on this surface. The UPD-Cu monolayer may be considered as a precursor layer on top of which the bulk phase is formed. The fact that this adlayer is "expanded" with respect to the interatomic spacing if a bulk Cu(111) surface, influences both the thermodynamics and kinetics of bulk copper deposition. Clearly, it would be thermodynamically unfavorable for bulk copper to develop epitaxially by a single monolayer-by-monolayer growth mode on top of this metal adlayer, since great strains would be expected to develop in such a film as a result of the unfavorable Cu-Cu spacing. This misfit at the interface, in turn, effects the growth kinetics of the bulk phase in different crystallographic directions, which is apparent from STM images presented in the following section.

In the case of Cu underpotential deposition on Au(111) from pure (halide free) sulphuric electrolytes, only two well-ordered commensurate metal adlayers are observed. The situation is quite different for the underpotential deposition of Pb on Ag(111). In this case, the metal-substrate misfit is in the opposite direction as that of Cu UPD on Au(111), with the Pb-Pb spacing in a Pb(111) surface (0.350 nm) being significantly greater than Ag-Ag (0.289 nm) in the Ag(111) substrate. In the underpotential range from 0 to 100 mV compressed and rotated hexagonal close-packed (hcp) Pb adlayers have been observed on Ag(111) by in situ STM [78,83]. Such an example is shown in Figure 3.5. In this case, the structure observed in the high-resolution STM image, has been interpreted as a direct imaging of Pb adatoms in the UPD adlayer. Such a conclusion is consistent with grazing incidence in situ X-ray diffraction measurements [128]. In the case of STM images such as Figure 3.3b, it is now clear that the strongly adsorbed sulphate is imaged and not the Cu adatoms. By contrast, Figure 3.5 has been interpreted as a direct imaging of Pb adatoms. In this case

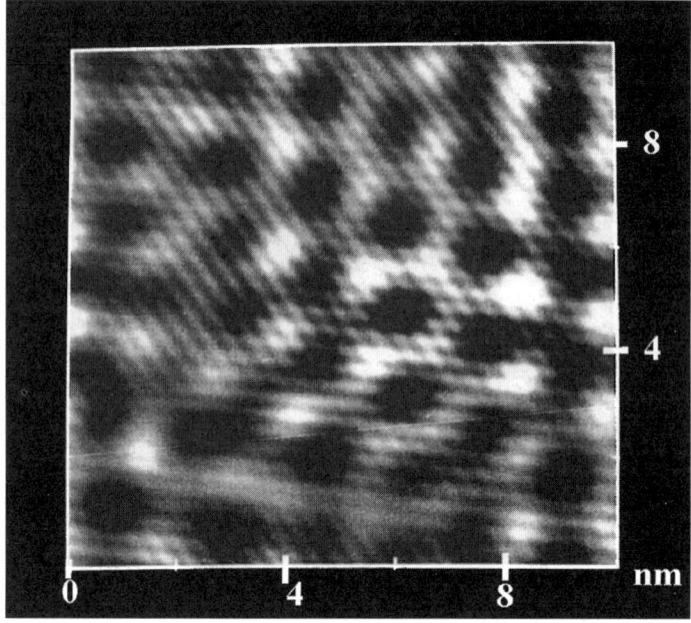

Figure 3.5 An in situ STM image of a compressed and rotated Pb adlayer on Ag(111), showing the moiré pattern of the Pb monolayer at E = 28 mV (vs. Pb/Pb^{2+}). Reprinted by permission of [78].

ClO_4^- is the anion present in the electrolyte, which is known to be a typically much weaker adsorbate than sulphate or bisulphate. It seems likely that this accounts for the difference.

The distinctive Moiré pattern apparent in Figure 3.5 had been taken as evidence for the compression and rotation of the hexagonal close-packed (hcp) Pb adlayer [78,83]. Such images are consistent with the results of in situ grazing incidence X-ray diffraction [128]. In addition, such X-ray diffraction measurements have shown that the Pb-Pb spacing for the hcp monolayer decreases linearly with electrode potential, in the range from +150 mV versus Pb/Pb^{2+} (Pb-Pb spacing 0.345 nm) to 0 mV (0.339 nm) [128]. The compression of these metal adlayers has marked consequences for the growth of the bulk phase. As for bulk deposition on Cu-monolayer covered Au(111), this significant misfit between the deposit and the monolayer-covered substrate effects both the thermodynamics and kinetics of the formation of the bulk phase, as discussed in the section on metal growth.

3.3 NUCLEATION

Following metal adatom formation and electron transfer the adatoms aggregate to form nuclei. Clusters larger than the critical nucleus size (N_c) stably evolve

3.3 NUCLEATION

into growth centers. As such, N_c is an important quantity in the thermodynamics and kinetics of the nucleation process. Clearly this process of nucleation plays a critical role in the evolution of the deposit, by determining the number density of growth centers and their distribution across the surface. It has been well accepted, at least since the early theories of Kossel and Stranski, that surface defects can play a determining role in nucleation and growth [129,130]. Defects, such as monoatomic steps and kink sites in steps, are certainly randomly distributed and can also be dynamic. It is here that local probe methods can provide the essential real-space information. By contrast, other structural techniques, such as the diffraction methods, require long-ranging periodicity and are hence not suited to the examination of individual defects. SPM also stands out from other microscopic methods, such as optical and electron microscopies. Clearly, SPM offers about a 10^4 greater lateral resolution than optical methods that have been used to examine metal deposition. Although transmission electron microscopy also offers atomic scale resolution of metallic deposits, it cannot be applied in situ and usually requires rather restrictive sample preparation.

Figure 3.6 shows the originally published in situ STM images of copper clusters that have formed during electrodeposition on Au(111) substrates [35].

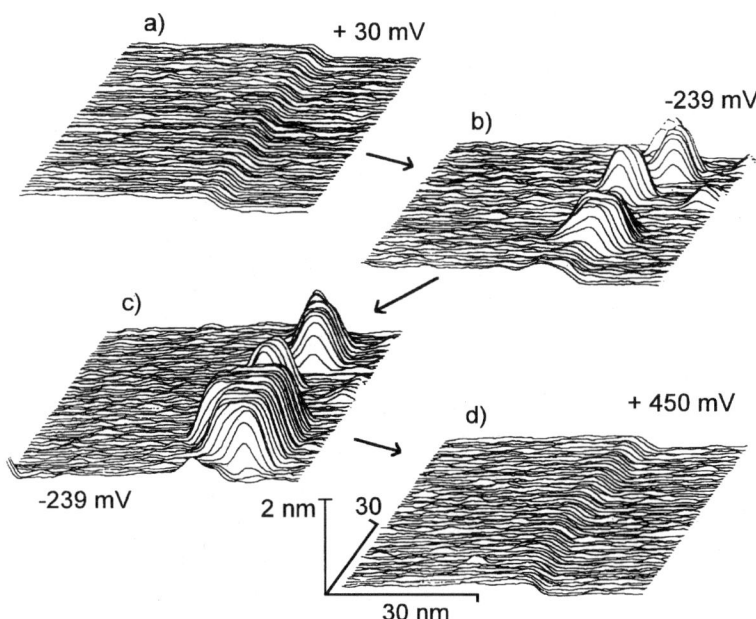

Figure 3.6 In situ STM images taken (a) before, (b,c) during and after bulk copper deposition on a Au(111) electrode in 0.1 M $HClO_4$ + 5 × 10^{-5} M $Cu(ClO_4)_2$. Tunnelling current, I_t = 2.5 nA and tip potential E_{tip} = +60 mV. Adapted from [35], with permission of Elsevier Science S.A. [154].

The top image shows the gold substrate at potentials positive of the Nernst potential for bulk copper deposition. This area of the surface is characterized by two atomically smooth terraces separated by a monoatomic high step edge. Upon stepping the electrode potential negative of the Nernst potential, distinctive copper clusters form at the step edges. By contrast, the terraces in Figure 3.6b remain free from copper clusters. The clusters are seen to grow from Figure 3.6b to the subsequent image in Figure 3.6c. These STM images are a visual verification of the important role surface defects can play in the initial stages of electrocrystallization. Clearly, they are in good accordance with the textbook model of Kossel and Stranski, which is depicted in Figure 3.7. Nucleation occurs preferentially at step edges, where an adatom is more high coordinated with the surface that an adatom on an atomically flat terrace. The energetics of copper nucleation on flat gold terraces are clearly less favorable, occurring only at longer times or higher overpotentials. Plate 3.1 shows a "top view" representation of Cu clusters electrodeposited on Au(111). Such STM images can be used to assess the distribution of growth centers and their morphological characteristics.

Images such as Figure 3.6 and Plate 3.1 enable the direct determination of the preferential sites for nucleation and their distribution across the surface. However, in situ has not been able to observe nuclei at their "conception." In fact, the metal clusters typically imaged in an in situ experiment far exceed the size of a critical nucleus. The number of metal adatoms in a critical cluster (N_c) can be estimated from analysis of conventional electrochemical current versus time transients, in the nucleation regime, using the relationship:

$$N_c = \frac{2.303kT}{ze} \frac{d \log J}{d|\eta|}$$

where J is the current density, and η is the overpotential [131,132]. For instance, a value of $N_c = 11$, for the overpotential deposition of Pb on Ag(111) from an acidic perchlorate electrolyte, has been obtained [131].

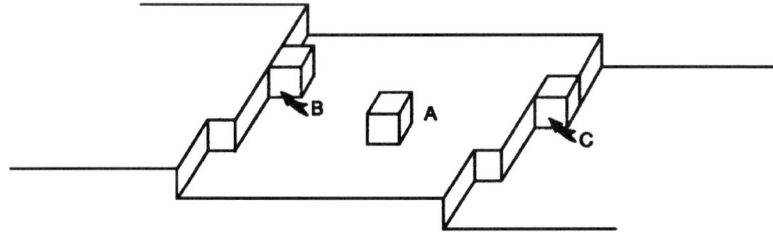

Figure 3.7 "Textbook" representation of the possible defect sites on an otherwise "perfect" single crystal surface. Incorporation of adatoms are shown at (A) terrace site, (B) step site, and (C) kink site.

3.3 NUCLEATION

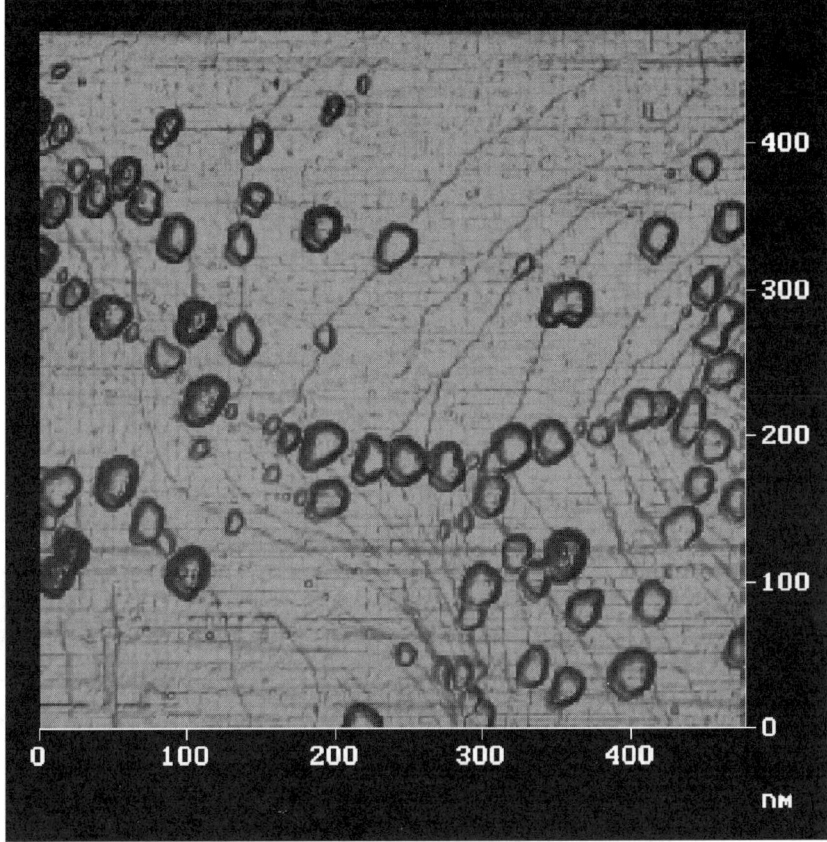

Plate 3.1 In situ STM image taken during the bulk electrodeposition of Cu on Au(111) at −200 mV, in 0.1 M H_2SO_4 + 1 mM $CuSO_4$. See color plate section.

The size of the smallest clusters typically imaged during a bulk metal deposition experiment typically exceeds this size. For instance, the smaller of the copper clusters, which has nucleated at the step in Figure 3.6, can be estimated to contain more than 500 metal atoms. These observations are consistent with the rapid kinetics of growth (or dissolution) of critical nuclei formed a significant overpotential. Clearly, the time frame of SPM imaging is too slow to observe this interesting phase of metal electrodeposition, and clusters are typically first observed at a much more advanced stage of their growth. By contrast, STM studies of metal growth at low temperatures in UHV have succeeded in imaging metal clusters near to the "critical" nucleus size [133].

The images of Figure 3.6 clearly show that step edges are significantly more favorable for copper nucleation than terraces. Following the Kossel-Stranski model, kinks in a step edge are expected to be lower energy nucleation sites than straight step edges. Unfortunately, the imaging of individual kink sites is impeded by the difficulty in obtaining *lateral* atomic resolution of step edges, particularly if the kink sites are mobile. However, comparisons of relatively straight with highly curved step edges are useful in this respect. A highly curved step edge, such as a small circular monoatomic high island of the substrate, is expected to possess a high kink site density. A comparison of straight and curved step edges has shown higher nuclei density on the curved steps. Indeed, in the case of highly curved steps a continuous rim of copper is formed along the monoatomic step edges [34].

SPM can also be used to analyze the effect of electrochemical conditions, such as overpotential or electrolyte composition, on the distribution and density of nuclei [36,49,51]. This analysis is illustrated in Figure 3.8, which shows topview images of the substrate surface prior to metal deposition (Figure 3.8a) and following potential steps to −200 and −400 mV, respectively [49]. The small raised patches in Figures 3.8b and 3.8c correspond to individual copper

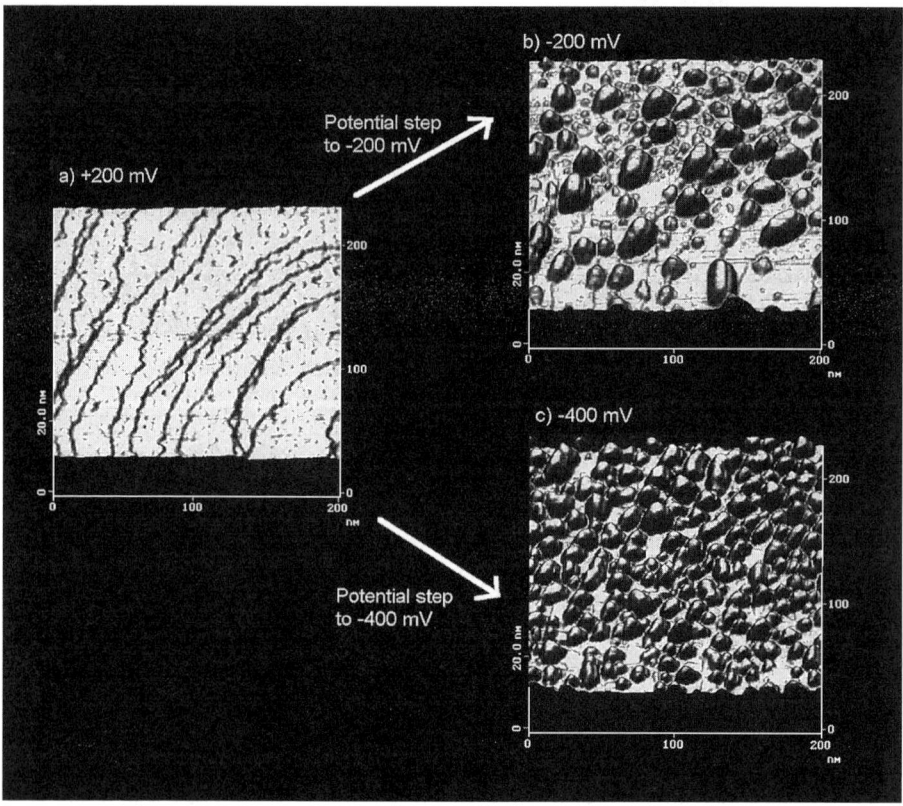

Figure 3.8 Topview images of an Au(111) surface taken by in situ STM, (a) prior to bulk copper deposition and (b) following potential steps to −200 mV and (c) −400 mV. The electrolyte: 0.1 M H_2SO_4 + 1 mM $CuSO_4$·; I_t = 5 nA and E_{tip} = +15 mV. Adapted from [49], with permission of Elsevier Science S.A. [154].

clusters. Magnification of the images shows, as before, that nucleation is predominantly at defects. About 85 readily distinguishable clusters (in a 240×240 nm area) are observed at the lower overpotential, while approximately 180 growth centers are seen at the higher overpotential. This result corresponds to nuclei densities of 1.5 and 3.1×10^{11} nuclei cm^{-2}, at the lower and higher overpotential, respectively. This observation of an increase in the number of growth centers with overpotential is consistent with the expectation of fundamental nucleation theories, since increasing the overpotential is equivalent to increasing the supersaturation.

Interestingly, the electrodeposition of Cu on Ag(111) shows a strikingly different behavior to that of Cu on Au(111) [27,28]. The two substrates, Ag(111) and Au(111), have similar lattice constants and as such form an interesting basis for comparison. In the case of Cu deposition on Au(111), nucleation at steps is observed by in situ STM, whereas Cu clusters on Ag(111) form predominantly on flat terraces, at lower overpotentials, with nucleation at steps being only generally observed at high overpotentials. Kolb and co-workers have proposed that this could arise from the very different potential of zero charge (p.z.c.) for the two substrates, of around +0.22 V (vs. Cu/Cu^{2+}) for Au(111) and −0.69 V (vs. Cu/Cu^{2+}) for Ag(111) [27,28]. Then, Cu deposition occurs on a negatively charged Au(111) surface, while the Ag(111) surface is positively charged for typical Cu deposition conditions. They have postulated that specific adsorption of sulphate on Ag(111) blocks the steps, leading to the unusual nucleation of bulk copper on terraces. This contrasts with the negatively charged Au surface, where anion adsorption is expected to be substantially lower and the "classical" nucleation at steps is observed. At higher overpotentials, nucleation at steps is observed on Ag(111), presumably arising from a lower degree of sulphate coverage at the more negative potential [27,28].

3.4 METAL GROWTH IN ELECTROCRYSTALLIZATION

Following nucleation the metal clusters grow by incorporation of metal adatoms at lattice sites of the deposit. This stage is of particular interest, because it leads to the development of crystallographic and morphological characteristics of the deposit. Of course, whenever there is growth of the metal electrodeposit, the system must be displaced from equilibrium, as reflected by the overpotential. Clearly we have a difficult situation to appraise due to the separate influences of the thermodynamics and kinetics of the electrocrystallization. Consequently, a competition arises between the drive of the system to reach its minimum energy configuration, as dictated by thermodynamics, and the limitations imposed by kinetics. The kinetic limitations may arise from, for instance, the rate of transport to growth centers or the rate of adatom formation or incorporation into lattice sties. Metal electrocrystallization is often performed far from equilibrium, and these kinetic limitations can dominate the growth morphology of the deposit. The final shape of the electrodeposit will depend on the relative rates of growth in different crystallographic directions. In the illustration in Figure 3.9, the slowest growing face dominates the external appearance, with the horizontal face growing itself out of existence [134].

Plate 3.2 shows copper clusters electrodeposited on a copper monolayer covered Au(111) substrate [52]. Such a growth morphology, involving the formation of three-dimensional islands on top two-dimensional adlayer may be termed Stranski-Krastanov growth. Other common growth modes, observed for other systems, are Volmer Weber growth (3-D deposit islands directly on top the substrate) and Frank-van-der-Merwe growth (layer-by-layer growth). However, strictly speaking these are thermodynamic models for the growth and they do not take into account kinetic factors, which may give rise to a certain growth morphology.

The clusters in Plate 3.2 have a distinct plateau-like morphology, with flat tops and steep edges. They are significantly broader than they are high (in this respect it should be noted that the z-axis is significantly expanded with respect to the x and y axes). The shape of these clusters has been rationalized by considering a significantly different rate of growth in the lateral and normal directions, with respect to the surface.

A model for the growth of the *plateau* crystallites is shown in Figure 3.10. For the purpose of this model three individual growth rates have been taken. The first is the lateral growth of the crystallite by copper atom incorporation at the edge of the crystallite, rate ν_1. The second is the formation of a two-dimensional copper nucleus on top of the crystallite, rate ν_2. The third is propagation of the two-dimensional nucleus across the upper face of the crystallite, increasing its height by a single monolayer, rate ν_3. Hence the rate of lateral growth is determined by ν_1, while the rate of vertical growth is determined by ν_2 and/or

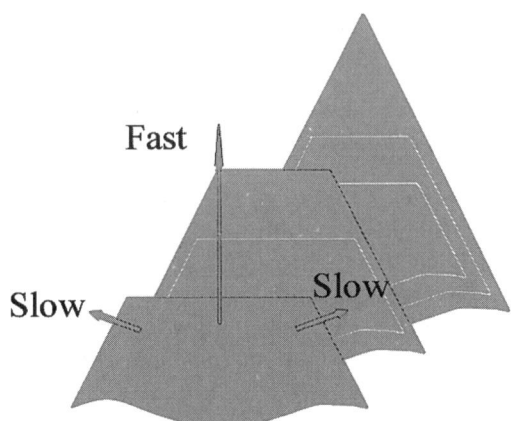

Figure 3.9 An illustration showing how the slowest growing face of a crystallite can dominate its external appearance. Adapted from [134], by permission of Oxford University Press.

3.4 METAL GROWTH IN ELECTROCRYSTALLIZATION 117

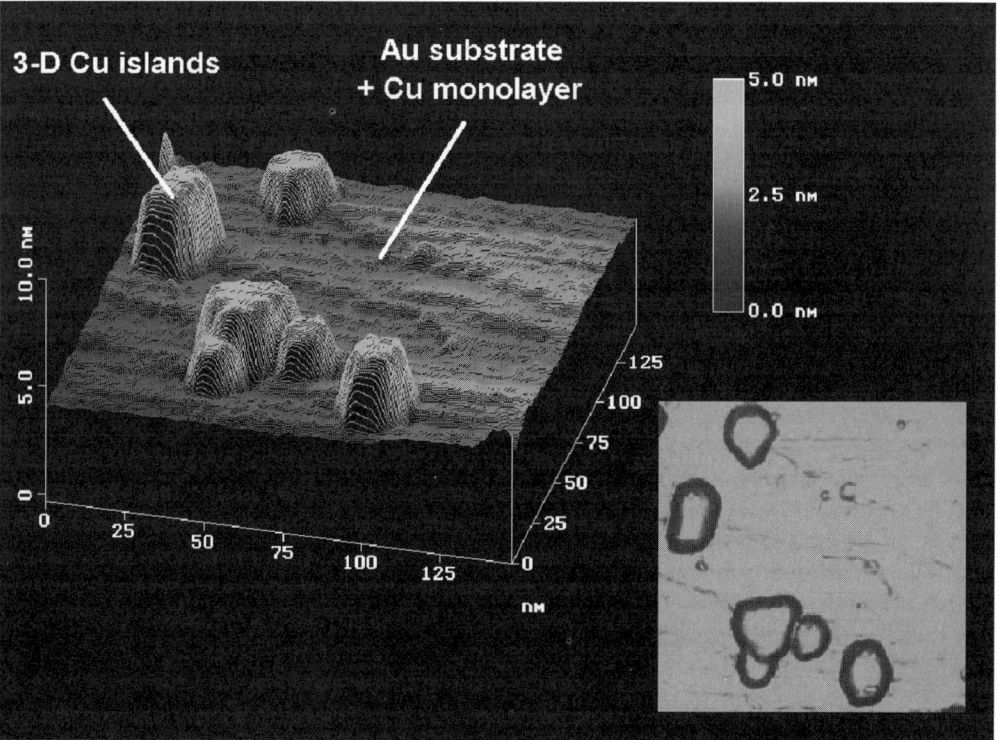

Plate 3.2 An in situ STM image of bulk copper deposition on a gold film at −200 mV (vs. Cu/Cu^{2+}) in 0.1 M H$_2$SO$_4$ + 1 mM CuSO$_4$. 3D and top view representations are shown. Notice the plateau form of the copper islands. Reprinted by permission of [52], with permission of [153]. See color plate section.

ν_3. In this model, the overall shape of the electrodeposited crystallite depends on the relative rates of the processes shown in Figure 3.10. Nichols et al. have analyzed the growth morphologies of a large number of copper clusters electrodeposited on Au(111), as observed in situ by STM during growth [49]. It once again illustrates the power of the local probe methods, which can be used to individually assess the growth behavior of single clusters. Emphasis was placed on analyzing the growth rates normal to the surface with respect to those lateral to the surface. The precise ratio of these two growth rates was found to be very sensitive to the electrochemical conditions, the electrolyte composition and the net rate of growth [49,36]. In any case, the lateral propagation of the clusters was found to be generally at least an order of magnitude faster than the growth normal to the surface [49,36]. This difference reflects a low rate of nucleation on top the atomically flat cluster, with ν_1 hence being significantly greater than ν_2. STM observations also indicate that nuclei propagate relatively rapidly across the top of the plateau crystallites once formed (i.e., $\nu_3 \gg \nu_2$) [51].

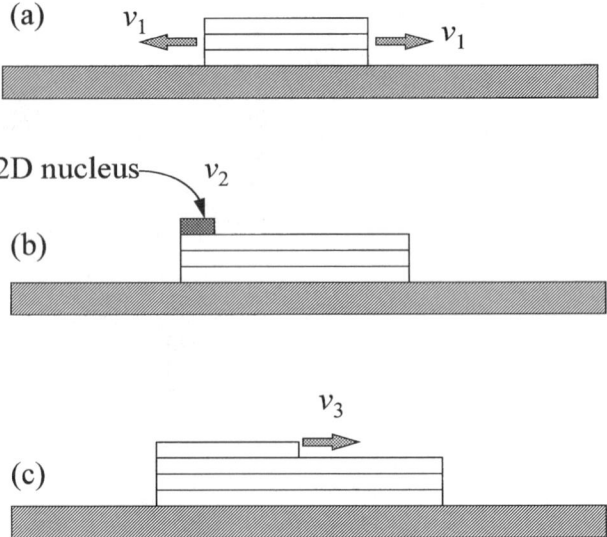

Figure 3.10 Growth model for the *"flat topped"* crystallites formed during Cu electrodeposition on Au(111) showing different growth parameters necessary to account for growth morphology: (a) lateral growth of the cluster by adatom incorporation at edge of the crystallite, rate v_1; (b) formation of a 2-D nucleus on top of cluster, rate v_2; (c) propagation of the 2-D nucleus across the upper face of the cluster, increasing its height by a single monolayer, rate v_3.

Kolb and co-workers have used another method for assessing the growth of electrodeposited metal clusters [46,28]. In this case, instead of observing the growth of copper clusters nucleated at step edges, copper clusters artificially nucleated on terraces were examined. A tip-induced cluster was formed on a terraced region of the surface and its growth was monitored by STM. Additionally, instead of imaging the whole growing cluster, a section through the cluster was monitored by scanning across a single line, which is readily experimentally achieved, by scanning only in the fast (x-) direction and disabling the slow scan (y-) direction. Such an x-t scan is presented in Figure 3.11, which shows the growth of a copper cluster on a copper monolayer covered Au(111) substrate [28]. The tip-induced cluster was generated at about t = 10 s and the growth of 14 layers was observed in this time frame. From this x-t scan different lateral growth rates were estimated for the 1st and 2nd layers (2 nm/s) and the 3rd to 14th layers (25–35 nm/s), respectively. They tentatively assigned these differences as arising from misfit strain [28]. Thus, the relatively slow spread of the initial layers may be explained by the significant lattice misfit of the copper deposit and the gold substrate.

As discussed, there is a low nucleation probability on top of the atomically flat top of clusters, such as those observed in Plate 3.2. From in situ STM, 2-D nucleation of these subsequently formed copper monolayers was seen to

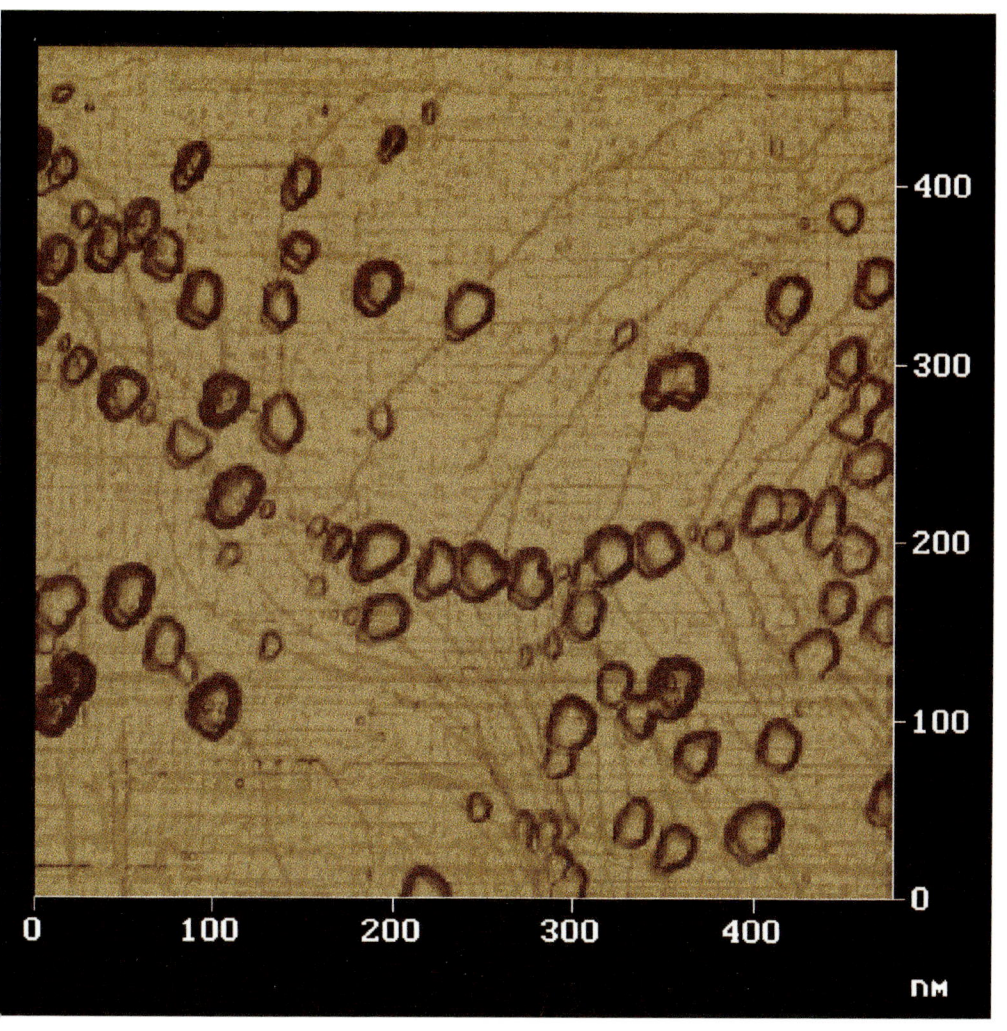

Plate 3.1 In situ STM image taken during the bulk electrodeposition of Cu on Au(111) at −200 mV, in 0.1 M H_2SO_4 + 1mM $CuSO_4$.

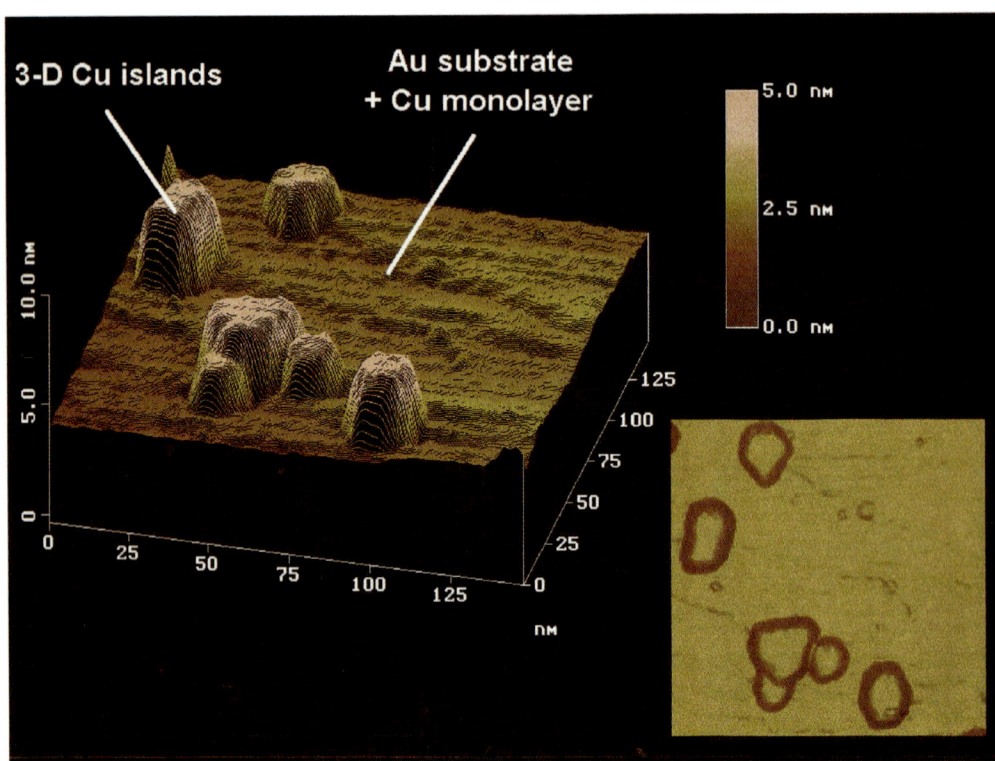

Plate 3.2 An in situ STM image of bulk copper deposition on a gold film at −200 mV (vs. Cu/Cu^{2+}) in 0.1 M H$_2$SO$_4$ + 1 mM CuSO$_4$. 3D and top view representations are shown. Notice the plateau form of the copper islands. Reprinted by permission of [52], with permission of [153].

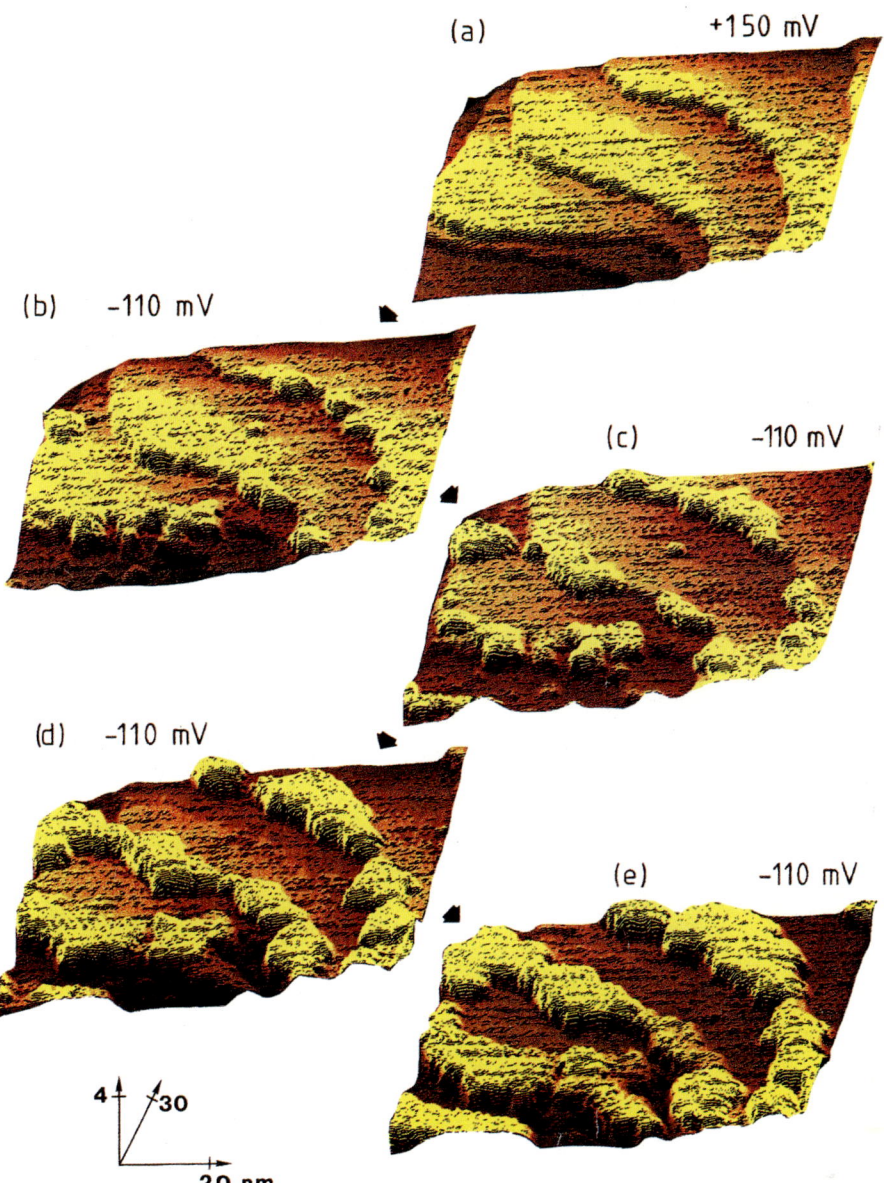

Plate 3.3 In situ STM images taken (a) before and (b) through (e) during bulk copper deposition onto Au(111) in 0.1 M H_2SO_4 + 1 mM $CuSO_4$. + 2.5 x 10^{-5} M crystal violet. I_t = 10 nA, U_{tip} = + 20 mV. Reprinted from [49] with kind permission of Elsevier Science S.A. [154].

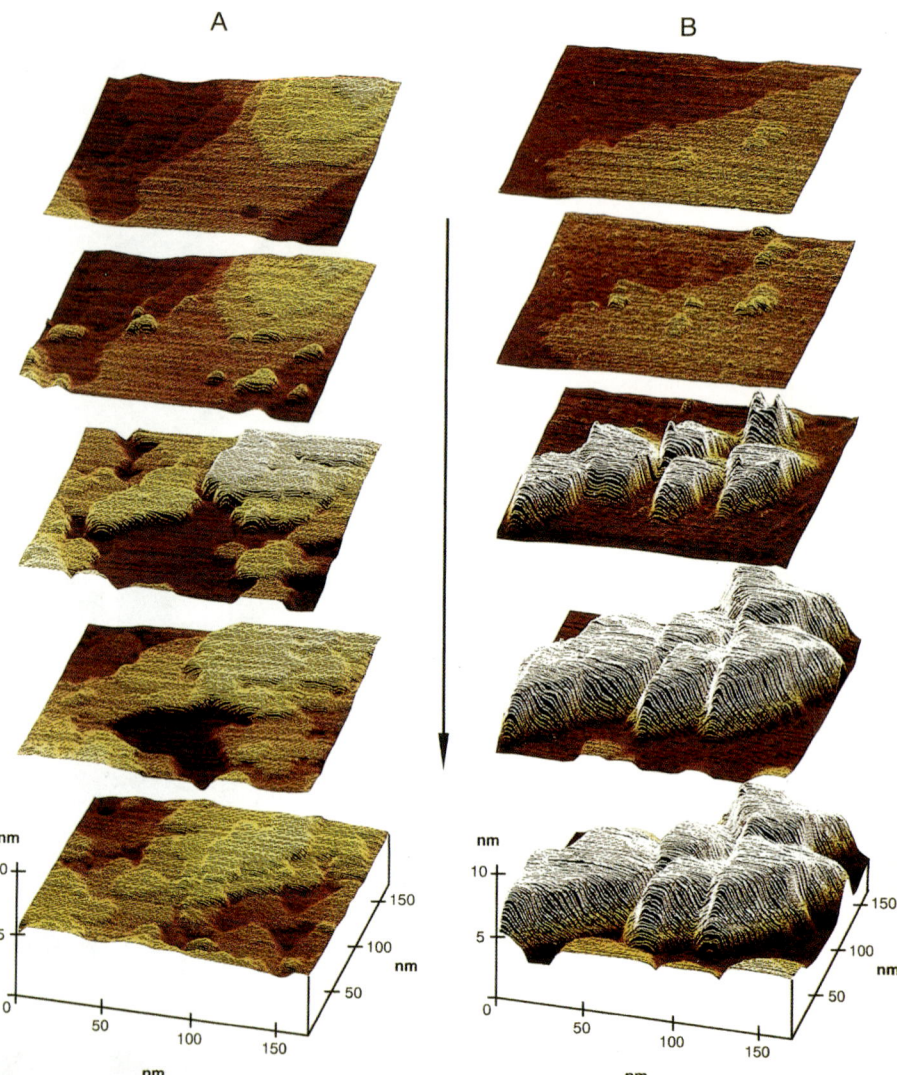

Plate 3.4 In situ STM images showing a direct comparison of the influence of two additives on the electrodeposition of copper on Au(111) from in 0.1 M H_2SO_4 + 1 mM $CuSO_4$. + (a) 10 mgl^{-1} benzothiazole derivative (b) 10 mgl^{-1} crystal violet [53]. Copyrighted and reprinted with the permission of SCANNING, and/or the Foundation for Advances of Medicine and Science (FAMS), Box 832, Mahwah, NJ 07430, USA.

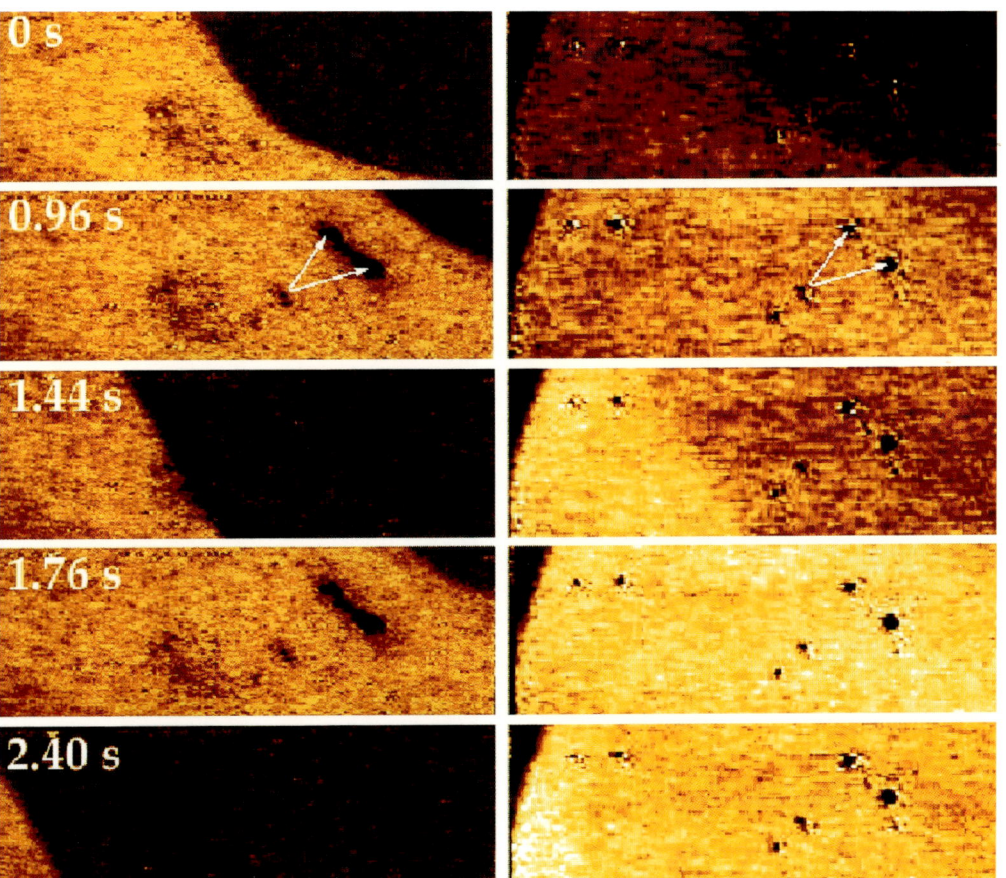

Plate 4.1 A single O front observed at a corresponding surface area simultaneously with EMSI and IR-imaging at p_{O_2} = 113 mbar and p_{CO} = 3 mbar, T = 453 K, imaged area for each frame is 3 x 1 mm²; times are indicated in the EMSI frames (lefthand side), the same two defects on the surface are marked by arrows in the frames for 0.96 s.

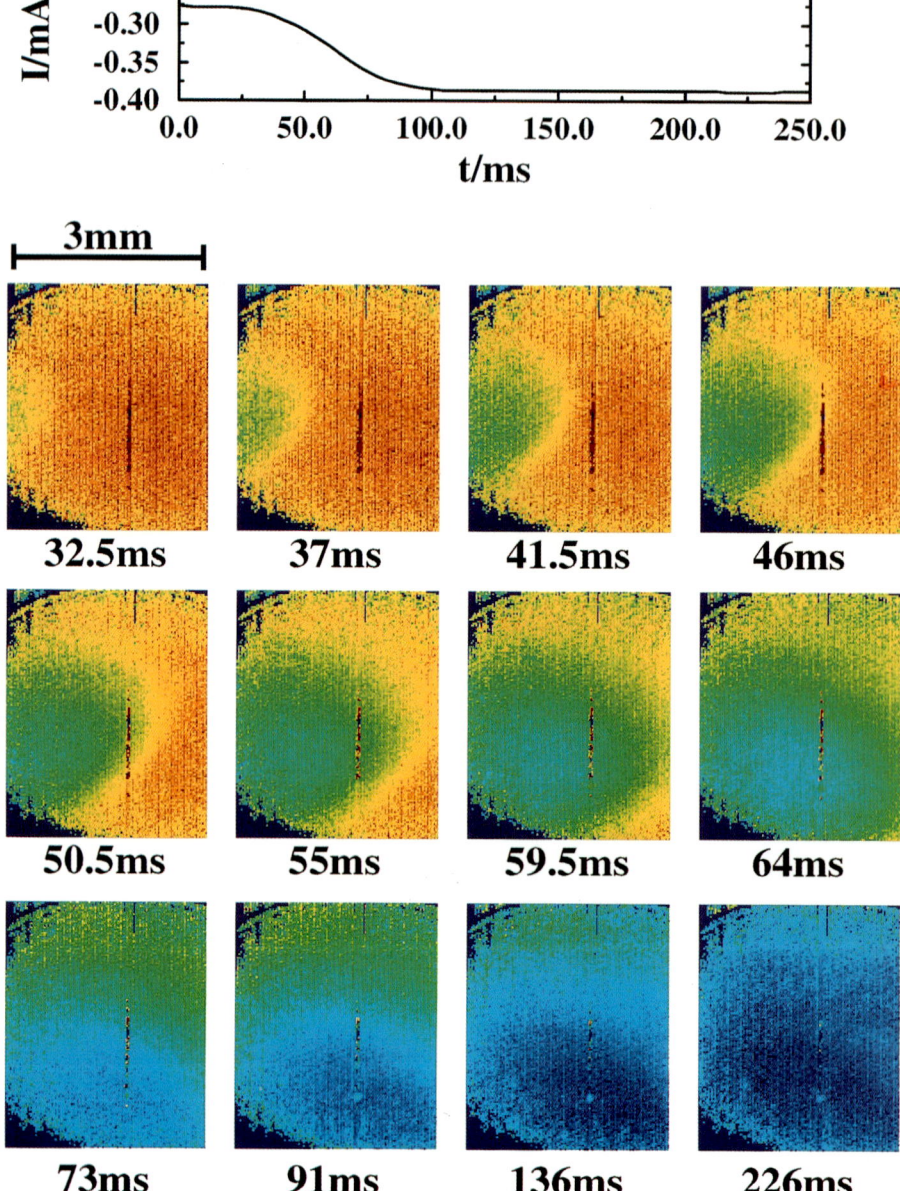

Plate 4.2 (a) Time trace of the global current during a transition in the bistable regime. (b) SPP microscope images of the electrode during this transition. Reproduced from Flätgen et al.[46]

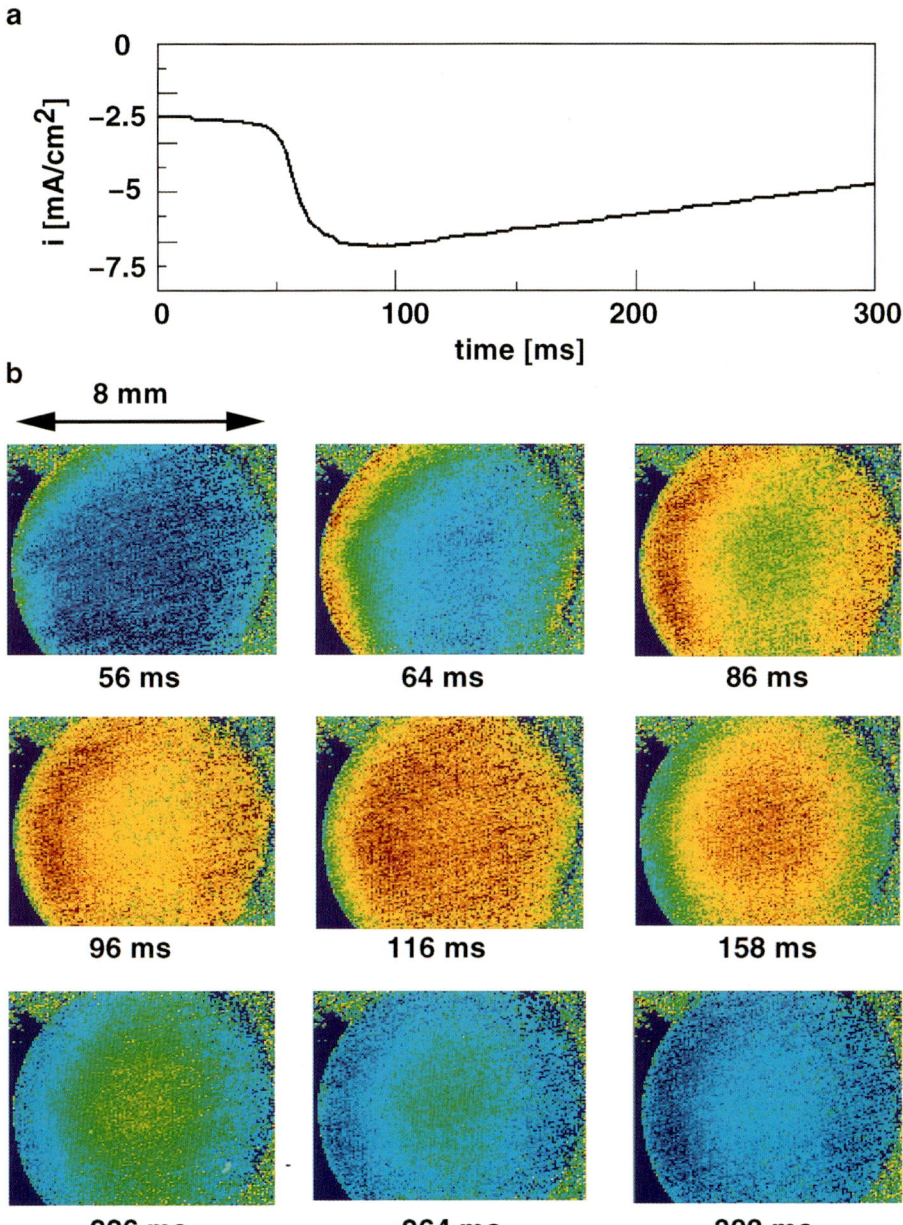

Plate 4.3 (a) Time trace of the global current during a transition in the bistable regime at parameter values closer to the oscillatory regime than in Plate 4.2. Note that the relaxation to the steady state is not completely shown. (b) SPP microscope images during this transition. (The color is reversed with respect to Plate 4.2 as the angle of incidence lay on the other side of the resonance minimum.) Reproduced from Flätgen et al.[160]

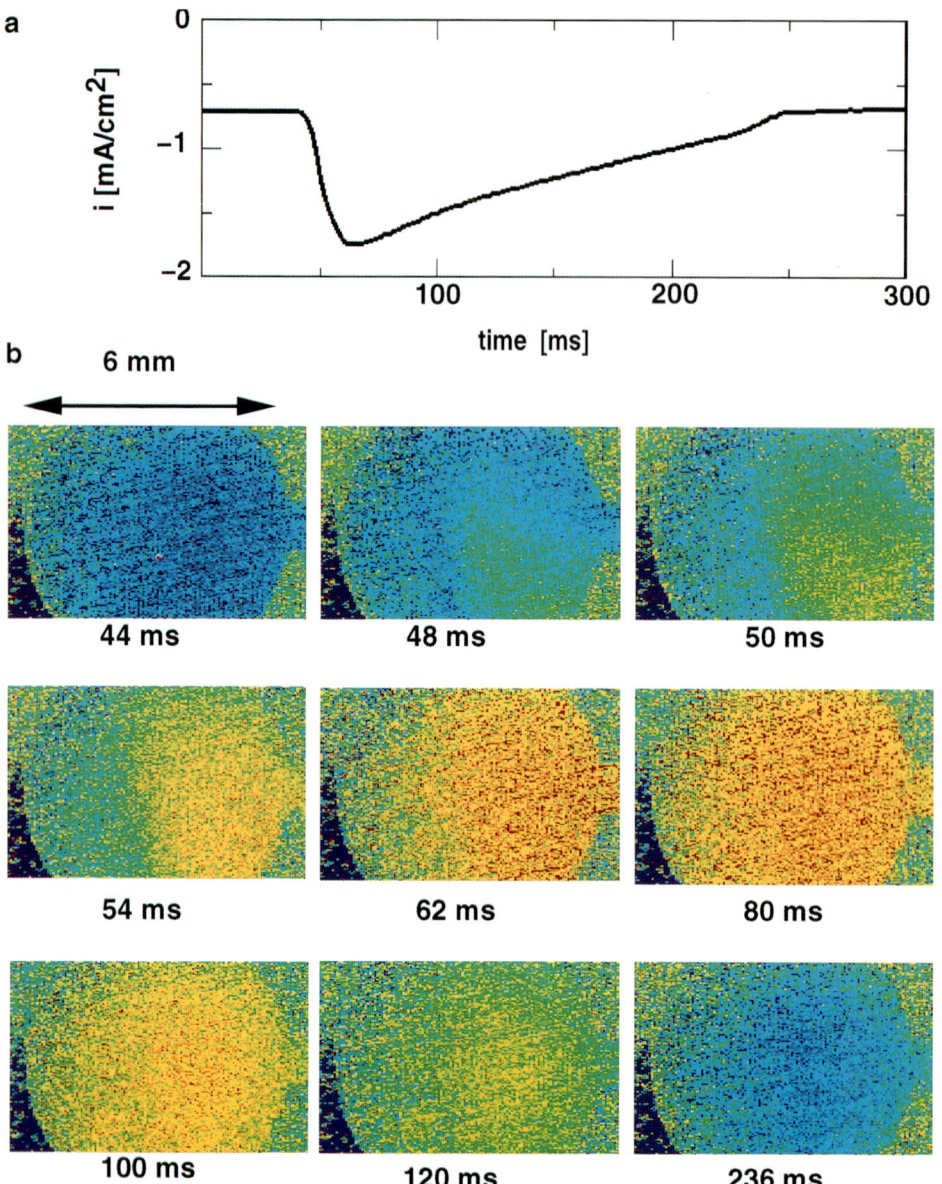

Plate 4.4 Time trace of the global current during an oscillation of 2 Hz frequency. (b) SPP microscope images during the oscillation. Reproduced from Flätgen et al.[160]

3.4 METAL GROWTH IN ELECTROCRYSTALLIZATION

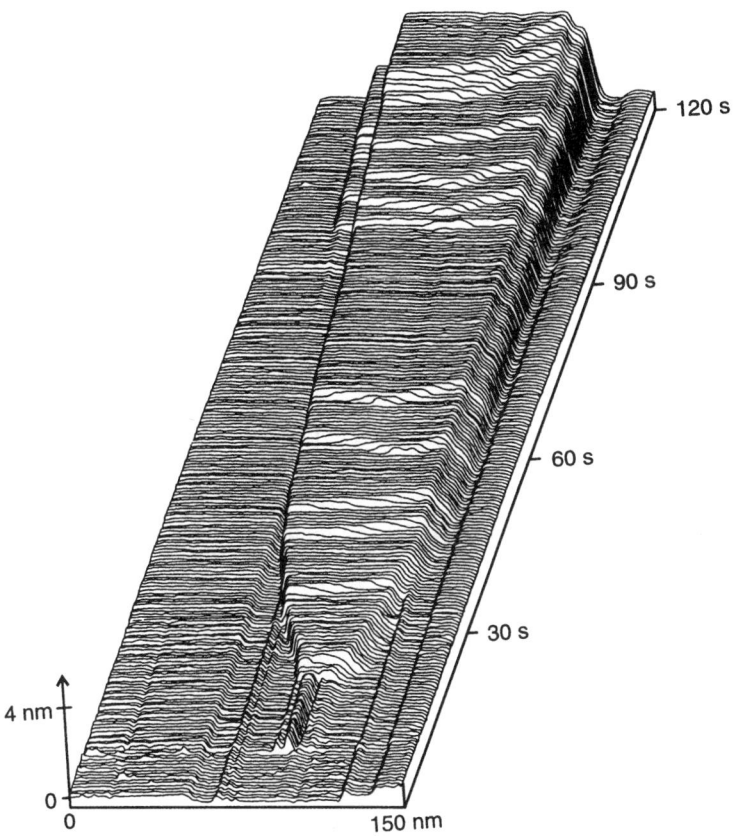

Figure 3.11 The growth of a Cu cluster on Au(111) monitored by recording cross sections (x-t scans) with STM. The nucleation of the cluster was tip induced, and growth was monitored at -180 mV vs. Cu/Cu^{2+}. Used with permission of [28].

occur at the edge of the upper face of the plateau clusters, as depicted in the model in Figure 3.10 [51]. The relatively low rate of vertical propagation of the crystallites points to these nucleation sites as significantly less favorable nucleation sites than monoatomic step edges. In addition, comparison may be made to this form of nucleation and the nucleation of metal clusters on the upper terrace side of the monoatomic step; a phenomenon observed in the presence of certain organic adsorbates [51].

Strain effects, arising from misfit, are of general significance in influencing the growth morphology of metal deposits on foreign metal substrates. A well-investigated example is the electrodeposition of Pb on Ag(111) [78–83,88,89,114,136–138]. In the section on metal adlayers, STM images of compressed and rotated Pb underpotential deposits on Ag(111) are shown. Bulk Pb deposition occurs on top of these adlayers, whose structure effects the growth of the bulk phase. The misfit between the bulk Pb deposit and the Pb monolayer on Ag(111) has been used to account for the 3-D growth of the bulk phase [78,83]. Since the deposition morphology is 3-D Pb crystallites on top of a 2-D Pb precursor layer, it has been designated as Stranski-Krastanov growth [83]. In addition, hexagonal-shaped crystallites have been observed by optical microscopy, rotated by 4.5° with respect to the major (110) substrate directions [114,136]. The metal monolayer is also rotated by 4.5° with respect to the (110) directions on the Ag(111) substrate, showing that the metal monolayer determines the epitaxial orientation of the bulk crystallites [78].

The structure of the Pb monolayer formed at underpotentials also has marked effects on the kinetics of bulk Pb electrocrystallization [114,136–138]. As discussed in the section on metal adlayers, the degree of compression of the Pb monolayer increases as the potential is decreased from about +150 mV to 0 mV (versus Pb/Pb^{2+}) [128]. The linear decrease in Pb-Pb spacing in this potential range has been related to a corresponding increase in the strain of the monolayer [114,138]. Further, it has been deduced that these compressed metal monolayers are slow to relax upon performing potentials steps from underpotentials to overpotentials where the bulk phase nucleates [114,136–138]. The kinetics of nucleation have been electrochemically determined by performing potential steps from various underpotentials, corresponding to Pb monolayers with differing degrees of strain, to a given overpotential [114,136–138]. From these measurements, the kinetics of the bulk phase formation were found to be slower on the more highly strained Pb monolayers formed at low underpotentials [114,136–138]. This has been related to a greater "difficulty" of forming the bulk Pb phase on more strained monolayers and once again emphasizes the significant role that the structure of the underpotential deposit can play in the development of the bulk phase.

Lorenz et al. have also presented STM images of the surface of growing Pb crystallites on Ag(111) and Ag(100) [79,82,83]. They found that the surface of such crystallites consists of atomically flat Pb(111) terraces separated by parallel monoatomic high steps [79,82]. They report that it indicates a spiral growth mechanism for the growth of Pb crystallites on Ag surfaces. Spiral growth originating from screw dislocations is of great importance in crystal growth, because it provides a mechanism for the sustained growth of the deposit without the need for nucleation of new growth centers. In the case of Cu deposition on Au(111), morphological features typical of spiral growth were not generally observed, which is presumably related to the surprisingly low screw dislocation density on well-prepared Au(111) surface [35,36]. However, on areas of the Au(111) surface where screw dislocations had been located, Cu clusters were seen by STM to effectively decorate the dislocation edge during electrodeposition [35].

3.5 THE INFLUENCE OF ORGANIC ADDITIVES

Organic compounds are commonly added in small quantities to metal plating baths. They are of particular importance to the metal plating industry, be-

3.5 THE INFLUENCE OF ORGANIC ADDITIVES

cause they are used to control the deposition process and ensure that the deposit has acceptable properties. However, only relatively recently has STM enabled a direct real-space and high-resolution visualization of how organic additives effect the growth of electrodeposits [36,41,46,49,51,52,53]. In this respect, most effort has been concentrated on examining the bulk Cu deposition on Au(111) model system, in the presence of organic additives such as crystal violet [36,46,49,51,53], BT-A, BT-B [36,51,53], and thiourea [28,41,46]. The structures of these additives are shown in Figure 3.12.

In the case of crystal violet and the benzothiazole derivatives, BT-A and BT-B, defects, such as monoatomic step edges in the substrate surface, are favorable sites for the nucleation of the bulk copper deposits [36,49,51]. As for additive free electrolytes, nucleation and growth on atomically flat terraces is significantly less favorable. This can be clearly seen in Plate 3.3 for crystal violet addition. Plate 3.3 shows the substrate surface at an underpotential. Following a potential step to −110 mV the bulk copper deposit can be seen to develop. A nucleation and initial growth at the monoatomic step edges is apparent, with the step edges effectively decorated with bulk copper as in Plate 3.3.

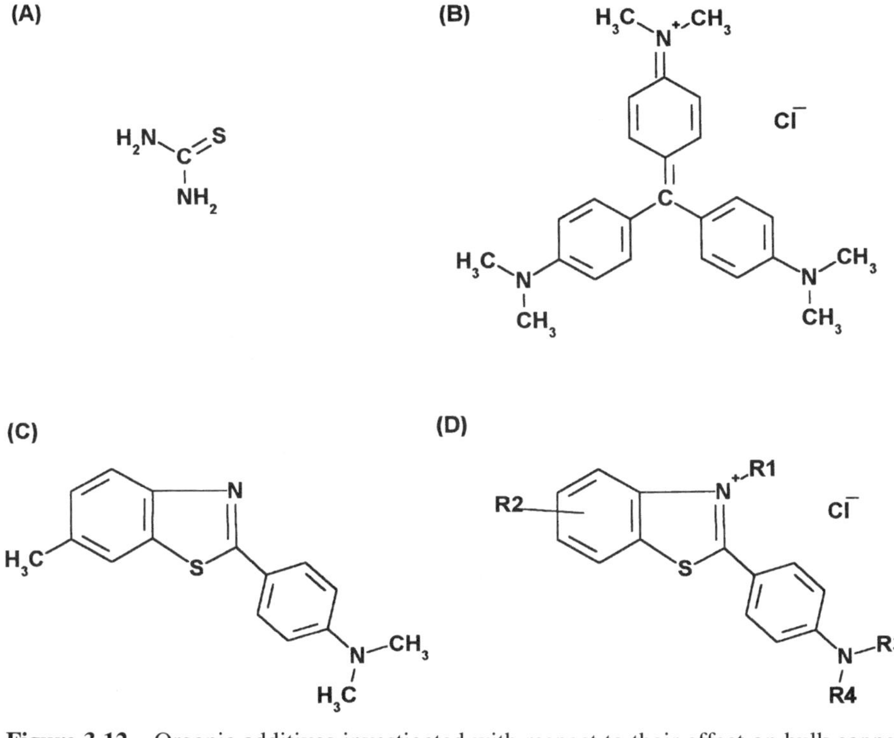

Figure 3.12 Organic additives investigated with respect to their effect on bulk copper electrodeposition on gold.

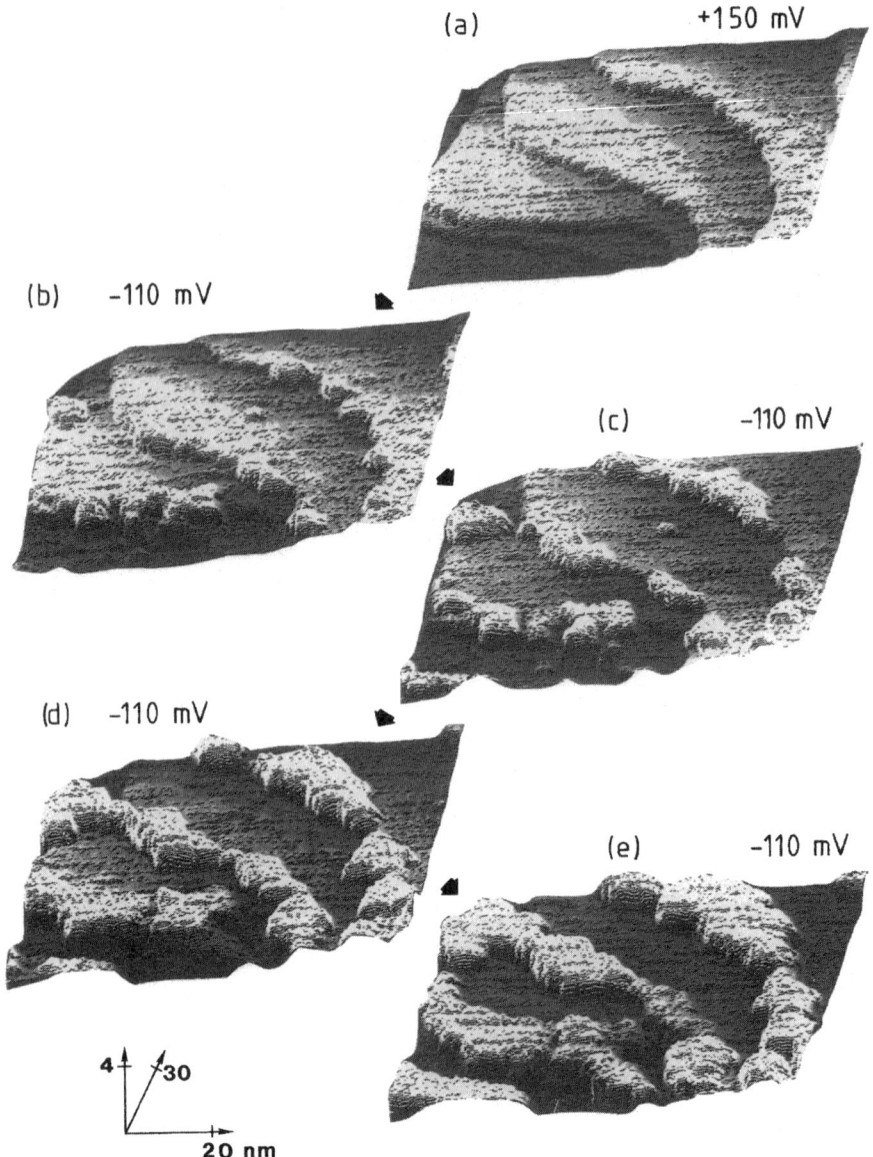

Plate 3.3 In situ STM images taken (a) before and (b) through (e) during bulk copper deposition onto Au(111) in 0.1 M H_2SO_4 + 1 mM $CuSO_4$. + 2.5 × 10^{-5} M crystal violet. I_t = 10 nA, U_{tip} = + 20 mV. Reprinted from [49] with kind permission of Elsevier Science S.A. [154]. See color plate section.

3.5 THE INFLUENCE OF ORGANIC ADDITIVES

Prior to in situ STM observation such as those in Plate 3.3, there has been speculation that such additives function by blocking defects, hence enhancing deposition on smoother regions of the surface, giving a more "balanced" metal deposition across the entire surface. Images such as those shown in Plate 3.3 show that this is not the case, with nucleation and growth occurring preferentially at defects [49]. With hindsight, it is not surprising that the surface defects remain the favorable nucleation sites, even with crystal violet or BT-addition, when one compares the size of such organic molecules with the size of, for instance, an atomic scale kink in a step edge. These additives are significantly larger than such a kink and the experimental evidence supports the hypothesis that they do not pack effectively enough around the defect in order to block nucleation [51].

A somewhat different behavior has been observed for Cu deposition on Au(111) in the presence of thiourea, which at first sight seems to contradict the preceding conclusions [28,41,46]. In this case, the nucleation and growth of 3-D crystallites *appears* to occur on flat terraced regions of the substrate. However, a closer examination of these terraced regions has revealed a 2-D island structure for the underpotential deposit on the "flat" terraces [28,41,46]. Nucleation and growth of the bulk deposit initiates at the rims of these islands, which may be likened to monoatomic high step edges. Once again, it illustrates the importance of considering the structure of the underpotential deposit when studying the growth of the bulk metal phase.

The influence of organic additives is first apparent upon examination of the growth morphology of the bulk Cu electrodeposit in the presence of additives [36,49,51]. In additive-free electrolytes the copper clusters show a three-dimensional growth, which leads to inhomogeneous films. Addition of crystal violet, BT-A or BT-B to acidic copper plating electrolytes has been shown to markedly influence the growth behavior of the copper deposit [36,49,51]. These additives suppress the vertical growth of the copper crystallites, so that they effectively spread parallel to the surface, hence exhibiting a quasi two-dimensional growth. Such growth behavior yields compact and well-defined copper deposits, which clearly illustrates the technological merits of using such additives in metal plating bath formulations [53].

The precise mechanism by which the additives suppress the vertical growth of the copper clusters in favor of their lateral growth cannot as yet be determined. However, from examination of the model in Figure 3.10, it is clear that such a growth morphology would occur if the rate of 2-D nucleation (ν_2) on top of the plateau crystallites is greatly suppressed. Some additives are more effective at inhibiting the vertical propagation of the copper crystallites in favor of their lateral spreading. For instance, crystal violet only effectively suppresses the vertical growth for very slow overall growth rates from dilute copper electrolytes [51]. BT-B is more effective than crystal violet and a comparison is shown in Plate 3.4 for 10^{-3} M $CuSO_4$ + 0.1 M H_2SO_4 with either (a) crystal violet or (b) BT-B addition [51,53]. As can be seen from the in situ STM images taken during the course of electrodeposition, uniform and compact copper films

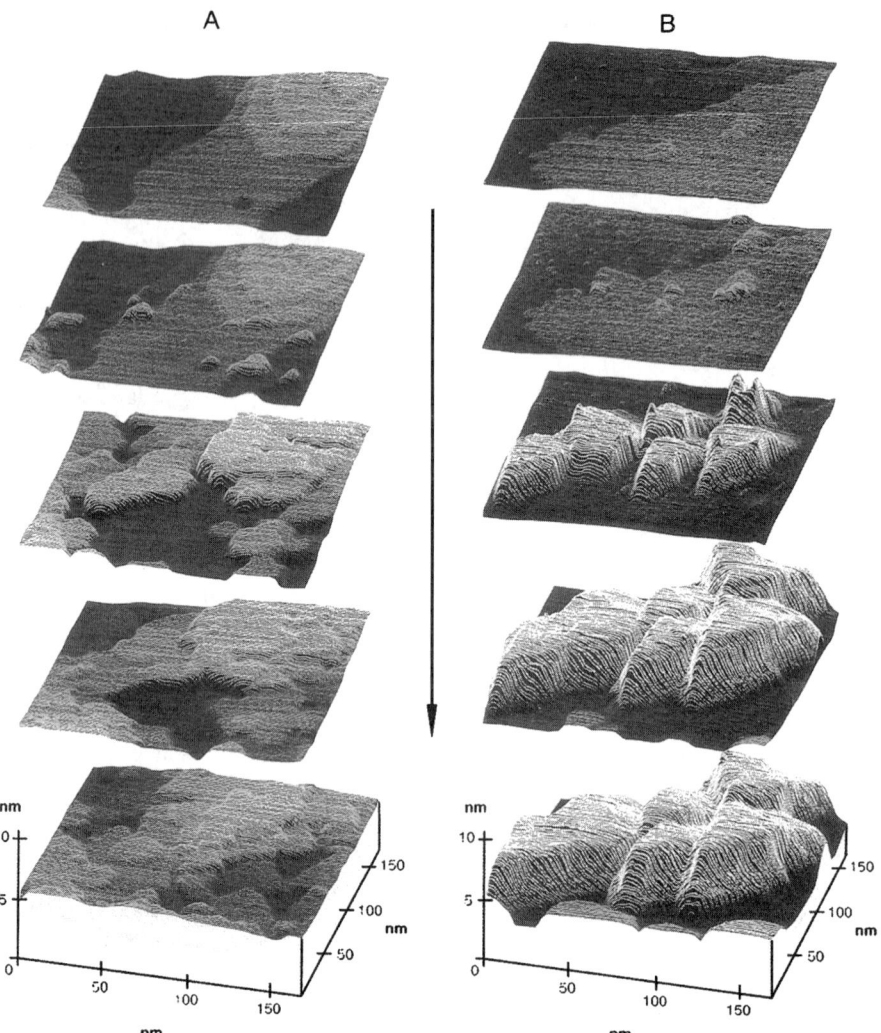

Plate 3.4 In situ STM images showing a direct comparison of the influence of two additives on the electrodeposition of copper on Au(111) from in 0.1 M H_2SO_4 + 1 mM $CuSO_4$. + (a) 10 mgl^{-1} benzothiazole derivative (b) 10 mgl^{-1} crystal violet [53]. Copyrighted and reprinted with the permission of SCANNING, and/or the Foundation for Advances of Medicine and Science (FAMS), Box 832, Mahwah, NJ 07430, USA. See color plate section.

are deposited in the presence of BT-B, while the copper film deposited in the presence of crystal violet is markedly inhomogeneous.

The growth of copper films deposited in the presence of various organic additives has been directly assessed using in situ STM. On the basis of these STM observations, additives have been ranked in terms of their ability to suppress

3.6 NANOTECHNOLOGY AND ELECTRODEPOSITION

the vertical growth of copper crystallites relative to their lateral spreading [36]:

$$\text{BT-B} > \text{BT-A} > \text{crystal violet} \gg \text{HSO}_4^- \text{ (additive free)}$$

In this list, BT-B is the additive that leads to the most uniform films under the greatest variety of electrodeposition conditions.

3.6 NANOTECHNOLOGY AND ELECTRODEPOSITION

A number of methods have been developed, combining scanning probe microscopy techniques and metal electrodeposition to enable the lateral structuring of metal electrodes with nano-scale metal features [139–147]. Penner et al. have pioneered this use of SPM to create highly localized nano-scale electrodeposits [139]. For instance, silver nano-disk structures have been electrodeposited onto graphite surfaces using STM [139–144]. This electrodeposition was achieved by applying a rapid high bias pulse between the tip and the graphite surface immersed in Ag^+ containing electrolytes [144]. A pulse width and duration of +6.0 V and 50 μsec was used to create stable silver structures directly under the tip, with a diameter of approximately 15–20 nm and height of 2–3 nm [140]. A two-step mechanism for the deposition of such structures is proposed. At short times the high bias voltage ruptures the graphite surface to create a monolayer deep pit, followed by nucleation and growth of silver at this artificially created defect [140,144]. Penner et al. have also succeeded in depositing copper nano-disks in the close vicinity of silver nano-disks. This deposition is done by using the aforementioned procedure to deposit the two silver disks on the right-hand side of Figure 3.13 for 0.5 mM AgF. The silver plating electrolyte in the STM cell was then changed for a copper-containing

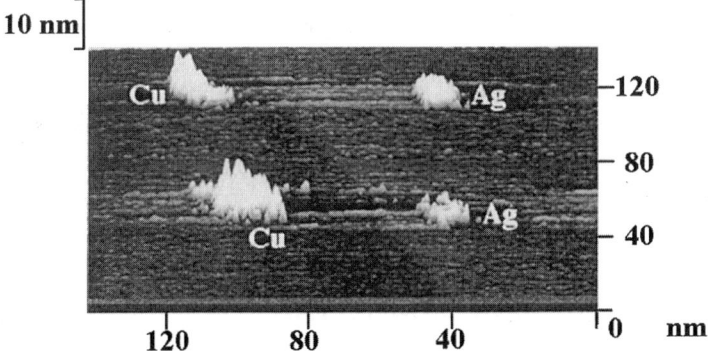

Figure 3.13 STM image of silver and copper nano-disk structures codeposited on graphite. Used with kind permission of the ACS [140].

electrolyte and then the two copper nano-disks were deposited, as seen on the left-hand side of the image. They then studied the behavior of this "nano-meter scale galvanic cell" [140].

Ullmann et al. have devised another method for creating STM tip-generated metal clusters on electrodes in electrochemical media [145,146]. In this method copper is predeposited onto the metallic STM tip, which is held at potentials slightly negative with respect to Cu/Cu^{2+}. The tip potential is held at this constant value and the sample potential is stepped from positive values to somewhat negative of the tip potential. During this potential excursion the tip-to-sample bias passes through zero [145,146]. In doing so a copper cluster is transferred from the tip to the sample, due perhaps to a fleeting contact between the tip and sample at zero bias. As with the Penner method, Ullmann could controllably deposit clusters at given surface locations. In addition, the Cu clusters could be erased by locating the tip over a cluster and applying an appropriate high positive tip bias [145,146]. They have recently extended the tip-induced cluster generation method to form Pd clusters [169,170].

A method utilizing in situ AFM for nanostructuring copper surfaces with electrodeposited Cu has been developed by Gewirth and LaGraff [147]. It was shown that enhanced electrochemical deposition of Cu can occur in the region of the surface where the AFM tip is scanning. In such experiments the copper single crystal electrodes were polarized at relatively low overpotentials, where the net overall rate of copper nucleation and growth across the entire surface was low. Enhanced deposition of copper was obtained in the scan area of the AFM tip by imaging with high tip-to-sample interaction forces. This enabled nano-scale copper features to be selectively grown in a given area of the surface. Two possible mechanisms for this enhanced metal deposition have been proposed. One suggestion is that the tip may create defects in the underlying copper substrate, hence enhancing the nucleation and subsequent growth of the copper deposit. The other suggestion is that the AFM tip removes the passivating oxide or hydroxide adlayer, which covers the copper substrate and results in a locally enhanced copper deposition rate [147].

These results clearly show that it is possible to create nanoscopic metallic electrodeposits and controllably position them on given areas of the electrode surface. Penner and co-workers have made further advances by examining the properties of individual metallic structures by STM, such as the discharge of the nanoscale battery shown in Figure 3.13 [140,144]. Kolb's group has also recently extended their tip-induced cluster generation method to form micrometer-size arrays of Cu nano-clusters [171,185,186].

3.7 IMAGING ORGANIC ADDITIVES

Only recently have adsorbed organic additives, which are of relevance to metal plating, been directly imaged by STM [148–150]. Tetramethylthiourea and crystal violet have both been imaged with submolecular resolution with STM [148–150]. In the case of crystal violet a number of unsuccessful previous attempts had been made to image this molecule adsorbed on Au(111) substrates in electrochemical environments. However, the adsorbate was successfully imaged when it was adsorbed on top of pre-adsorbed iodine adlayers on the Au(111) substrate [150]. Under such conditions, well-defined adsorbate superstructures could be observed and the characteristic symmetry of the individual crystal violet molecules could be recognized. In the case of tetramethylthiourea (TMTU), the molecule directly adsorbed on the metallic surface could be imaged [148,149].

In situ imaging of organic molecules, of relevance to metal plating as well as many other technologically important processes, such as corrosion or electrocatalysis, is still at an early stage. However, rapid progress is being made, for instance, the recent STM imaging of the dynamics of organic molecule adsorption and desorption from electrode surfaces [149]. It remains to be established how far the scope of imaging adsorbed organic molecules on electrode surfaces can be extended. In this respect, an imaging of appropriate organic adsorbates under conditions relevant to metal deposition, would help in constructing molecular and atomic scale mechanisms of the important influence that additives can have on metal electrocrystallization.

3.8 CONCLUSIONS AND OUTLOOK

This brief review has attempted to illustrate, through a number of chosen examples, the use of SPM in investigating metal electrocrystallization. SPM has provided a direct visualization of electrodeposition processes down to single atom resolution, as well as offering visualization of length scales in excess of 100 μm. As such SPM allows contact to be directly established between microscopic models and resulting macroscopic features of the electrodeposit.

SPM has so far provided insights into the atomic structure of surfaces, the distribution and nature of surface defects, underpotential metal deposits, nucleation centers, the evolution of deposit morphology and the influence of metal plating bath additivs. Recent advances have been made, with STM providing a submolecular resolution of certain adsorbed organic additives, which offers hope that microscopic mechanisms of additive action can be constructed [149–150]. Other recent advances have been the use of in situ STM to assess more complex underpotential deposition, metal monolayer formation and ECALE systems including studies of Cd, Te, CdTe, GaAs, Se, and Hg deposition on gold single crystals (see Table 3.1). The range of well-defined single crystal surfaces that can now be prepared for in situ STM studies has been extended, with noteworthy recent additions including Co and Ni single crystals [157,206]. There have been a number of detailed studies on the electrodeposition of Ni on Au(111), including assessment of the nucleation sites, monolayer structure, the transition to multilayer growth and the effect of sulfamate adlayers on the nickel monolayer formation [75,190–191]. Another recent advance is the application of STM to study in situ metal deposition from room temperature molten salts [168]. Probe microscopy has also been used to follow electroless deposition processes [183]. There have also been recent articles on the electrodeposition of metals on semiconductor surfaces [58,104,182,184,196,203]. Clearly these recently published papers show the diversity of electrodeposition problems to which electrochemical-STM (and AFM) is being successfully applied.

Another important factor in the growth of metal deposits, which has been discussed here, is the influence of strain at the metal–foreign substrate interface arising from misfit between the deposit and substrate lattice. In this respect, in situ SPM offers the chance to assess the influence of this misfit by monitoring changes in the lattice spacing or structure on top of metal clusters or ultra-thin metal films as they develop on foreign substrates. Several groups have recently made progress with such measurements. For instance, Dietterle et al. have observed by STM that Cu initially deposits (up to 8 layers) pseudomorphically on Ag(100) in the thermodynamically less stable bcc form [164]. They have observed that a transition occurs during deposition of the ninth layer, which results in a relaxation in the lattice parameter. This leads to the surface becoming buckled with the driving force being presumably strain relief [164]. Another recent advance of relevance here, is the development of STM methods for in situ monitoring of minute deformations of electrode surfaces, arising from changes in surface stress [151,152,176]. These techniques have enabled stress in electrodeposited metal films to be assessed during their evolution [176]. A recent study describes the quantification of such stresses as a function of deposit thickness and microscopic structure for copper deposits on Au(111) [176].

In conclusion, interest is generally moving from using SPM to investigate purely structural aspects of the metal–electrode interface to using in situ SPM to directly visualize dynamic processes, such as metal nucleation and growth. In this respect, the use of SPM to investigate kinetic mechanisms of processes at the atomic scale, is of fundamental importance. This growing scope of application of SPM will ensure that SPM continues to be a technique of great relevance for investigating electrochemical interfaces well into the next millennium.

ACKNOWLEDGMENTS

I would like to thank Dr. Wolfgang Haiss for many helpful discussions and proofreading of this manuscript. I also gratefully acknowledge Professor Jacek Lipkowski for his guidance in the preparation of this manuscript. Support from the Scanning Probe Microscopy Initiative of EPSRC is also gratefully acknowledged.

REFERENCES

1. D. Pletcher and F. C. Walsh, Industrial Electrochemistry, Second Edition, 1900 Blackie Academic, London.
2. J. A. Harrison and H. R. Thirsk, in A. J. Bard (ed.), Electroanalytical Chemistry, Vol. 5, Marcel Dekker, London, 1971, p. 67.
3. J. P. Vandereerden, M. A. H. Mickers, J. W. Gerritsen, and M. H. J. Hottenhuis, *Electrochimica Acta*, 1989, Vol. 34, No. 8, pp. 1141–1145.
4. N. Breuer, U. Stimming, and R. Vogel, *Electrochima Acta*, 1995, Vol. 40, No. 10, pp. 1401–1409.
5. S. G. Garcia, D. Salinas, C. Mayer, J. R. Vilche, H. J. Pauling, S. Vinzelberg, G. Staikov, and W. J. Lorenz, *Surface Science*, 1994, Vol. 316, No. 1–2, pp. 143–156.
6. K. Ogaki and K. Itaya, *Electrochima Acta*, 1995, Vol. 40, No. 10, pp. 1249–1257.
7. S. G. Corcoran, G. S. Chakarova, and K. Sieradski, *Journal of Electroanalytical Chemistry*, 1994, Vol. 377, No. 1–2, pp. 85–90.
8. S. Sugita, T. Abe, and K. Itaya, *Journal of Physical Chemistry*, 1993, Vol. 97, No. 34, pp. 8780–8785.
9. T. Hachiya and K. Itaya, Ultramicroscopy, 1992, Vol. 42, No. Pta, pp. 445–452.
10. C. H. Chen, S. M. Vesecky, and A. A. Gewirth, *Journal of the American Chemical Society*, 1992, Vol. 114, No. 2., pp. 451–458.
11. P. Mrozek, Y. E. Sung, M. Han, M. Gamboaaldeco, A. Wieckowski, C. H. Chen, and A. A. Gewirth, *Electrochima Acta*, 1995, Vol. 40, No. 1, pp. 17–28.
12. K. Endo, Y. Sugawara, S. Mishima, T. Okada, and S. Morita, *Japanese Journal of Applied Physics, Part 1—Regular Papers, Short Notes & Review Papers*, 1991, Vol. 30, No. 10, pp. 2592–2593.
13. S. G. Corcoran, G. S. Chakarova, and K. Sieradzki, *Physical Review Letters*, 1993, Vol. 71, No. 10, pp. 1585–1588.
14. E. E. Mola, A. G. Appignanessi, J. Vicente, L. Vazquez, R. C. Salvarezza, and A. J. Arvia, Surface Review and Letters, 1995, Vol. 2, No. 4, pp. 489–494.
15. R. T. Potzschke, C. A. Gervasi, S. Vinzelberg, G. Staikov, and W. J. Lorenz, *Electrochimica Acta*, 1995, Vol. 40, No. 10, pp. 1469–1474.
16. R. S. Robinson, *Journal of Vacuum Science & Technology, A—Vacuum Surfaces and Films*, 1990, Vol. 8, No. 1, pp. 511–514.
17. R. Sonnenfeld and B. C. Schardt, Appl. Phys. Lett., 49 (1986) 1172.
18. K. Itaya and E. Tomita, *Surf. Sci.*, 201 (1988) L507.
19. R. S. Robinson, *J. Electroanal. Chem.*, 136 (1989) 584.
20. K. Kowal, L. Xie, R. Huq, and G. C. Farrington, *Journal of the Electrochemical Society*, 1994, Vol. 141, No. 1, pp. 116–122.
21. N. Shinotsuka, K. Sashikata, and K. Itaya, *Surface Science*, 1995, Vol. 335, No. 1–3, pp. 75–82.
22. N. Kimizuka and K. Itaya, Faraday Discussions, 1992, No. 94, pp. 117–126.
23. J. Schneir, V. Elings, and P. K. Hansma, *J. Electrochem. Soc.*, 135 (1988) 2774.

24. C-H. Chen and A. A. Gewirth, *J. Am. Chem. Soc.*, 114 (1992) 5439–5440.
25. C-H. Chen, K. D. Kepler, A. A. Gewirth, B. M. Ocko, and J. Wang, *Journal of Physical Chemistry*, 1993, Vol. 97, No. 28, pp. 7290–7294.
26. M. H. Ge and A. A. Gewirth, *Surface Science*, 1995, Vol. 234, No. 2–3, pp. 140–148.
27. M. Dietterle, T. Will, and D. M. Kolb, *Surface Science*, 1995, Vol. 342, No. 1–3, pp. 29–37.
28. T. Will, M. Dietterle, D. M. Kolb, in: Nanoscale Probes of the Solid–Liquid Interface, Nato ASI, Vol. E 288, p. 137, ed. By A. A. Gewirth and H. Siegenthaler, Kluver, Dordrecht, 1995.
29. J. E. T. Andersen and P. Moller, *Journal of the Electrochemical Society*, 1995, Vol. 142, No. 7, pp. 2225–2232.
30. J. E. T. Andersen, G. Bechnielsen, and P. Moller, Surface & Coatings Technology, 1994, Vol. 70, No. 1, pp. 87–95.
31. F. A. Moller, O. M. Magnussen, and R. J. Behm, Physical Review B—Condensed Matter, 1995, Vol. 51, No. 4, pp. 2484–2490.
32. O. M. Magnussen, J. Hotlos, G. Bettel, D. M. Kolb, and R. J. Behm, *Journal of Vacuum Science & Technology B*, 1991, Vol. 9, No. 2, pp. 969–975.
33. O. M. Magnussen, J. Hotlos, R. J. Nichols, D. M. Kolb, and R. J. Behm, Physical Review Letterrs, 1990, Vol. 64, No. 24, pp. 2929–2932.
34. N. Batina, D. M. Kolb, and R. J. Nichols, Langmuir, 1992, Vol. 8, No. 10, pp. 2572–2576.
35. R. J. Nichols, D. M. Kolb, and R. J. Behm, *Journal of Electroanalytical Chemistry*, 1991, Vol. 313, No. 1–2, pp. 109–119.
36. R. J. Nichols in Nanoscale Probes of the Solid–Liquid Interface, NATO ASI, Vol. E 228, p. 163, ed. by A. A. Gewirth and H. Siegenthaler, Kluver, Dordrecht, 1995.
37. N. Ikemiya, S. Miyaoka, and S. Hara, *Surface Science*, 1995, Vol. 327, No. 3, pp. 261–273.
38. F. Moller, O. M. Magnussen, and R. J. Behm, *Electrochima Acta*, 1995, Vol. 40, No. 10, pp. 1259–1265.
39. W. Haiss, D. Lackey, J. K. Sass, H. Meyer, and R. J. Nichols, *Chemical Physics Letters*, 1992, Vol. 200, No. 4, pp. 343–349.
40. T. Hachiya, H. Honbo, and K. Itaya, *Journal of Electroanalytical Chemistry*, 1991, Vol. 315, No. 1–2, pp. 275–291.
41. M. H. Holzle, C. W. Apsel, T. Will, and D. M. Kolb, *Journal of the Electrochemical Society*, 1995, Vol. 142, No. 11, pp. 3741–3749.
42. M. P. Green and K. J. Hanson, *Journal of Vacuum Science & Technology A—Vacuum Surfaces and Films*, 1992, Vol. 10, No. 5, pp. 3012–3018.
43. J. Hotlos, O. M. Magnussen, and R. J. Behm, *Surface Science*, 1995, Vol. 335, No. 1–3, pp. 129–144.
44. H. Matsumoto, I. Oda, J. Inukai, and M. Ito, *Journal of Electroanalytical Chemistry*, 1993, Vol. 356, No. 1–2, pp. 275–280.
45. D. M. Kolb, R. J. Nichols, R. J. Behm, in Electrified Interfaces in Physics, Chemistry and Biology, ed. R. Guidelli, NATO ASI Series C (Kluwer, Dordrecht, 1992), p. 275.

46. N. Batina, T. Will, and D. M. Kolb, Faraday Discussions, 1992, No. 94, pp. 93–106.
47. S. Manne, P. K. Hansma, J. Massie, V. B. Elings, and A. A. Gewirth, *Science*, 1991, Vol. 251, No. 4990, pp. 183–186.
48. N. Ikemiya, S. Miyaoka, and S. Hara, *Surface Science*, 1994, Vol. 311, No. 1–2, pp. L641–L648.
49. R. J. Nichols, W. Beckmann, H. Meyer, N. Batina, and D. M. Kolb, *Journal of Electroanalytical Chemistry*, 1992, Vol. 330, No. 1–2, pp. 381–394.
50. U. Stimming, R. Vogel, D. M. Kolb, and T. Will, *Journal of Power Sources*, 1993, Vol. 43, No. 1–3, pp. 169–180.
51. R. J. Nichols, C. E. Bach, and H. Meyer, *Berichte der Bunsen Gesellschaft für Physikalische Chemie*, 1993, Vol. 97, No. 8, pp. 1012–1020.
52. R. J. Nichols, E. Bunge, H. Meyer, and H. Baumgärtel, *Surface Science*, 1995, Vol. 335, No. 1–3, pp. 110–119.
53. R. J. Nichols, D. Schröer, and H. Meyer, Scanning Vol. 15, (1993) 266–273.
54. X. G. G. Zhang and U. Stimming, *Journal of Electroanalytical Chemistry and Interfacial Electrochemistry*, 1990, Vol. 291, No. 1–2, pp. 273–279.
55. X. Zhang and U. Stimming, *Corros. Sci.*, 30 (1990), 951.
56. R. Srinivasan, P. Gopalan, *Surface Science*, 1995, Vol. 338, No. 1–3, pp. 31–40.
57. N. Breuer, U. Stimming, and R. Vogel, *Surface & Coatings Technology*, 1994, Vol. 67, No. 3, pp. 145–149.
58. M. Koinuma and K. Uosaki, *Electrochima Acta*, 1995, Vol. 40, No. 10, pp. 1345–1351.
59. R. J. Nichols, D. Schroer, and H. Meyer, *Electrochima Acta*, 1995, Vol. 40, No. 10, pp. 1479–1485.
60. K. Uosaki and H. Kita, *Journal of Electroanalytical Chemistry and Interfacial Electrochemistry*, 1989, Vol. 259, No. 1–2, pp. 301–308.
61. F. R. F. Fan and A. J. Bard, *Journal of the Electrochemical Society*, 1989, Vol. 136, No. 11, pp. 3216–3222.
62. K. Uosaki and H. Kita, *J. Vac. Sci. Technol. A*, 8 (1990) 520.
63. H. Matsumoto, J. Inukai, and M. Ito, *Journal of Electroanalytical Chemistry*, 1994, Vol. 379, No. 1–2, pp. 223–231.
64. R. M. Rynders and R. C. Alkire, *Journal of the Electrochemical Society*, 1994, Vol. 141, No. 5, pp. 1166–1173.
65. W. U. Schmidt and R. C. Alkire, *Journal of the Electrochemical Society*, 1994, Vol. 141, No. 7, pp. L85–L87.
66. G. Beitel, O. M. Magnussen, and R. J. Behm, *Surface Science*, 1995, Vol. 336, No. 1–2, pp. 19–26.
67. M. J. Armstrong and R. H. Muller, *Journal of the Electrochemical Society*, 1991, Vol. 138, No. 8, pp. 2303–2307.
68. K. Sashikata, N. Furuya, and K. Itaya, *Journal of Electroanalytical Chemistry*, 1991, Vol. 316, No. 1–2, pp. 361–368.
69. Y. Shingaya, H. Matsumoto, H. Ogasawara, and M. Ito, *Surface Science*, 1995, Vol. 335, No. 1–2, pp. 23–31.

70. T. Abe, G. M. Swain, K. Sashikata, and K. Itaya, *Journal of Electroanalytical Chemistry*, 1995, Vol. 382, No. 1–2, pp. 73–83.
71. M. Wunsche, R. J. Nichols, R. Schumacher, W. Beckmann, and H. Meyer, *Electrochima Acta*, 1993, Vol. 38, No. 5, pp. 647–652.
72. X. M. Yang, K. Tonami, L. A. Nagahara, K. Hashimoto, Y. Wei, and A. Fujishima, Chemistry Letters, 1994, No. 11, pp. 2059–2062.
73. C. H. Chen and A. A. Gewirth, *Ultramicroscopy*, 1992, Vol. 42, No. Pta, pp. 437–444.
74. N. Koura, S. Tamura, M. Yoshikawa, *Denki Kag.*, 1995, Vol. 63, No. 7, pp. 623–628.
75. F. A. Möller, O. M. Magnussen, and R. J. Behm, *Phys. Rev. Lett.*, 1996, Vol. 77, No. 15, pp. 3165–3168.
76. Z. F. chen, J. Li, and E. K. Wang, *Journal of Electroanalytical Chemistry*, 1994, Vol. 373, No. 1–2, pp. 83–87.
77. R. Christoph, H. Siegenthaler, H. Rohrer, and H. Wiese, *Electrochima Acta*, 1989, Vol. 34, No. 8, pp. 83–87.
78. W. Obretenov, U. Schmidt, W. J. Lorenz, G. Staikov, E. Budevski, D. Carnal, U. Muller, H. Siegenthaler, and E. Schmidt, *Journal of the Electrochemical Society*, 1993, Vol. 140, No. 3, pp. 692–703.
79. W. Obretenov, U. Schmidt, W. J. Lorenz, G. Staikov, E. Budevski, D. Carnal, U. Muller, H. Siegenthaler, and E. Schmidt, Faraday Discussions, 1992, No. 94, pp. 107–116.
80. W. J. Lorenz, L. M. Gassa, U. Schmidt, W. Obretenov, G. Staikov, V. Bostanov, and E. Budevski, *Electrochima Acta*, 1992, Vol. 37, No. 12, pp. 2173–3178.
81. G. Staikov, K. Juttner, W. J. Lorenz, and E. Budevski, *Electrochima Acta*, 1994, Vol. 39, No. 8–9, pp. 1019–1029.
82. G. Staikov and W. J. Lorenz, in Nanoscale Probes of the Solid–Liquid Interface, Nato Asi, Vol. E 288, p. 215, ed. by A. A. Gewirth and H. Siegenthaler, Kluver, Dordrecht, 1995.
83. W. J. Lorenz and G. Staikov, *Surface Science*, 1995, Vol. 335, No. 1–3, pp. 32–43.
84. U. Muller, D. Carnal, H. Siegenthaler, E. Schmidt, W. J. Lorenz, W. Obretenov, U. Schmidt, G. Staikov, and E. Budevski, Physical Review B—Condensed Matter, 1992, Vol. 46, No. 19, pp. 12899–12901.
85. D. Carnal, P. I. Oden, U. Muller, E. Schmidt, and H. Siegenthaler, *Electrochima Acta*, 1995, Vol. 40, No. 10, pp. 1223–1235.
86. J. Halbritter, G. Repphun, S. Vinzelberg, G. Staikov, and W. J. Lorenz, *Electrochima Acta*, 1995, Vol. 40, No. 10, pp. 1223–1235.
87. D. Carnal, U. Muller, and H. Siegenthaler, *Journal De Physique* Iv, 1994, Vol. 4, No. C1, pp. 297–302.
88. G. Staikov, E. Budevski, W. Obretenov, and W. J. Lorenz, *Journal of Electroanalytical Chemistry*, 1993, Vol. 349, No. 1–2, pp. 225–233.
90. M. Binggeli, D. Carnal, R. Nyffenegger, H. Siegenthaler, R. Christoph, and H. Rohrer, *Journal of Vacuum Science & Technology B*, 1991, Vol. 9, No. 4, pp. 1985–1992.

91. M. P. Green, K. J. Hanson, R. Carr, and I. Lindau, *Journal of the Electrochemical Society*, 1990, Vol. 137, No. 11, pp. 3493–3498.
92. M. P. Green, M. Richter, X. Xing, D. Scherson, K. J. Hanson, P. N. Ross, R. Carr, and I. Lindau, *Journal of Microscopy*, Oxford, 1988, Vol. 152, pp. 823–829.
93. M. P. Green, K. J. Hanson, D. A. Scherson, X. Xing, M. Richter, P. N. Ross, R. Carr, and I. Lindau, *Journal of Physical Chemistry*, 1989, Vol. 93, No. 6, pp. 2181–2184.
94. M. P. Green and K. J. Hanson, *Surface Science*, 1991, Vol. 259, No. 3, pp. L743–L749.
95. C. H. Chen, N. Washburn, and A. A. Gewirth, *Journal of Physical Chemistry*, 1993, Vol. 97, No. 38, pp. 9754–9760.
96. M. Aindow and J. P. G. Farr, *Transactions of the Institute of Metal Finishing*, 1992, Vol. 70, No. 4, pp. 171–176.
97. M. Szklarczyk and J. O. Bockris, *Journal of the Electrochemical Society*, 1990, Vol. 137, No. 2, pp. 452–457.
98. A. Gonzalezmartin, R. C. Bhardwaj, and J. O. Bockris, *Journal of Applied Electrochemistry*, 1993, Vol. 23, No. 6, pp. 531–546.
99. R. R. Adzic, J. Wang, C. M. Vitus, and B. M. Ocko, *Surface Science*, 1993, Vol. 293, No. 3, pp. L876–L883.
100. M. Baldauf and D. M. Kolb, *Electrochima Acta*, 1993, Vol. 38, No. 15, pp. 2145–2153.
101. X. Q. Tong, M. Aindow, and J. P. G. Farr, *Journal of Electroanalytical Chemistry*, 1995, Vol. 395, No. 1–2, pp. 117–126.
102. K. Itaya and S. Sugawara, *Chem. Lett.*, (1987) 1927.
103. N. Breuer, U. Stimming, and R. Vogel, in Nanoscale Probes of the Solid–Liquid Interface, Nato ASI, Vol. E 288, p. 121, ed. by A. A. Gewirth and H. Siegenthaler, Kluver, Dordrecht, 1995.
104. S. Eriksson, P. Carlsson, B. Holmstrom, and K. Uosaki, *Journal of Electroanalytical Chemistry*, 1992, Vol. 337, No. 1–2, pp. 217–227.
105. W. Polewska, J. X. Wang, B. M. Ocko, and R. R. Adzic, *Journal of Electroanalytical Chemistry*, 1994, Vol. 376, No. 1–2, pp. 41–47.
106. S. Tamura, N. Koura, and D. Kagaku, 1994, Vol. 62, No. 6, pp. 483–488.
107. N. Koura and S. Tamura, *Denki Kag.*, 1992, Vol. 60, No. 7, pp. 662–663.
108. T. R. I. Cataldi, I. G. Blackham, G. A. D. Briggs, J. B. Pethica, and H. A. O. Hill, *J. Electroanal. Chem.*, 290 (1990) 1.
109. Nanoscale Probes of the Solid–Liquid Interface, Nato ASI, Vol. E 228, ed. by A. A. Gewirth and H. Siegenthaler, Kluver, Dordrecht, 1995.
110. C. E. Bach, R. J. Nichols, W. Beckmann, H. Meyer, A. Schulte, J. O. Besenhard, and P. D. Jannakoudakis, *J. Electrochem. Soc.*, 140 (1993) 1281.
111. W. Haiss, D. Lackey, J. K. Sass, and K. H. Besocke, *J. Chem. Phys.*, 95 (1991) 2193.
112. W. Haiss, PhD dissertation, Technical University of Berlin, 1994.
113. D. M. Kolb, in Advances in *Electrochemistry and Electrochemical Engineering*, Vol. 11, eds. H. Gerischer and C. W. Tobias (Wiley, New York, 1978) p. 125.

114. K. Jüttner and W. J. Lorenz, *Z. Phys. Chem. NF*, 122 (1980) 163.
115. Y. Nakai, M. S. Zei, D. M. Kolb, and G. Lempfuhl, *Ber. Bunsenges. Phys. Chem.*, 88 (1984) 340.
116. D. M. Kolb, *Ber. Bunsenges. Phys. Chem.*, 92 (1933) 1175.
117. M. S. Zei, G. Qiao, G. Lempfuhl, and D. M. Kolb, *Ber. Bunsenges. Phys. Chem.*, 91 (1987) 349.
118. Z. Shi and J. Lipkowski, *J. Electroanal. Chem.*, 365 (1994) 303.
119. Z. Shi and J. Lipkowski, *J. Electroanal. Chem.*, 364 (1994) 289.
120. M. F. Toney, J. N. Howard, J. Richer, G. L. Borges, J. G. Gordon, O. R. Melroy, D. Yee, and L. B. Sorensen, *Phys. Rev. Lett.*, 75 No. 24 (1995) 4472.
121. J. G. Gordon, O. R. Melroy, and M. F. Toney, *Electrochima Acta*, Vol. 40, No. 1 (1995), 3–8.
122. L. Blum, H. D. Abruña, J. White, J. G. Gorden II, G. L. Borges, M. G. Samant, and O. R. Melroy, *J. Chem. Phys.*, 85 (1986) 6732.
123. O. R. Melroy, M. G. Samant, G. L. Borges, J. G. Gorden II, J. H. White, M. J. Abbarelli, M. McMillan, and H. Abruña, Langmuir 4 (1988) 728.
124. A. Tadjeddine, D. Guay, M. Ladouceur, and G. Tourillon, *Phys. Rev. Lett.*, 66 (1991) 2235.
125. G. Tourillon, D. Guay, and A. Tadjeddine, *J. Electroanal. Chem.*, 289 (1990) 263.
126. P. Zelenay, L. M. Rice-Jackson, and A. Wieckowski, Surf. Sci., 256 (1991) 253.
127. D. A. Huckaby and L. Blum, *J. Electroanal. Chem.*, 315 (1991) 255.
128. M. F. Toney, J. G. Gordon, M. G. Samant, G. L. Borges, O. R. Melroy, D. Yee, and L. B. Sorensen, *J. Phys. Chem.*, 99 (1995) 4733.
129. H. Fischer, *Electrolytischer Abscheidung und Electrokristallisation von Metallen*, Springer Verlag 1954.
130. I. N. Stranski, *Z. Phys. Chem.*, 136, (1928) 259.
131. H. Bort, K. Jüttner, W. J. Lorenz, G. Staikov, and E. Budevski, *Electrochim. Acta*, 28 (1983) 985.
132. D. Kaschiev, *J. Chem. Phys.*, 76 (1982) 5098.
133. H. Brune, H. Roder, C. Boragno, and K. Kern, *Phys. Rev. Lett.*, 73 No. 14 (1994) 1955.
134. P. W. Atkins, Physical Chemistry, 5th Edition, Oxford University Press, U.K.
135. R. Koch, *Journal of Physics—Condensed Matter*, Vol. 6, No. 45 (1994) pp. 9519–9550.
136. W. J. Lorenz, E. Schmidt, G. Staikov, and H. Bort, *Faraday Symposium of the Chemical Society*, No. 12 (1977) 14.
137. H. Bort, K. Jüttner, W. J. Lorenz, G. Staikov, and E. Budevski, *Electrochim. Acta*, 28 (1983) 985.
138. K. Jüttner, W. J. Lorenz, G. Staikov, and E. Budevski, *Electrochim. Acta*, 23 (1978) 741.
139. R. M. Penner, M. J. Heben, T. L. Longin, and N. S. Lewis, Science, 1990, Vol. 250, No. 4948, pp. 1118–1121.
140. W. J. Li, J. A. Virtanen, and R. M. Penner, *Journal of Physical Chemistry*, 1992, Vol. 96, No. 16, pp. 6529–6532.

141. W. J. Li, J. A. Virtanen, and R. M. Penner, *Langmuir*, 1995, Vol. 11, No. 11, pp. 4361–4365.
142. W. J. Li, J. A. Virtanen, and R. M. Penner, *Journal of Physical Chemistry*, 1994, Vol. 98, No. 45, pp. 11751–11755.
143. W. J. Li, J. A. Virtanen, and R. M. Penner, *Applied Physics Letters*, 1992, Vol. 60, No. 10, pp. 1181–1183.
144. W. Li, T. Doung, J. A. Virtanen, and R. M. Penner, in Nanoscale Probes of the Solid–Liquid Interface, Nato ASI, Vol. E 288, p. 183, ed. by A. A. Gewirth and H. Siegenthaler, Kluver, Dordrecht, 1995.
145. R. Ullmann, T. Will, and D. M. Kolb, *Chemical Physics Letters*, 1993, Vol. 209, No. 3, pp. 238–242.
146. R. Ullmann, T. Will, and D. M. Kolb, Berichte Der Bunsen-Gesellschaft-Physical Chemistry, 1995, Vol. 99, No. 11, pp. 1414–1420.
147. J. R. Lagraff and A. A. Gewirth, *Journal of Physical Chemistry*, 1995, Vol. 99, No. 24, pp. 10009–10018.
148. E. Bunge, R. J. Nichols, H. Baumgärtel, and H. Meyer, *Ber. Bunsenges. Phys. Chem.*, 99 (1995) 1243–1246.
149. E. Bunge, H. Meyer, H. Baumgärtel, B. Roelfs, and R. J. Nichols, *Langmuir*, Vol. 12, No. 12 (1996) 3060–3066.
150. N. Batina, M. Kunitake, and K. Itaya, *J. Electroanal. Chem.*, Vol. 405 (1996) 245–250.
151. H. Ibach, *J. Vac. Sci. Technol.*, A12, 1994, 2240.
152. W. Haiss and J. K. Sass, *J. of Electroanal. Chem.*, 1995, Vol. 386, No. 1–2, pp. 267–270.
153. Reprinted from [52] with kind permission of Elsevier Science—NL, Sara Burgerhartstraat 25, 1055 KV Amsterdam, The Netherlands.
154. Reprinted with kind permission from Elsevier S.A., P.O. Box 564, 1001 Lausanne, Switzerland.
155. J. E. T. Andersen, G. BechNielsen, P. Moller, and J. C. Reeve, *Journal of Applied Electrochemistry*, 1996, Vol. 26, No. 2, pp. 161–170.
156. J. E. T. Andersen and P. Moller, *Surface and Coatings Technology*, 1997, Vol. 89, No. 1–2, pp. 1–9.
157. S. Ando, T. Suzuki, and K. Itaya, *Journal of Electroanalytical Chemistry*, 1997, Vol. 431, No. 2, pp. 277–284.
158. J. C. Bondos, A. A. Gewirth, and R. G. Nuzzo, *Journal of Physical Chemistry*, 1996, Vol. 100, No. 21, pp. 8617–8620.
159. T. A. Brunt, T. Rayment, S. J. Oshea, and M. E. Welland, *Langmuir*, 1996, Vol. 12, No. 24, pp. 5942–5946.
160. J. L. Bubendorff, L. Cagnon, V. CostaKieling, J. P. Bucher, and P. Allongue, *Surface Science*, 1997, Vol. 384, No. 1-3, pp. L836–L843.
161. O. Cavalleri, S. E. Gilbert, and K. Kern, *Surface Science*, 1997, Vol. 377, No. 1-3, pp. 931–936.
162. A. I. Danilov, J. E. T. Andersen, E. B. Molodkina, Y. M. Polukarov, P. Moller, and J. Ulstrup, *Electrochima Acta*, 1998, Vol. 43, No. 7, pp. 733–741.
163. W. L. Desimone and J. J. Breen, *Langmuir*, 1995, Vol. 11, No. 11, pp. 4428–4432.

164. M. Dietterle, T. Will, and D. M. Kolb, *Surface Science*, 1998, Vol. 396, No. 1-3, pp. 189–197.
165. J. Divisek, B. Steffen, U. Stimming, and W. Schmickler, *Journal of Electroanalytical Chemistry*, 1997, Vol. 440, No. 1-2, pp. 169–172.
166. E. D. Eliadis, R. G. Nuzzo, A. A. Gewirth, and R. C. Alkire, *Journal of the Electrochemical Society*, 1997, Vol. 144, No. 1, pp. 96–105.
167. E. D. Eliadis and R. C. Alkire, *Journal of the Electrochemical Society*, 1998, Vol. 145, No. 4, pp. 1218–1226.
168. F. Endres, W. Freyland, and B. Gilbert, *Berichte Der Bunsen-Gesellschaft-Physical Chemistry Chemical Physics*, 1997, Vol. 101, No. 7, pp. 1075–1077.
169. G. E. Engelmann, J. C. Ziegler, and D. M. Kolb, *Journal of the Electrochemical Society*, 1998, Vol. 145, No. 8, p. 2970.
170. G. E. Engelmann, J. C. Ziegler, and D. M. Kolb, *Journal of the Electrochemical Society*, 1998, Vol. 145, No. 3, pp. L33–L35.
171. G. E. Engelmann, J. C. Ziegler, and D. M. Kolb, *Surface Science*, 1998, Vol. 401, No. 2, pp. L420–L424.
172. M. L. Foresti, G. Pezzatini, M. Cavallini, G. Aloisi, M. Innocenti, and R. Guidelli, *Journal of Physical Chemistry* B, 1998, Vol. 102, No. 38, pp. 7413–7420.
173. S. G. Garcia, D. E. Salinas, and C. E. Mayer, *Anales De La Asociacion Quimica Argentina*, 1996, Vol. 84, No. 6, pp. 597–605.
174. S. Garcia, D. Salinas, C. Mayer, E. Schmidt, G. Staikov, and W. J. Lorenz, *Electrochima Acta*, 1998, Vol. 43, No. 19-20, pp. 3007–3019.
175. A. Gichuhi, B. E. Boone, U. Demir, and C. Shannon, *Journal of Physical Chemistry* B, 1998, Vol. 102, No. 34, pp. 6499–6506.
176. W. Haiss, R. J. Nichols and J. K. Sass, *Surface Science*, 1997, Vol. 388, No. 1-3, pp. 141–149.
177. B. E. Hayden and I. S. Nandhakumar, *Journal of Physical Chemistry* B, 1997, Vol. 101, No. 39, pp. 7751–7757.
178. B. E. Hayden and I. S. Nandhakumar, *Journal of Physical Chemistry* B, 1998, Vol. 102, No. 25, pp. 4897–4905.
179. B. M. Huang, T. E. Lister, and J. L. Stickney, *Surface Science*, 1997, Vol. 392, No. 1-3, pp. 27–43.
180. J. Inukai, S. Sugita and K. Itaya, *Journal of Electroanalytical Chemistry*, 1996, Vol. 403, No. 1-2, pp. 159–168.
181. J. Inukai, Y. Osawa, M. Wakisaka, K. Sashikata, Y. G. Kim, and K. Itaya, *Journal of Physical Chemistry* B, 1998, Vol. 102, No. 18, pp. 3498–3505.
182. C. Kaneshiro and T. Okumura, Physica B, 1996, Vol. 227, No. 1-4, pp. 271–275.
183. H. Kind, A. M. Bittner, O. Cavalleri, K. Kern, and T. Greber, *Journal of Physical Chemistry* B, 1998, Vol. 102, No. 39, pp. 7582–7589.
184. M. Koinuma and K. Uosaki, *Journal of Electroanalytical Chemistry*, 1996, Vol. 409, No. 1-2, pp. 45–50.
185. D. M. Kolb, R. Ullmann, and T. Will, *Science*, 1997, Vol. 275, No. 5303, pp. 1097–1099.
186. D. M. Kolb, R. Ullmann, and J. C. Ziegler, *Electrochima Acta*, 1998, Vol. 43, No. 19-20, pp. 2751–2760.

187. A. Lachenwitzer, M. R. Vogt, O. M. Magnussen, and R. J. Behm, *Surface Science*, 1997, Vol. 382, No. 1-3, pp. 107–115.
188. T. E. Lister and J. L. Stickney, *Journal of Physical Chemistry*, 1996, Vol. 100, No. 50, pp. 19568–19576.
189. H. Martin, P. Carro, A. H. Creus, S. Gonzalez, R. C. Salvarezz, and A. J. Arvia, *Langmuir*, 1997, Vol. 13, No. 1, pp. 100–110.
190. F. A. Moller, O. M. Magnussen, and R. J. Behm, *Physical Review Letters*, 1996, Vol. 77, No. 26, pp. 5249–5252.
191. F. A. Moller, J. Kintrup, A. Lachenwitzer, O. M. Magnussen, and R. J. Behm, Physical Review B–Condensed Matter, 1997, Vol. 56, No. 19, pp. 12506–12518.
192. J. Morales, S. M. Krijer, P. Esparza, S. Gonzalez, L. Vazquez, R. C. Salvarezza, and A. J. Arivia, *Langmuir*, 1996, Vol. 12, No. 4, pp. 1068–1077.
193. H. Naohara, S. Ye, and K. Uosaki, *Journal of Physical Chemistry* B, 1998, Vol. 102, No. 22, pp. 4366–4373.
194. I. Oda, Y. Shingaya, H. Matsumoto, and M. Ito, *Journal of Electroanalytical Chemistry*, 1996, Vol. 409, No. 1-2, pp. 95–101.
195. D. Oyamatsu, M. Nishizawa, S. Kuwabata, and H. Yoneyama, *Langmuir*, 1998, Vol. 14, No. 12, pp. 3298–3302.
196. B. Rashkova, B. Guel, R. T. Potzschke, G. Staikov, and W. J. Lorenz, *Electrochima Acta*, 1998, Vol. 43, No. 19-20, pp. 3021–3028.
197. J. Sackmann, A. Bunk, R. T. Potzschke, G. Staikov, and W. J. Lorenz, *Electrochima Acta*, 1998, Vol. 43, No. 19-20, pp. 2863–2873.
198. W. U. Schmidt, R. C. Alkire, and A. A. Gewirth, *Journal of the Electrochemical Society*, 1996, Vol. 143, No. 10, pp. 3122–3132.
199. U. Schmidt, S. Vinzelberg, and G. Staikov, *Surface Science*, 1996, Vol. 348, No. 3, pp. 361–279.
200. D. W. Suggs and J. L. Stickney, *Surface Science*, 1993, Vol. 290, No. 3, pp. 375–387.
201. X. Q. Tong, M. Aindow, and J. P. G. Farr, Institute of Physics Conference Series, 1995, Vol. 147, No. 259–262.
202. K. Uosaki, S. Ye, H. Naohara, Y. Oda, T. Haba and T. Kondo, *Journal of Physical Chemistry* B, 1997, Vol. 101, No. 38, pp. 7566–7572.
203. K. Uosaki, T. Kondo, M. Koinuma, K. Tamura, and H. Oyanagi, *Applied Surface Science*, 1997, Vol. 121, No. 102-106.
204. I. Villegas and J. L. Stickney, *Journal of Vacuum Science & Technology A—Vacuum Surfaces and Films*, 1992, Vol. 10, No. 5, pp. 3032–3038.
205. B. U. Yoon, K. C. Cho, and H. Kim, *Analytical Sciences*, 1996, Vol. 12, No. 2, pp. 321–326.
206. T. Suzuki, T. Yamada, and K. Itaya, *J. Phys. Chem.*, 1996, Vol. 100, pp. 8954–8961.

4

IMAGING OF REACTION FRONTS AT SURFACES AND INTERFACES

HARM HINRICH ROTERMUND, KATHARINA KRISCHER, AND BRUNO PETTINGER

Fritz-Haber-Institut der Max-Planck-Gesellschaft Faradayweg 4-6, D-14195 Berlin, Germany

4.1 INTRODUCTION

4.1.1 Origin and General Importance of Reaction Fronts

In surface science as well as in electrochemistry the concept of homogeneous surfaces and interfaces has been employed for smooth or crystalline substrates for a long time. However, already during the last decades, but certainly during the last few years, evidence has been accumulated which shows that surfaces and interfaces of single crystalline samples are in reality rather inhomogeneous. Due to the breakdown of the bulk lattice structure at the phase boundary solid/gas phase or solid/electrolyte, the surfaces tend to reconstruct or microfacet, an effect that can be supported or even induced by adsorption. In addition, on a microscopic scale, there are always defects like steps, kinks, or screw dislocations.

The overstructures are naturally organized in domains with sizes in the range of a few angstroms up to several μm. Thus, the patterns on the surface—today easily monitored by STM or AFM—resemble a patchwork rather than a uniform surface structure. Evidence showing that the degree of coverage can dif-

---

Imaging of Surfaces and Interfaces (Frontiers of Electrochemistry, Volume 5).
Edited by Jacek Lipkowski and Philip N. Ross.
ISBN 0-471-24672-7 © 1999 Wiley-VCH, Inc.

fer at various overstructures has also been accumulated. Moreover, reactions may introduce specific overstructures and even the formation of microfacets large enough to macroscopically roughen the surface. Consequently, the surface chemistry occurs locally at different rates. These local parameters are connected with more global parameters such as pressure, temperature, or concentration and electrode potential. As a result, the surface/interfacial chemistry generally possesses a rich dynamical and spatial structure on all scales. Most spectacular in this respect is the formation and propagation of reaction fronts.

The understanding of these spatio-temporal structures on the microscopic and macroscopic scale is certainly a key point in fundamental and applied science and technology. In this chapter, we review various techniques for imaging different properties of catalytic surfaces and electrodes. We will demonstrate how the measurement of work function, coverage, or potential across the electrode interface can be used to investigate reaction fronts at interfaces.

The subject of chemical reaction fronts exerts a continual fascination even after at least twenty years from the beginning of their intense study. The reason for this continued interest is manifold: Firstly, the existence of a reaction front seems to contradict scientific intuition. A simple example of a reaction front is the propagation of a state of high reactivity into a state of low reactivity, the interface between the two states possessing a constant shape [1]. Obviously, if the reaction rate of the two states is different, then so is the concentration of the reacting species. Hence, there exists a concentration gradient across the interface which leads to diffusion, and naively thinking, one would expect that diffusion broadens the interfacial region, ending eventually in a flat, homogeneous state. By now it is well known that a spatially inhomogeneous structure can be maintained during its propagation if the reaction mechanism exhibits certain features (e.g. an autocatalytic step). This fact can be rationalized using a simple picture. Consider a situation in which the concentration of the autocatalytic species and the reaction rate is high and the opposite situation, where both are low. In such cases, the autocatalytic species will diffuse from the state of high reactivity to the state of low reactivity, increasing there its concentration. If the concentration in this *"passive"* state exceeds a certain threshold, the autocatalytic reaction will drive the system locally into the high reactivity state. Thus, this *"active"* state spreads while keeping the shape of the interface constant. This front is the result of the interplay of an autocatalytic reaction and diffusion.

A second reason for the continued interest in reaction fronts is that systems exhibiting autocatalytic reactions generally form a large variety of patterns [2]. One prominent example, which we consider in Section 4.2 in some detail, is the CO oxidation on Pt single crystal surfaces. In the 1970s and early 1980s it was recognized that not only do patterns in disciplines as different as astrophysics, biology, geology, physics, and chemistry look alike, but they are also described by equations exhibiting the same mathematical structure [3,4]. Hence, from a study of the formation of spirals in one particular system, for example, it is possible to understand how spirals are formed in general. As chemical systems

4.1 INTRODUCTION

are comparatively easy to handle, they proved to be important model systems for the understanding of pattern formation in non-equilibrium systems.

Finally, there is a third reason for the unbroken interest in reaction fronts: Along with a deeper understanding of the mechanisms that lead to instabilities and spatial structures in chemical systems, it was recognized how easily temporal oscillations or spatial structures can form in catalytic reactions. In other words, dynamical instabilities are widespread in surface or electrochemical reactions [5–9].

Experimentally, it is in general easy to study the temporal behavior of an average (or global) quantity, such as the overall reaction rate, the global current density or an average concentration. By contrast, the visualization of spatial patterns in heterogeneously catalyzed reactions or in electrochemical systems requires more sophisticated techniques. At the gas/solid interface and, at least in the case of electrocatalytic reactions, also at the liquid/solid interface, the concentration patterns are formed by adsorbates, and hence the coverages range between an empty surface and one monolayer. In exothermic reactions the concentration patterns are accompanied by temperature patterns, at least at pressures above 1 mbar or at supported catalysts. In electrochemical systems, different concentrations (i.e., reaction rates) are accompanied by different potential drops across the interface (except in a few, rare cases), and hence concentration patterns are accompanied by potential patterns. In order to obtain an image of the patterns forming at interfaces, methods that resolve either spatial differences of concentrations in the submonolayer region, different temperatures or potentials have to be employed. Furthermore, the methods must not interfere with the dynamics. In this context the free accessibility of an electrode or a catalyst by the reactants is the most critical part.

The necessary spatial resolution depends on the dominant spatial coupling mechanism, which determines the characteristic length of the patterns. Roughly speaking, the faster information between neighboring parts of the catalyst or electrode is exchanged with respect to the characteristic time of the corresponding reaction, the larger is the characteristic length. In electrochemical systems the dominant spatial coupling occurs via the electric field, which adjusts rapidly to changes at the electrode, and the observed characteristic length of patterns is on the order of centimeters. In contrast, in heterogeneous gas phase reactions at low pressures, the dominant spatial coupling is diffusion of the adsorbates, with a typical diffusion length on the order of 1 μm. When thermal conductivity dominates spatial coupling in catalyzed systems, characteristic lengths are on the order of 1 mm. Also, the necessary temporal resolution might vary orders of magnitude in different systems. In electrochemical systems, time constants are on the average much smaller than those in heterogeneous catalysis. Typical oscillation frequencies in the former systems range between 10^{-2} s to 10 s, which restricts the use of scanning methods and methods requiring long sampling times to a few systems. With typical oscillation periods between 1 sec and several minutes, characteristic time scales in heterogeneous catalysis are considerably longer. Nevertheless, they still may limit the use of scanning techniques.

4.1.2 Imaging in Heterogeneous Catalysis

The early observations of reaction fronts in heterogeneous catalysis constituted observations of temperature waves and were recorded by visual or infrared thermography. This method was introduced by Barelko and coworkers [10–12] and subsequently refined by several other groups [13–17]. Nowadays, temperature differences of less than 0.05 K can be probed in real time, and a spatial resolution of about 100 μm^2 is achieved. The use of several thermocouples is technically easier but not equivalent, since the spatial resolution is worse and there is the danger of interference with the reaction dynamics (see for example references in [5]).

In the following examples, we will focus on the CO oxidation at platinum single crystals. Since this reaction is highly exothermic (with the potential to explode!), beginning at pressures of 10^{-2} mbar O_2, the sample temperature increases several degrees when reacting at 220°C. This increase is generated only on the reactive, or oxygen-covered regions, and gives rise, at least for very thin probes, to temperature patterns, observable with a sensitive IR camera. One problem to overcome is the very low emissivity for IR radiation of Pt surfaces, which have a typical value of about 0.02. In other words, Pt emits IR radiation only with 2% intensity compared to a black body at the same temperature. Therefore, the experiment has to be configured in such a way that no objects close to the sample possess a much higher emissivity. Then it is even possible to observe changes in an IR image, while a clean surface is exposed to CO, presumably due to modification of the IR emissivity of Pt.

Temperature sensitive methods are no longer applicable when studying reactions under low pressures and, hence, nearly isothermal reaction conditions. The first evidence of the propagation of chemical waves on single crystal surfaces was obtained by Cox et al. [18], using a scanning LEED (low energy electron diffraction) technique, where the LEED beam was deflected across the surface with a pair of Helmholtz coils and the intensity variations of the diffraction spots were monitored. The resolution in these experiments was about 1 mm laterally and about a minute for time resolution, too rough to observe sophisticated pattern formation. Nevertheless, these observations of broad waves inspired the further development of tools and techniques that are capable of imaging reaction fronts at surfaces and interfaces.

The first of these techniques was scanning photoemission microscopy (SPM) [19,20]. Here, a UV light source is focused into a spot of about 0.5 μm in diameter, and the total yield of photoemitted electrons is then measured with a channeltron. As the spot is scanned across the surface, an image of the local work function arises. With this method it was possible to image slow-moving reaction diffusion fronts during the CO oxidation on Pt(100) directly [21]. However, the temporal resolution of this scanning method is still rather restricted: With all improvements implemented, it allows only up to one image per second, due to signal-to-noise considerations. In many situations, such as the imaging of reaction fronts on Pt(110), this rate is not sufficient.

4.1 INTRODUCTION

Imaging the photoelectrons with electron microscope optics solves the problem of the low temporal resolution and is also superior to SPM in other ways. This method has become known as photoemission electron microscopy (PEEM) [22,23]. In comparison to SPM, PEEM's greatest advantage is certainly that the images are processed in parallel, yielding a visible picture on a phosphorescent screen. It means that the temporal resolution is only dependent on the video equipment available. In addition, the spatial resolution is enhanced: it is about 0.2 μm in the instruments used nowadays. (Note, however, that in principle a PEEM can be developed into an instrument with a resolution in the nm region [24]. It will remain a "soft" probe in terms of radiation damage due to the energy of the photons, even though the intensities in real-time imaging have to be very high.) Further, the storage of the data on video tape as opposed to a computer allows the recording of much longer sequences. When the PEEM was first employed in 1990, it led to a breakthrough in the understanding of pattern formation under isothermal conditions.

A technique that in some way combines elements of the first and last mentioned methods (i.e., LEED and PEEM) is low-energy electron microscopy (LEEM), with a spatial resolution of 100–1000 Å and a sensitivity to the surface structure as well as topography [25]. Examples where pattern formation was observed with LEEM can be found in Rausenberger et al. and Rose et al [26–28].

In the early 1990s several other microscopic techniques were developed, which either aimed to enhance further contrast or spatial resolution, such as the mirror electron microscopy (MEM) [29,30], or to cover experimental conditions that were not accessible with the previously mentioned techniques. In this context reflection electron microscopy (REM) [31] has to be mentioned, making it possible to study reaction-induced substrate changes that occur during oscillatory reactions on small Pt single crystal spheres (diameter ~ 0.2 mm) [32] (see also chapter 3 in this volume).

All of these methods that probe the concentration field, directly or indirectly via structural changes of the catalyst, can only be applied under low pressures ($p < 10^{-3}$ mbar) where temperature variations are negligibly small. At atmospheric pressure, on the other hand, only temperature patterns could be made visible. Two new optical methods, ellipso-microscopy for surface imaging (EMSI) and reflectance anisotropy microscopy (RAM), overcome the restriction of low pressures. Both exploit reflectivity changes of the surface upon adsorption of molecules or atoms, though under different angles of incidence [33,34]. They thus provide a promising possibility to bridge the pressure gap between the "pure" surface science approach with its ultra high vacuum (UHV) conditions to the ordinary atmospheric environment of chemical reactions. Therefore, it has become possible to investigate the transition from diffusional coupling through the adsorbates, which is responsible for the occurrence of patterns at low pressure, to temperature coupling, the dominant coupling mechanism at high pressure. Moreover, EMSI and RAM can be operated in parallel with an IR camera, enabling the simultaneous monitoring of temperature and concentration fields.

The last class of microscopic techniques available for the study of reaction fronts enables studies at the microscopic level with a resolution of 10–20 Å. These are field ion microscopy (FIM) and field electron microscopy (FEM) [35–37]. Together with these techniques a group of methods has been developed or adapted during the last decade, allowing the in situ study of wave phenomena on catalytic surfaces at virtually all length scales and under various experimental conditions.

4.1.3 Reaction Fronts in Electrochemistry

At the solid/liquid interface the situation is quite different. Although the history of electrochemical waves is much older than the one of concentration or temperature waves in heterogeneous catalysis, there are comparatively few reports on reaction fronts. As already mentioned, pattern formation in electrochemical systems is almost always connected with potential patterns, and the first investigations of electrochemical fronts were carried out with two potential probes placed at some distance from each other but close to the electrode surface (see for example the articles by Bonhoeffer and Franck [38–40] and by Suzuki [41] for a review on early Japanese articles). Though the use of potential probes has some limitations that restrict its applicability to one-dimensional geometries, it is still an important tool for the study of waves in an electrochemical environment. Observations of patterns on two-dimensional electrodes had been long restricted to metal dissolution [42–44]. In these systems, regions of different reactivity are connected with distinct thicknesses of salt or oxide layers that are so pronounced that they possess visible contrast in the reflectivity. Thus, waves can be directly recorded with a video camera. In addition, the differences in the refractive index of the concentration boundary layer become so large that profiles of the concentration boundary layer during the oscillatory Fe dissolution could be obtained using holographic microphotography [45].

However, the kinetics of metal dissolution reactions are very complex. From the viewpoint of nonlinear dynamics, several phenomena make the understanding of the primary mechanisms giving rise to the observed wave patterns more difficult: the continuous change of the electrode surface, the formation of thick passive films (which requires a three-dimensional description), and the large density gradients between the electrode surface and bulk resulting in convection. From this standpoint, the introduction of surface plasmon (SP) microscopy, which was the first method to enable the observation of waves in electrocatalytic reactions at two-dimensional electrodes, was a major breakthrough [46]. Surface plasmon microscopy exploits the dependence of the resonance condition for SP excitation on the potential drop across the double layer, as studied by Tadjeddine, Kolb, and others in the early 1980s [47,48]. Spatial resolution is achieved by illuminating a major part of the electrode with a broadened laser beam and imaging the irradiated electrode onto a screen. An experimental setup with a freely accessible electrode can be easily realized, such that the dynamics, which crucially depend on the mass transport of the reactants and products, are undisturbed. Though this approach can be extended to other optical methods, such as electroreflectivity measurements, at the moment surface plasmon microscopy remains the only alternative to potential probe measurements when studying spatio-temporal dynamics in noncorrosive systems.

4.1.4 Organization of the Paper

Section 4.2 deals with methods that visualize waves on catalytic surfaces at the gas/solid interface. As described previously, there exists a whole hierarchy of techniques yielding spatial resolutions from atomic to macroscopic scale. Obviously, we cannot cover all of them in the present article. We chose to discuss SPM, PEEM, EMSI/RAM, and IR imaging in more detail as, on the one hand, most of the experiments on reaction fronts were carried out with these techniques; on the other hand, these experiments are also the ones with which we are most familiar. They are subsequently discussed in 4.2.2 to 4.2.5. In the introductory section of 4.2 the most common mechanisms leading to dynamic instabilities in surface reactions are summarized first, then followed by a detailed description of the oscillatory mechanism of the CO oxidation on Pt(110), the example we refer to throughout chapter 4.2.

Section 4.3 deals with imaging of reaction fronts at the solid/liquid interface. In 4.3.1 we briefly survey when electrochemical reactions might exhibit bistability or oscillations in the global behavior, which is a necessary condition for electrochemical waves to occur. These considerations are followed in 4.3.2 by a discussion on the use of potential probes for the study of electrochemical waves. Potential probes have been widely used in the context of corrosion. Here we consider only those additional aspects that have to be taken into account when studying electrochemical waves. In 4.3.3 a thorough discussion of the physical foundations, the experimental setup, and the applicability of surface plasmon microscopy to the imaging of electrochemical reaction fronts is given.

4.2 IMAGING REACTION FRONTS IN HETEROGENEOUS CATALYSIS

4.2.1 Introduction

In the last section we emphasized that temporal oscillations and spatial structures are widespread phenomena in catalytic surface reactions. Before we discuss the methods that can be used to image these spatio-temporal patterns, it is worthwhile to review the two most important mechanisms giving rise to dynamic instabilities in surface reactions.

The first one applies to non-isothermal systems. It is well known that any activated exothermic reaction in an open system has the potential for exhibiting bistable or oscillatory behavior as well as developing spatial patterns [49]. Of course, whether any of these features are in fact observed depends on the operative reaction parameters. Because the crucial effects are thermokinetic in nature, the main features of these reactions, independent of their detailed mechanism, can already be understood when taking into account the temperature variations resulting from the heat balance equation and a single "chemical" variable. Simulations of spatio-temporal phenomena described by such a general two-variable thermokinetic model are reviewed in Volodin et al. [50]. Experimental examples of patterns in exothermic reactions are the oxidation of ammonia and hydrocarbons on Pt [11,12,51,52] or hydrogen oxidation on Ni [53].

For the majority of isothermal catalytic surface reactions, the conditions for the occurrence of dynamic instabilities are also fulfilled. With one possible exception [54], all surface reactions proceed via the Langmuir-Hinshelwood mechanism (i.e., all gases have to adsorb at the surface before reacting). Feinberg and Terman [55] proved that, whenever the site requirements for the adsorption of the gases involved is different, bistable behavior occurs in some parameter regime. Consider CO oxidation, for example. CO requires one adsorption site, whereas O_2 adsorbs dissociatively if two adjacent adsorption sites are available. Hence, the bistability observed in the CO oxidation can be traced back to the Langmuir-Hinshelwood mechanism.

For oscillations to occur, different site requirements of the species are a sufficient condition if at least three adsorbates are involved [56]. In binary surface reactions an additional feedback must exist, which can have quite different origins [56]. An overview of different models describing surface reactions can be found in Schüth et al. [5]. Prominent examples of feedback that lead to oscillatory behavior are:

1. The conversion of one of the species into another one with a different reactivity; it is, for example, the case in the oxide model suggested by Sales, Turner, and Maple [57], which describes CO oxidation at high pressure on a supported Pt/SiO_2 catalyst where chemisorbed oxygen reacts to give a less reactive oxide [58].

2. The formation of an equilibrium between chemisorbed oxygen and subsurface oxygen, where the latter reduces the reactivity of the catalyst (as was found during CO oxidation on Pd single crystal surface) [59,60].

3. The modification of the reactivity of the catalyst by an adsorption-induced change in surface structure (e.g., in reconstruction). This "reconstruction model" describes the oscillations during CO oxidation on Pt(100) and Pt(110) [61,62].

As we focus mainly on the latter system, we describe the oscillation mechanism of the reconstruction model in more detail.

It has been known for a long time [63] that CO oxidation on Pt single crystal surfaces proceeds via the Langmuir-Hinshelwood mechanism. Hence, both of the reactants have to adsorb on the surface first, oxygen will do so dissociatively, resulting in fairly strongly bound atoms at the surface, while CO stays

4.2 IMAGING REACTION FRONTS IN HETEROGENEOUS CATALYSIS

intact as a molecule and may diffuse around on the surface. If a CO molecule encounters an O atom, they will spontaneously form CO_2, which then desorbs from the surface, leaving two empty sites ready for the next reactants. If the partial pressure of CO is too high, the surface will be covered completely by CO. And, therefore, no empty sites will be available for any O_2 adsorption. This situation is illustrated in Figure 4.1a by a simplified sketch. The surface is then in the poisoned or nonreactive state. The impinging O_2 molecules can only adsorb dissociatively on a Pt surface at temperatures above 150 K. To do so, they need at least two empty surface sites next to each other. Too much

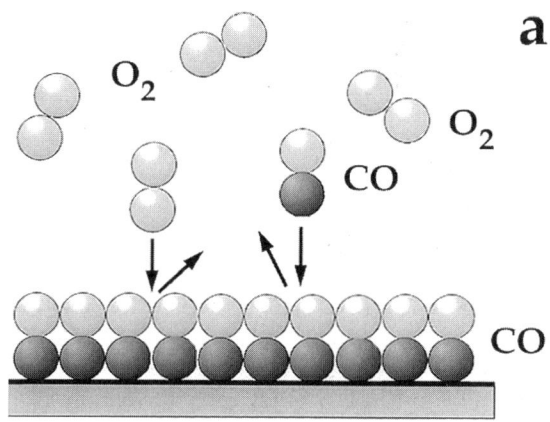

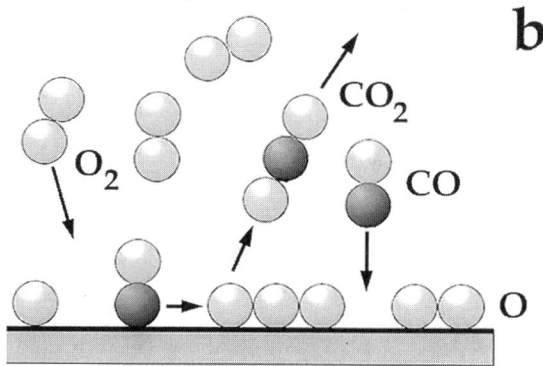

Figure 4.1. A simplified sketch illustrating (a) the poisoned or nonreactive state or (b) the reactive state on a Pt surface. The CO molecules are imaged as gray (C) and bright (O) balls clinched together, while O_2 is shown by two bright balls. The longer arrows indicate movements of gas phase molecules, the short and horizontal one in (b) represents surface diffusion of CO.

O_2, on the other hand, will not poison the surface, since O atoms can only cover up to 0.7 monolayers (ML) of the surface leaving plenty of empty sites for CO adsorption, as indicated in Figure 4.1b. The main reason is the low diffusivity of O atoms, which remain at their original adsorption sites. The poisoning of a Pt surface by too much CO, as previously explained, is a consequence of the different site requirements of CO and oxygen. Hence, it is also the explanation for the well-known strong hysteresis for the production rate of CO_2 versus partial pressure of CO at a constant temperature and a constant partial pressure of O_2. Starting from a small partial pressure of CO, a linear rise for the CO_2 signal with increasing CO is found, which then, at a certain CO/O_2 ratio, decreases rapidly. When finally the CO partial pressure is lowered again, it has to be well below the value for the inhibition of the reaction to start CO_2 production again.

Under appropriate and constant partial pressures of CO and O_2 at a given temperature, commonly 400–550 K, the reaction rate exhibits strong temporal oscillations that were first described for polycrystalline platinum at atmospheric pressures [64] and later thoroughly investigated for single crystal surfaces by Ertl and coworkers under well-defined UHV conditions [65–67]. The model proposed for the Pt(110) surface for these oscillations is based on a surface phase transition, due to adsorption [62]. For instance, a clean Pt(110) surface is reconstructed in a missing row manner and exhibits a 1×2 LEED pattern. Adsorption of 0.5 monolayers (ML) of CO lifts the reconstruction completely, [68] and a volume-like surface structure showing a 1×1 LEED pattern is established. Between 0.2 and 0.5 ML of CO, the fraction of the surface that is 1×1 increases monotonically. The key point is that this nonreconstructed surface phase now has a 50% higher sticking probability for O_2 than the reconstructed 1×2 phase, which is portrayed in Figure 4.2, where an oscillation cycle on Pt(110) is shown. If the partial pressures for the reactants are chosen in such a way that, starting with a clean surface, the adsorption of CO completely dominates, then the lifting of the reconstruction is likely. The sudden increase of O_2 uptake due to the increase of the O_2 sticking coefficient, which is a consequence of this phase transition, will momentarily stop any further increase of the CO coverage. Most of the adsorbed CO will then be consumed by the reaction and leave the surface as CO_2. As soon as the coverage of CO drops again below about 0.5 ML, the reconstruction of the surface sets in, lowering the sticking probability for O_2 to the old value. Now the game can start all over again!

On Pd surfaces the formation of subsurface oxygen [59,60], which is an oxygen species just underneath the first metal layer, is the process that acts as the slowly changing variable, as previously mentioned. The formation of a subsurface oxygen species has also been proven for Pt(110) and Pt(100) surfaces [69–71] under specific circumstances, but without a continuously running reaction. Some pattern phenomena during CO oxidation on Pt present clear signs of a subsurface oxygen formation [72], where the latter represents one topic of the current investigations. The mechanism giving rise to bistable or oscillatory behavior as illustrated is determined by the kinetics of the reaction. As was

4.2 IMAGING REACTION FRONTS IN HETEROGENEOUS CATALYSIS

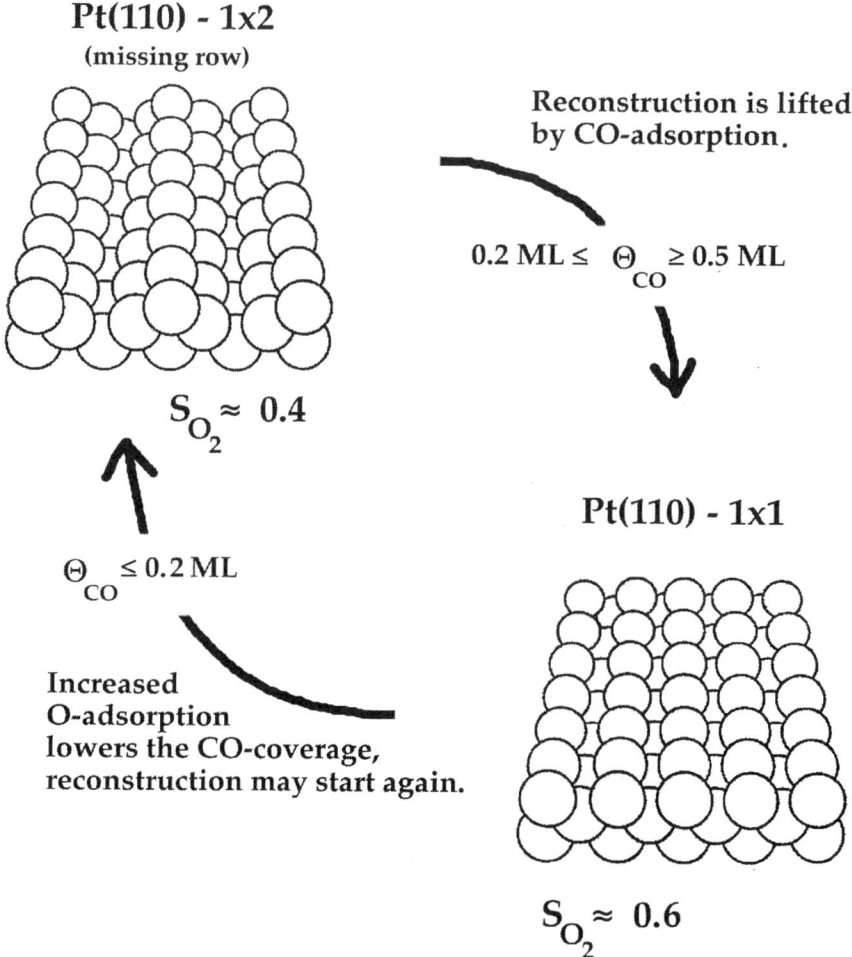

Figure 4.2. The oscillation cycle for the CO oxidation on Pt(110) illuminating the reconstruction model.

pointed out in 4.1.1, spatial patterns may develop if such an autocatalytic reaction is coupled to diffusion of the reactants. Alternatively, the role of diffusion can be assumed by any spatial coupling mechanism that allows one part of the system, in our case the catalyst, to inform the neighboring parts of the system about a changing variable, such as its coverage or temperature. Consider CO oxidation at low pressures, for example. In this case, the spatial communication is due to diffusion of CO molecules, and the formation of reaction fronts can best be envisaged as follows. In the bistable regime of CO oxidation we can imagine a situation in which the surface is in the CO covered state. Due to fluctuations or a defect with an enhanced oxygen sticking probability, a small portion of the surface might become the high reactive state with a low CO coverage. The gradient in the CO coverage causes CO molecules to diffuse from the poisoned to the reactive state. A characteristic property of bistable systems is that a transition from the less stable of the two steady states to the globally steady state can be induced by a small perturbation of the state. Hence, depending on the parameters, even a small increase of the number of free surface sites, resulting from a decrease in the CO coverage due to diffusion, may enhance the oxygen adsorption enough, in order to autocatalytically remove the CO molecules and thus to drive the system locally into the reactive steady state. In other words, diffusion of the CO molecules triggers a transition from one steady state to the other one close to the interface of the two regions of different reactivity, thereby enlarging the originally small island. While this process continues, a reaction front, forming the interface between the two states, propagates across the surface until the whole catalyst has acquired the high reactivity state. Likewise, we can imagine the development of a reaction front that mediates the transition from the state of high reactivity to the state of low reactivity, starting from a mainly O covered surface with a small portion in the low reactivity state. In this case, diffusion of CO molecules from the "nonreactive island" increases the CO coverage in its neighborhood, leading to an autocatalytic blocking of the free surface sites by CO and thus to a spreading of the CO covered state.

Normally, depending on the parameters, one of the two locally stable states in the bistable regime is globally stable, and only transitions from the metastable to the globally stable state can be realized; that is, for a given set of parameters only one of the two types of fronts described previously exists. However, the same interplay between the reaction kinetics and diffusion is responsible for the occurrence of more complex patterns such as spirals, solitary waves, or trains of pulses. In these cases, the reaction kinetics are more complex (e.g., double metastable, oscillatory or excitable). Double metastable characterizes a bistable system in which none of the two locally stable steady states is globally stable, and hence both types of fronts discussed can occur at the same set of parameters. In an excitable medium, there exists only one stable steady state. However, a local perturbation of the system can trigger a transition to a quasistationary state (e.g., the low reactivity branch if the stable steady state has a high reactivity), which then slowly evolves in time and eventually undergoes a transition back to the "original state." A typical pattern that is found in the excitable regime is a spiral wave.

In short, reactions determine the kinetics, whereas the coupling mechanisms (originating from diffusion, global gas phase, or temperature) discussed in the introduction, affect the type of pattern formation on a mesoscopic to macroscopic scale.

After this brief introduction to a specific reaction, namely CO oxidation on Pt(110), which serves as a model reaction, the different methods for imaging surface reactions will now be described in some detail.

4.2.2 Scanning Photoemission Microscope (SPM)

A simple and inexpensive setup to image surface reactions is an SPM. A sketch of it is reproduced in Figure 4.3. The UV light from a 30 W deuterium discharge lamp is monochromated by a high intensity grating monochromator with a blaze at 200 nm. A reduced picture of the exit slit is then imaged via a lens and a reflection objective (Schwarzschild objective) onto the sample. The Schwarzschild objective is located on the inside of an inward oriented viewing port with a quartz window, allowing a 20 mm working distance. This leaves enough space for installing a channeltron to the side of the sample, which collects the photoemitted electrons. The focus of the photon beam is scanned across the surface by moving the sample holder in x- and y-directions using stepping motors with 0.5 μm step width. At each point of the scan, the total yield of the photoelectrons is collected and stored on a computer, which then assembles the final image. For higher resolution, a small aperture down to 20 μm diameter can be used instead of the exit slit of the monochromator, thereby achieving spot sizes of about 0.5 μm. By moving a mirror into the optical axis, an ocular allows the optical inspection of the surface and the alignment of it, using an external light source for illumination. With this arrangement the exciting wavelength could be tuned, and by plotting the root of the photoemission intensity versus the photon energy in a so-called Fowler plot [73], the absolute work function of the surface could be determined as accurately as a couple of meV [20]. For standard pattern imaging just the "white" light of the discharge lamp was used to get an improved signal to noise ratio. Since the count rate of the channeltron is basically limited by its own time resolution to about 6 $\times 10^5$ cts/s, the scan speed may be varied from several hours to some seconds, depending on the pixel field.

The first SPM was able to measure the local work function ϕ with detail that had seldom been seen before. As an example, an image of a polycrystalline Pt foil is shown in Figure 4.4, revealing the whole variety of work functions governing a polycrystalline probe. Presumably, not only different work functions of the (110), (100), and (111) grains contribute to the constrast, but also their slightly different orientations with respect to the surface normal. The imaged work functions significantly vary for the different planes of Pt, the Pt(111) as the most densely packed plane has a ϕ of 5.94 eV, the Pt(100) has a $\phi = 5.84$ eV, and the most open plane, i.e., Pt(110) a $\phi = 5.67$ eV. The picture was taken in about 4 h, using an image format of 512 $\times$ 512 pixels allowing a resolution per pixel of 8 μm.

Another example observed with the SPM is the pattern formation during CO oxidation on Pt(100) as illustrated in Figure 4.5. Before the gases were admitted to the UHV chamber, the image was of course uniformly bright, indicating a single crystal surface with a constant work function. During the reaction all areas covered with oxygen are then imaged as dark regions, while the CO covered regions show up as bright spots. The sequence is imaged at 60 s time intervals, showing a slowly growing CO area at the upper left part (a–c). Suddenly, this spot changes from bright to dark between the frames c and d.

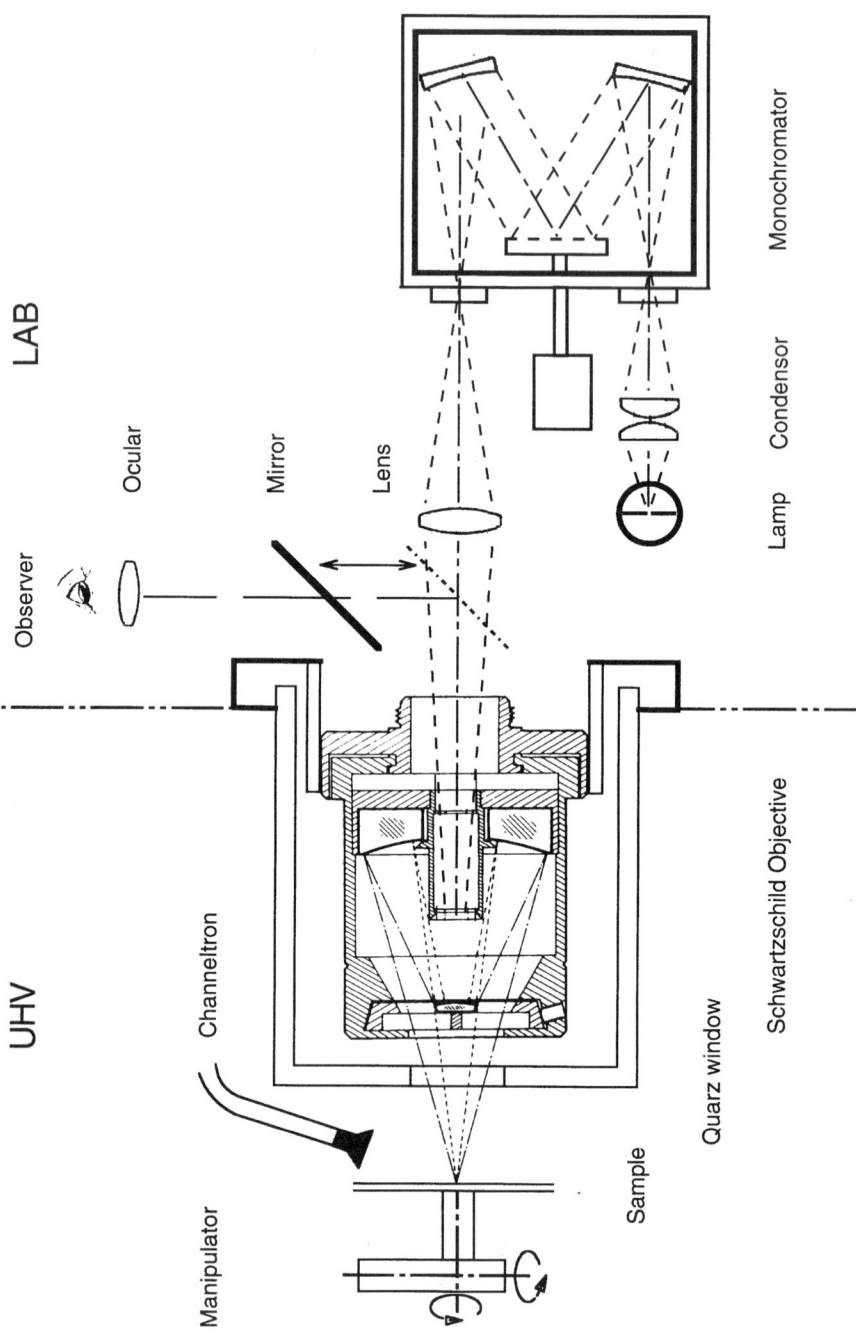

Figure 4.3. Scanning Photoemission Microscope. Reproduced from Rotermund et al. [20].

4.2 IMAGING REACTION FRONTS IN HETEROGENEOUS CATALYSIS

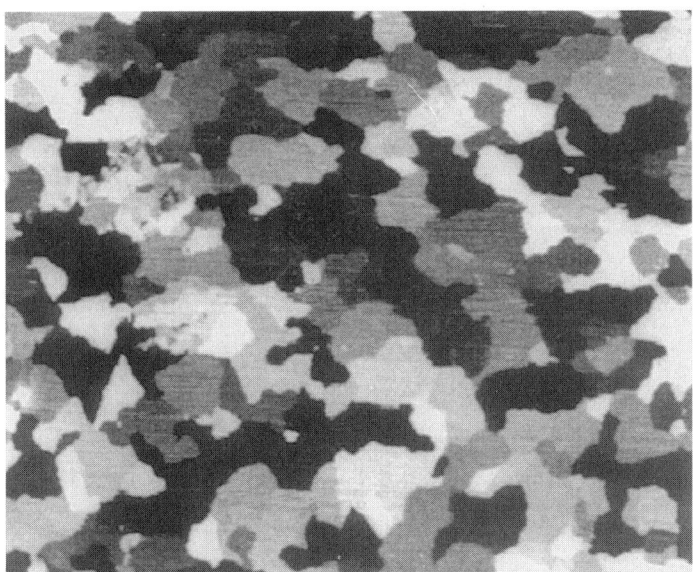

Figure 4.4. A SPM image of a clean 3.8×4.4 mm^2 section of a polycrystalline Pt foil. Reproduced from Rotermund et al. [20].

The preceding example clearly showed the need for improving the time resolution of the SPM. One attempt to achieve this was, instead of moving the complete manipulator stage back and forth, to rotate the sample holder periodically about 2 degrees in either direction, which translated into a horizontal motion of ±2 mm, due to the off axis mount of the sample. By turning the sample rather than moving the complete manipulator stage, the acquisition time for an image was reduced to some seconds, but this was still not enough for fast moving reaction fronts.

4.2.3 Photoemission Electron Microscope (PEEM)

In order to improve the time resolution, a PEEM was designed and built at the Fritz-Haber-Institute (FHI). A sketch of the PEEM [23] is shown in Figure 4.6. Electrostatic lenses have been used to simplify the design and construction of the PEEM, assuring easy operation under strict UHV conditions. It was built as a "snap on" instrument, fitting onto any standard LEED or Auger port of a UHV vessel. A base pressure of 2×10^{-11} mbar in the UHV-chamber guarantees optimal preparation conditions for the crystals. The sample is set at or near ground potential, which is a rather uncommon feature in electron microscopy. The first lens, the objective or so-called cathode lens, is situated at a distance of 4 mm from the sample, across which an accelerating potential of up to 20 kV is applied. The intermediate lens further magnifies the image, thereby reaching a total magnification between 100× and 1000×, allowing a field of view between

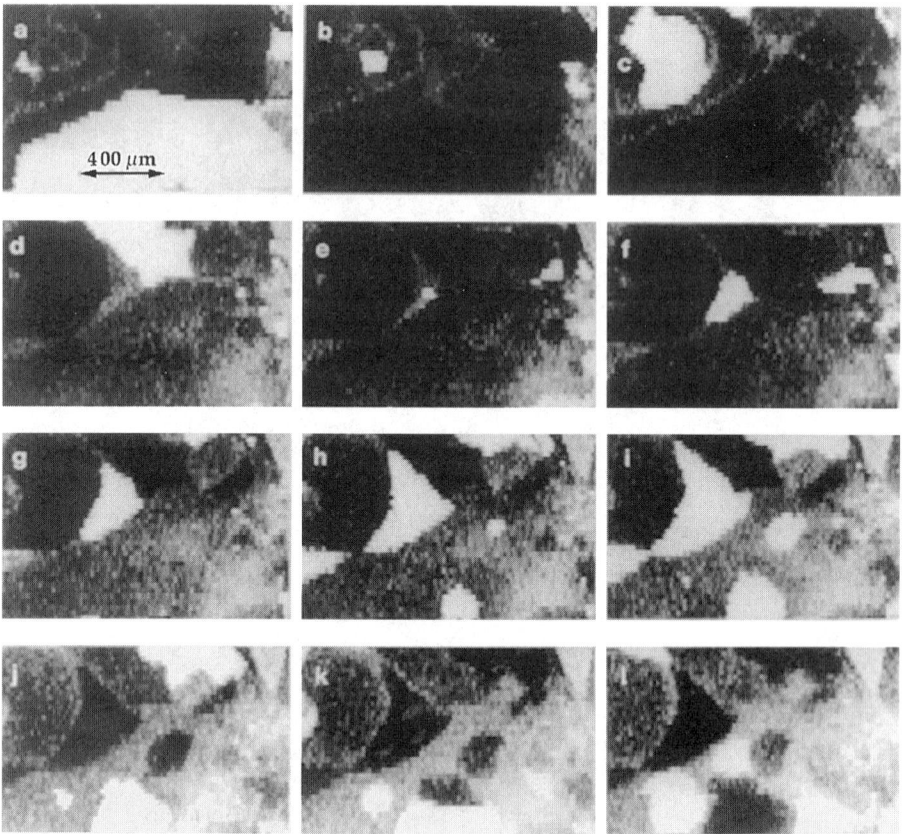

Figure 4.5. Sequence of SPM frames at 60 s time intervals imaging a 1.5 × 1 mm^2 area of a Pt(100) single crystal exhibiting pattern formation during CO oxidation: p_{O_2} = 9 × 10^{-5} mbar and p_{CO} = 8 × 10^{-6} mbar, T = 480 K. Reproduced from Rotermund et al. [21].

400 μm and 40 μm in diameter. The third lens decelerates the photoelectrons to about 1250 eV, allowing better yield on the channelplate, which is located directly in front of a phosphorous screen. At the focal points, two apertures are incorporated, allowing a pressure difference of about 10^{-4} mbar between sample and channelplate when the differential pumping is used. The total pressure for the reaction can be raised to about 10^{-3} mbar.

A rather striking example of PEEM results is shown in Figure 4.7. Here two different phenomena are illustrated, both being dominant features in pattern formation during surface reactions at low pressures for the reactants. In this sequence, a so-called target pattern evolves slowly and is governed by diffusion coupled reaction fronts. The centers of the target patterns are presumably pinned to defects, which act as small perturbations, initializing the elliptically distorted circular waves. These "pacemakers" perform independently of each other.

4.2 IMAGING REACTION FRONTS IN HETEROGENEOUS CATALYSIS

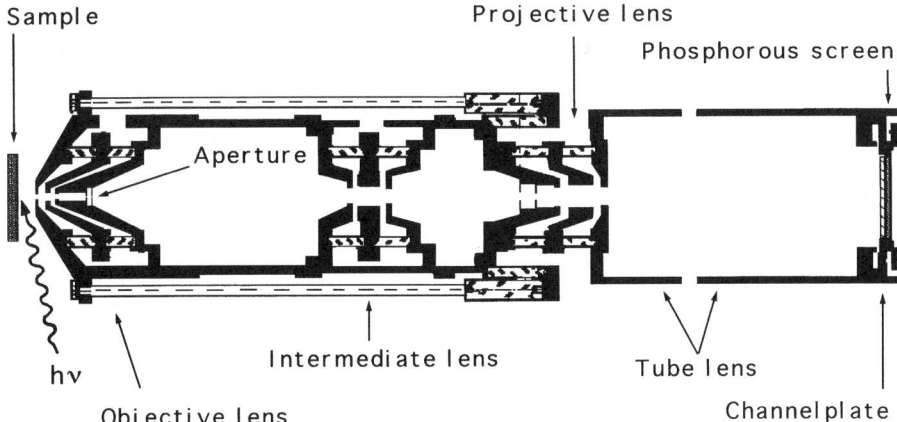

Figure 4.6. Photoemission Electron Microscope (PEEM). UV radiation from deuterium discharge lamp. Reproduced from von Oertzen [154].

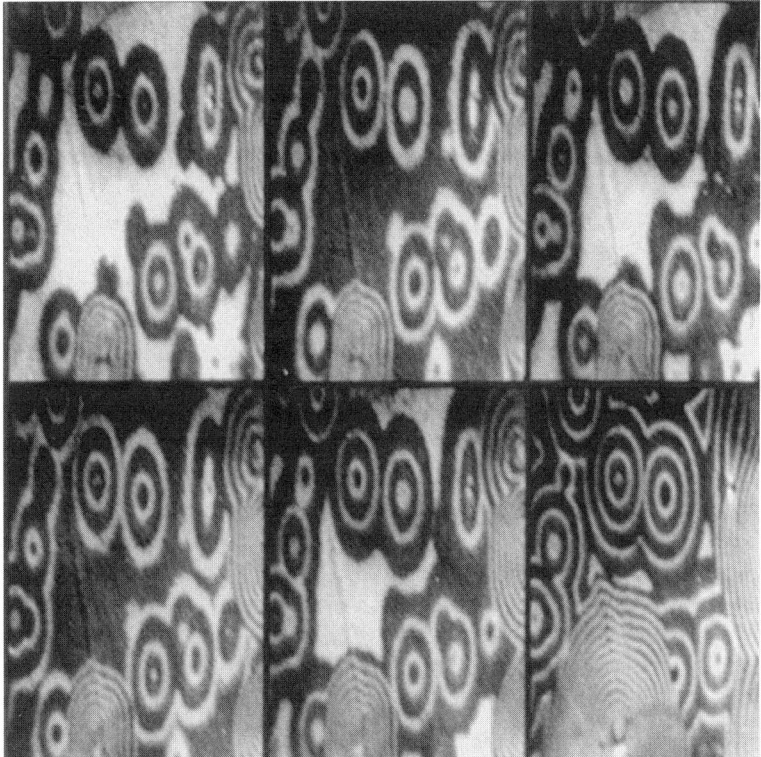

Figure 4.7. A sequence of PEEM images at 4.1 s time intervals (between the last two it is 30 s) imaging a 200 × 300 μm^2 area of a Pt(110) single crystal exhibiting target pattern formation during CO oxidation: $p_{O_2} = 3.2 \times 10^{-4}$ mbar and $p_{CO} = 3 \times 10^{-5}$ mbar, T = 427 K. Reproduced from Jakubith et al. [155].

In contrast, the much faster changing of the "background" from dark to bright in the same sequence is most likely a manifestation of global coupling through the gas phase. Normally, a single information transfer process dominates.

An example where defects and their sizes play an important role is reproduced in Figure 4.8. It shows a broad spectrum of spirals corresponding to diffusion controlled reaction fronts. The core size of the central spiral with a long wavelength is 25 μm × 14 μm. The short wavelength spiral annihilating the central spiral rotates around a smaller core of about 5 μm × 3 μm. The rotation direction appears to be random, two spirals rotate clockwise and two counter-clockwise.

If the defect from which the core of a spiral originates is large enough, several spirals frequently accommodate themselves to it. In Figure 4.9 twenty spirals are all pinned to one large defect.

While macroscopic defects on the surface act as pinning centers for the spirals and govern their rotation frequency and wavelength, such defects do not affect the appearance of "standing waves." The name "standing waves" was chosen because of the similarities with mechanical waves where some points remain unchanged with time. One half-cycle of a standing wave sequence is

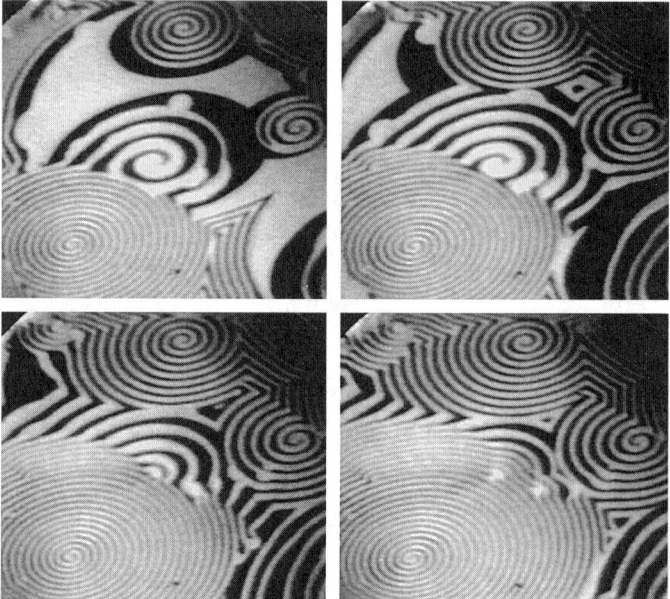

Figure 4.8. Temporal evolution of a population of spirals with strongly differing rotation periods and wavelengths at T = 448 K, p_{O_2} = 4 × 10^{-4} mbar and p_{CO} = 4.3 × 10^{-5} mbar. Each frame is 440 × 400 μm^2. Time interval between successive picture is 30 s. Reproduced from Nettesheim et al. [75].

Figure 4.9. 20 spirals rotating around a large defect, $p_{CO_2} = 4 \times 10^{-4}$ mbar and $p_{CO} = 5.6 \times 10^{-5}$ mbar, T = 463 K, image size 146×124 μm^2. Reproduced from Nettescheim et al. [75].

imaged in Figure 4.10. Information transfer here is controlled via the gas phase, using the consumption of CO to achieve global coupling. The rate of product formation, measured with a mass spectrometer fixed on the CO_2 signal, shows periodic oscillations during these experiments.

During the catalytic oxidation of CO on Pt surfaces, the parameter regions in which pattern formation has been observed can be separated into the following areas.

For a fixed partial pressure of oxygen and a constant temperature, reaction fronts occur when starting from low CO partial pressures with island growth in the so-called double metastable state. In this state, both CO as well as O covered areas appear to be bistable for the same obviously constant set of control parameters. By slowly increasing the CO pressure at the border of the excitable state, both dominant features of those states, namely island growth and spirals, might appear for the same set of parameters, thereby giving rise to particularly artistic images as depicted in Figure 4.11. The interchanging of the different patterns may be due to small periodic perturbations of the control parameters. Simulations of simple reaction-diffusion equations under bistable conditions have been successful in modeling this kind of pattern formation [74], which could be best characterized as a pulsating spiral-front hybrid.

Depending on the sample temperature, further pattern features for CO oxidation can be found. For the Pt(110) plane, which offers the largest variety of pattern formation, an experimental bifurcation diagram, showing all the types of patterns encountered by changing the temperature and CO partial pressure at one fixed partial pressure of O_2, is reproduced in Figure 4.12. Based on

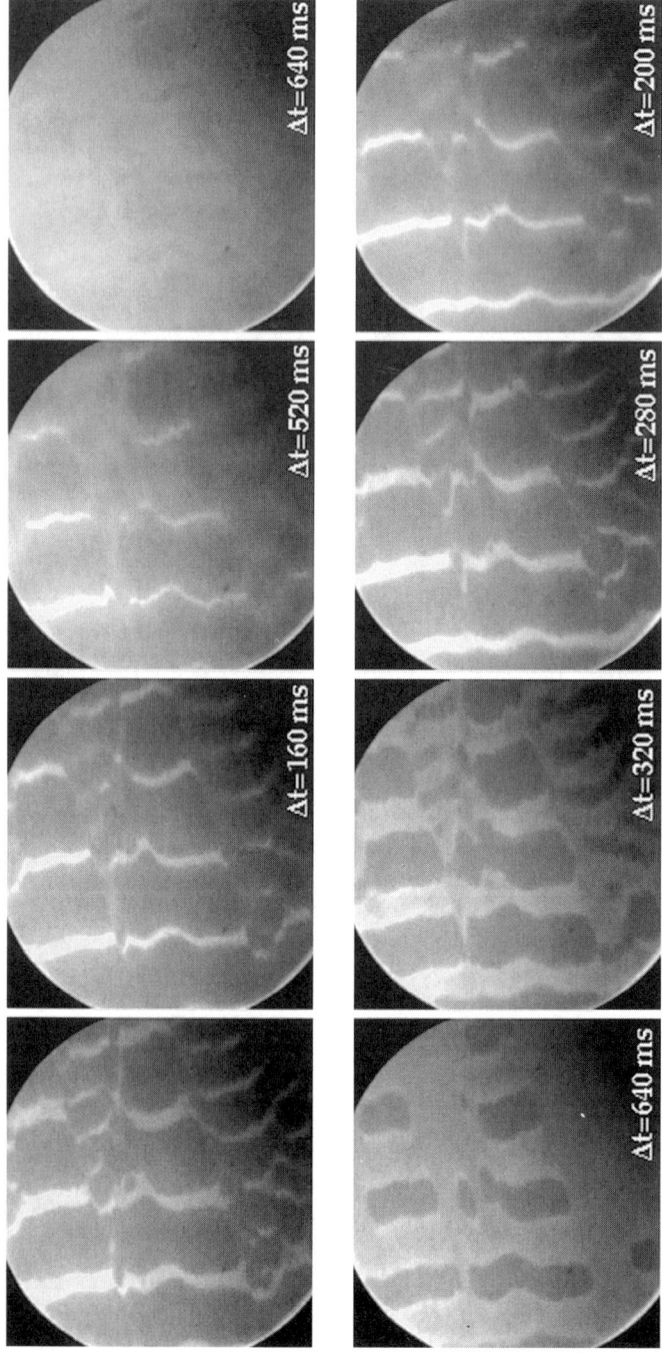

Figure 4.10. Standing waves during CO oxidation on Pt(110): $p_{O_2} = 4 \times 10^{-4}$ mbar and $p_C = 1.7 \times 10^{-4}$ mbar, T = 550 K, image diameter 500 μm.

4.2 IMAGING REACTION FRONTS IN HETEROGENEOUS CATALYSIS

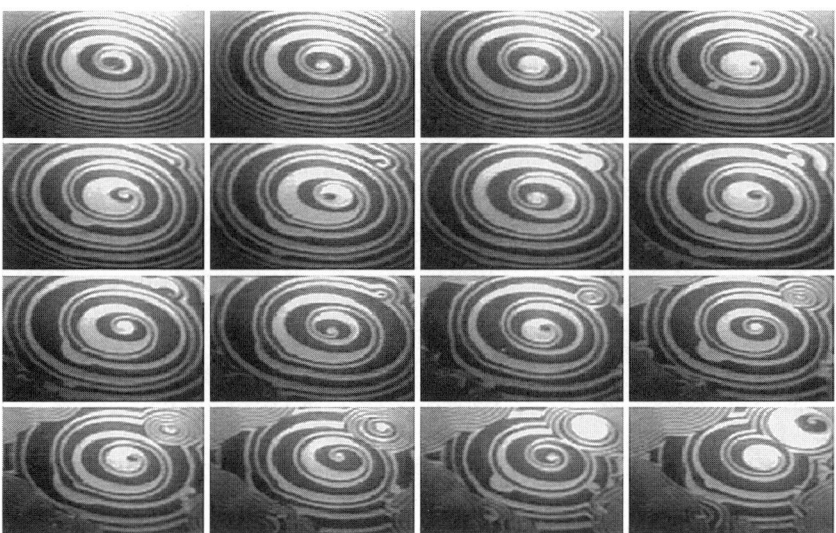

Figure 4.11. Intermittent growth of spirals and islands during CO oxidation on Pt(110). $p_{O_2} = 4 \times 10^{-4}$ mbar and $p_{CO} = 5.6 \times 10^{-5}$ mbar, T = 463 K, time interval 5 s, image size 260 × 160 μm^2. Reproduced from Nettesheim [156].

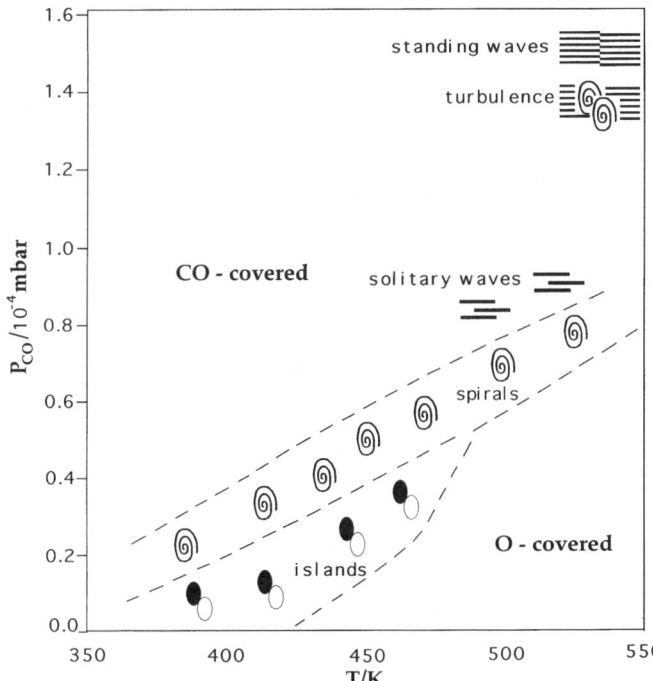

Figure 4.12. Experimental bifurcation diagram for pattern formation on Pt(110) at $p_{O_2} = 4 \times 10^{-4}$ mbar.

Nettesheim et al. [75], the range of CO partial pressure has been extended to include standing waves and turbulence as well. The solitary waves, exhibiting a constant profile and velocity, are indicated just above the formation region of spirals at about 500 K in Figure 4.12. These waves exhibit some unexpected soliton-like behavior [76] in the sense that in a few cases two colliding solitary waves both survive the impact and keep on traveling after their collision. This unforeseen behavior may be related to a locally changed sticking probability for O_2 [77].

Different reactions, such as $NO + CO$, $NO + H_2$, $H_2 + CO$, $H_2 + O_2$, exhibit various patterns on different catalytic surfaces. A parameter-dependent anisotropy of front propagation in the $H_2 + O_2$ reaction was found on Rh(110) [78], while the same reaction on Pt(100) did not show macroscopic pattern formation [37]. The catalytic reduction of NO with NH_3 on Pt(100) displays regular and irregular spatial patterns [79], while rather unusual patterns were observed for $NO + H_2$ on Rh(110) [80,81]. Depending on the control parameters, a variety of different chemical wave patterns (e.g., elliptical- and rectangular-shaped target patterns or traveling wave fragments) were found. The variations in the shape of the wave patterns are attributed to the presence of adsorbate-induced reconstructions with different anisotropy caused by O and N atoms. Cellular structures were found in catalytic reactions with global coupling [30].

One common reason for the complexity of pattern formation is the interaction of various reaction-diffusion fronts due to the extended nature of the surface. It is rarely possible to observe single waves without any collision. In an effort to isolate reaction waves, masks with defined geometric structures having special catalytic properties were deposited onto a Pt surface using microlithography. The boundary conditions can be, on the one hand, of a "no-flux" situation, using catalytically inert materials like Ti, or, on the other hand, of a "flux" type when material with a slightly different catalytic activity like Pd forms the outside boundaries. A completely new avenue for pattern formation is opened due to the special geometric structures surrounding the active Pt areas. As an example, a kind of a chemical reaction clock is shown in Figure 4.13.

Here, in a set of concentric Pt rings separated by catalytically inactive TiO rings, a single CO pulse rotates counterclockwise along the oxygen covered Pt surface. This pulse is quite stable compared to waves on a free surface. Even changes of +−5% of the CO partial pressure had no significant influence on the form and traveling speed of the pulse. The elongation of this pulse, when reaching the upper left or the lower right area of the ring, is due to the anisotropy of the CO diffusion. Along the fast [001] direction the front end of the pulse speeds up only to be slowed down again when the [1$\bar{1}$0] direction is reached. Of course, this anisotropy of the CO diffusion [82] is the reason for the distortions of patterns, like the target pattern in Figure 4.7. In simulations, this anisotropy is usually avoided by rescaling the space according to the differences of the diffusion coefficient. A similar approach for modeling the details of Figure 4.13 would have deformed the rings, and therefore a successful simulation had to use different diffusion constants along the two axes for the CO diffusion [83].

4.2 IMAGING REACTION FRONTS IN HETEROGENEOUS CATALYSIS

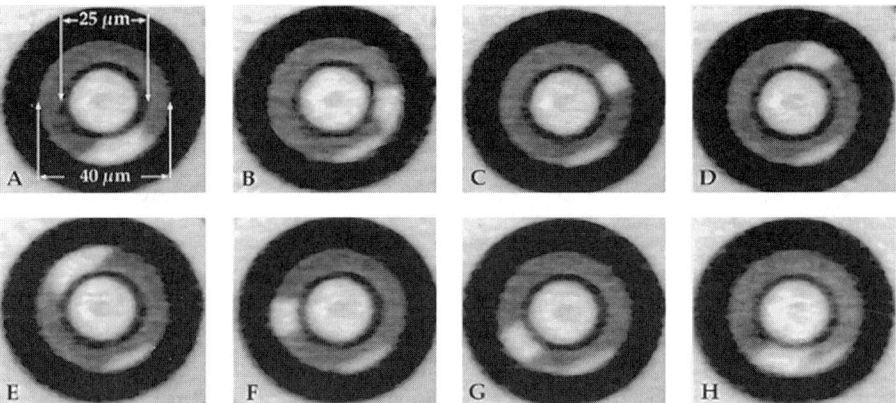

Figure 4.13. Single CO pulse rotating in a Pt(110) ring, time difference between frames 5 s, $p_{O_2} = 4 \times 10^{-4}$ mbar and $p_{CO} = 5.6 \times 10^{-5}$ mbar, T = 446 K. Reproduced from Graham et al. [157].

Another example for this kind of experiment shows a Pd structure on a Pt(110) surface, now both areas being catalytically active. In Figure 4.14 the Princeton shield consists of Pt(110) surrounded by amorphous Pd. In the shield, a single target pattern evolves (upper right-hand corner), and its oxygen waves move through the channels of the structure. They appear dark due to the strong

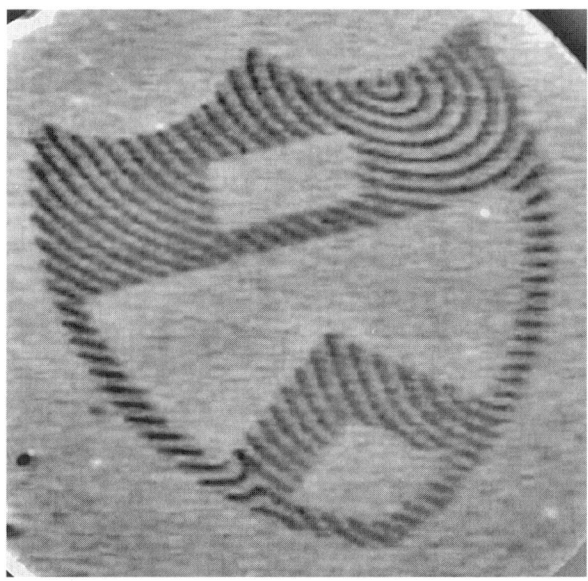

Figure 4.14. Oxygen waves traveling around a Princeton shield consisting of pure Pt(110) surrounded by many layers of Pd. $p_{O_2} = 4 \times 10^{-4}$ mbar and $p_{CO} = 3.5 \times 10^{-5}$ mbar, T = 434 K, image height is 350 μm.

increase of the work function, while the CO covered parts remain gray. The brightness of the surrounding Pd areas indicates that they remain completely O covered. Collision behavior between chemical waves, as seen in the lower part of the image, are now easily observable and can be much better controlled.

4.2.4 Ellipso-Microscope for Surface Imaging (EMSI) and Reflection Anisotropy Microscope (RAM)

The two novel techniques, EMSI and RAM, are based on changes of the polarization of light being reflected from a surface. The well-established technique of ellipsometry has even been used as an imaging method in a few experiments for characterizing structures on surfaces with thicknesses greater than nm [84,85].

In a simple extension, imaging ellipsometry has been used for the first time under vacuum conditions to allow real-time imaging of dynamic phenomena in distributions of submonolayer quantities of atoms adsorbed on a metal surface [33]. Although used more as a qualitative tool, quantitative information about the observed systems is in principle obtainable from local measurements of the usual ellipsometric parameters Δ and Ψ. Conventionally, Δ defines the change in phase difference that occurs upon reflection, while tan Ψ is the relative amplitude attenuation coefficient [86]. For practical purposes an image integrated over several seconds of a well-defined layer of adsorbed CO or O is used as a background picture, which is then subtracted from the following images by a real-time video processor (Hamamatsu Argus 20).

Comparable sensitivity in surface imaging is achieved if the beam is reflected near normal incidence. In this mode the optical reflectivity along the two nonequivalent directions of an anisotropic surface is probed. This new method was therefore denoted as Reflection Anisotropy Microscopy. It requires an anisotropic surface, such as the (110) plane of a fcc crystal, the (azimuthal) polarization angle of the incident light is then adjusted between the two principal axes. The contrast is due to the effect that the anisotropy of the reflectivity is often changed by the presence of a submonolayer coverage of adsorbates, for example due to a surface reconstruction or the formation of an overlayer with a unit cell of different symmetry.

The change in contrast for EMSI can be directly attributed to the change of the surface coverage from CO to oxygen, thereby changing the optical properties. In RAM an effect of structure is observed, as can be seen by turning the sample around the axis of its surface normal. For EMSI the contrast is independent of the sample azimuth, while for RAM the contrast disappears if the [001] or [1$\bar{1}$0] directions of the surface are parallel to the polarization plane of the analyzer. The contrast in RAM is presumably due to the occurrence of a phase transition from the 1 × 1 structure (stable when CO covered) to the 1 × 2 "missing row" structure (stable when oxygen covered). This change of anisotropy might be enhanced by the anisotropy of adsorption sites of the involved species.

The experimental arrangements for EMSI and RAM are sketched schematically in Figure 4.15. In this setup both can be applied simultaneously, using a

4.2 IMAGING REACTION FRONTS IN HETEROGENEOUS CATALYSIS

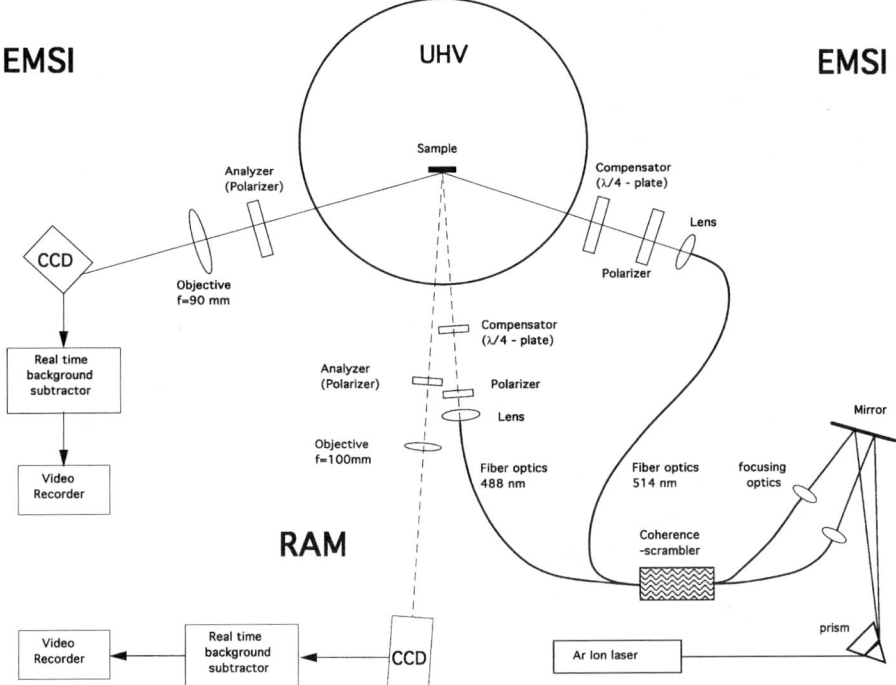

Figure 4.15. A schematic drawing of the parallel experimental setup for EMSI and RAM. Reproduced from Rotermund [158].

multi-line 5 W Ar ion laser, separating its two main wavelengths (488 nm for the RAM, dashed line and 514 nm for EMSI, solid line) via a prism and several mirrors (for clarity these are not shown in the figure). The coherence of each wavelength is first reduced by passing them through two vibrating multi-mode optical fibers. The same spot of the sample is then illuminated and imaged via objectives onto two CCD cameras. Due to the large angle of incidence, the EMSI image is significantly distorted. By tilting the CCD camera accordingly, this distortion is reduced, and a final compensation is performed via image processing on a computer later. For both techniques the incident light is elliptically polarized such that after reflection only linearly polarized light remains. This can be extinguished by an appropriate setting of the analyzer giving a homogeneous nearly dark image from a uniform surface. Local changes, due to a coverage or structural non-uniformity, may then be visible. In the EMSI image regional deviations of the ellipsometric parameters (i.e., the complex refractive index of the metal or the thickness of adsorbate layers) appear as brighter areas. For RAM, regions of different reflection anisotropy are contrasted. The data are preprocessed in real time by subtracting the background image with two Hamamatsu Argus 20 black boxes. They are then simultaneously stored on two independent S-VHS video recorders, both recording the same time code on their audio tracks for proper time alignment later.

The fields of view can be adjusted, ranging from several hundred μm to many mm in diameter. The lateral resolution is in principal diffraction limited at about one μm. To improve the resolution, a much shorter wavelength would have to be used, which has the additional advantage of raising the contrast in a RAM image considerably, at least for CO oxidation on Pt(110) [87]. Unfortunately, the optical alignment would be much more difficult, since the light would not be visible to the human eye and a special and expensive CCD camera would be required. The time resolution is again restricted by the video system, but in principle holds more potential for EMSI and RAM compared to a PEEM, since increasing the signal intensity is always possible just by using stronger lasers, while for a PEEM a much more intense light source such as a UV laser would create space charging effects, which in turn reduces the resolution of a photoelectron image significantly. Of course, utilizing a higher than standard video time resolution would require a special camera and a fast computer with large storage capabilities, especially if longer sequences with an appropriate lateral resolution are to be measured.

In Figure 4.16 two sequential RAM images are reproduced, which were recorded at a partial pressure of $p_{O_2} = 1 \times 10^{-2}$ mbar, a region that is not accessible with a PEEM instrument. At this pressure, an already noticeable convection cooling through the gas phase (about -10 K at 527 K) might be compensated by an increase in temperature due to the reaction energy; the CO oxidation is strongly exothermic and has the potential even to explode at high pressures.

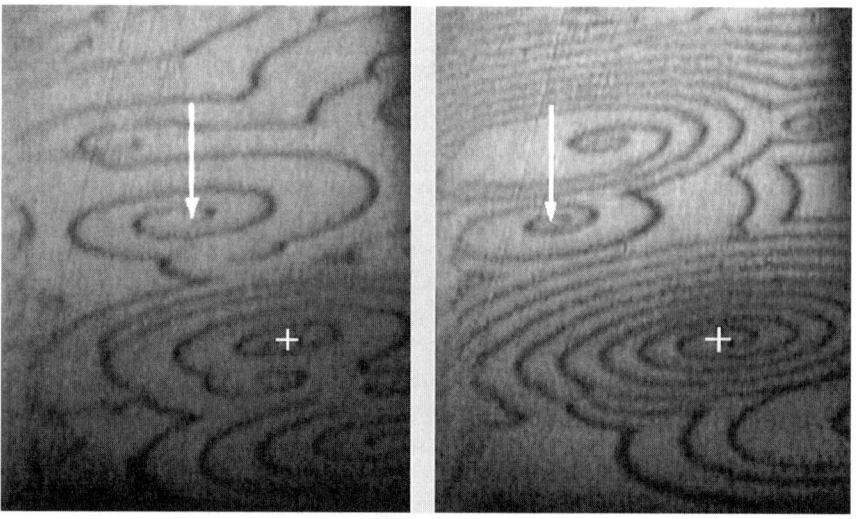

Figure 4.16. RAM images observed on Pt(110), 20 s apart and at $p_{O_2} = 1 \times 10^{-2}$ mbar and $p_{CO} = 1.1 \times 10^{-3}$ mbar, T = 473 K, imaged area is 0.9×1.1 mm^2.

Pattern formation is now only observable, when the temperature of the sample is stabilized by external regulated heating. In this case, for pressures below 1 mbar, a common pattern is the formation of spirals as seen in Figure 4.16. Interestingly, the spiral near the center (marked by an arrow) has moved about 130 μm during the 20 s between the images, while the lower more developed spiral is pinned to one location (marked by a cross). The more common features are pinned spirals, but small fluctuations in one control parameter can lead to freely meandering spirals. In the latter case, the temperature might have been slightly oscillating due to heat generation and dissipation.

On the Pt(110) plane, spirals, target patterns, and solitary waves have been discovered in the excitable region and, within the oscillatory section, standing waves and period doubling into chaotic patterns could be observed. Contrary to that, on the Pt(100) plane, only one type of pattern formation has been found exhibiting a so-called double metastable region up to 4×10^{-4} mbar, detected with PEEM [88]. The same has also been found for total pressures up to 10^{-2} mbar using EMSI.

Figure 4.17 shows a series of EMSI images illustrating pattern formation during the CO oxidation on Pt(100). The series spans a total time of only 0.88 s, in equal intervals of 80 ms. Nevertheless, each frame is substantially

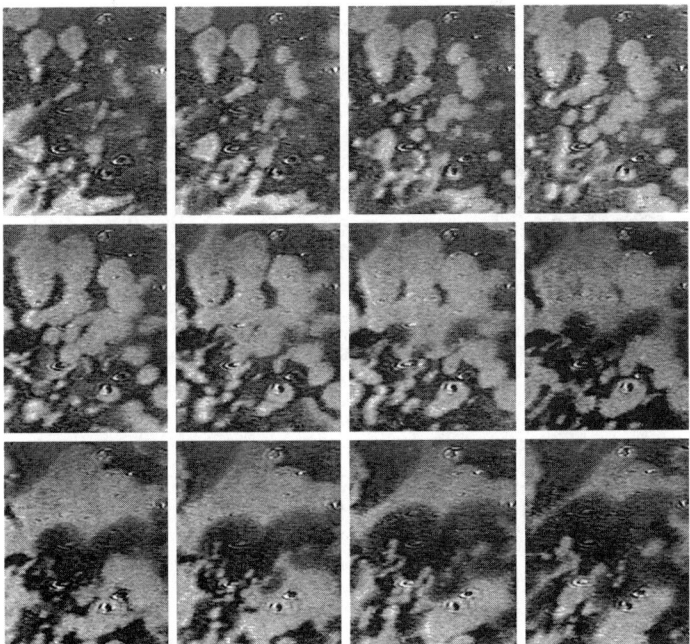

Figure 4.17. EMSI images observed on Pt(100), 80 ms apart, bright areas CO covered, dark ones oxygen covered, $p_{O_2} = 2 \times 10^{-3}$ mbar and $p_{CO} = 1.5 \times 10^{-4}$ mbar, T = 510 K, imaged area is 0.63×0.8 mm^2.

different from the next. In the first row the surface is dominated by CO islands growing with front speeds of about 90 μm/s. In the last row, oxygen island growth becomes dominant. These fronts move with speeds of about 440 μm/s, roughly 5 times faster than the CO-fronts, which continue to grow. It can be expected that at higher pressures the front speeds will increase further so that video cameras with a high speed shutter and/or operating at much increased frames/s will have to be used.

By simultaneous application of the two optical imaging methods EMSI and RAM, two different aspects of pattern formation can be imaged for CO oxidation on Pt(110). This imaging is illustrated in Figure 4.18, which shows the growth of O islands on a CO covered surface. The change in the adsorbate coverage is accompanied by a change in the surface structure. As on Pt(110), CO lifts the 1×2 reconstruction into a 1×1 phase, while O covered areas will stabilize a 1×2 structure. About 9 s before the first image of Figure 4.18, the valve controlling CO into the chamber is quickly closed. The surface stays CO poisoned for several seconds. The first EMSI image (t = 0 s) is arbitrarily chosen, when bright oxygen islands begin to grow at defects on the surface, the temporarily and spatially corresponding RAM image does not show any islands yet. In the RAM series they only start to appear in the second image, 1.24 s later, while the EMSI they have already grown in size. In the RAM they appear in the opposite contrast, namely black, due to different extinction settings of the polarizer and compensator. Even in the last image the dark areas can still be seen with RAM, while with EMSI the surface is already completely covered by oxygen.

Most of the experiments described so far were performed in a so-called continuously stirred tank reactor (CSTR), where the reactants as well as the product were steadily pumped out of the reaction chamber. In an experiment at 1 bar, the chamber was filled with pure oxygen with no pumps running. Then the partial pressure for CO was slowly increased so that a solitary reaction front could be observed. In Figure 4.19 the result of such an experiment is represented. The figure shows the simultaneously recorded RAM and EMSI images for the same surface area.

Clearly a delay of about 1 s between the EMSI and RAM is observed, most easily seen at the 2 s and the 3 s images, where a dominant scratch on the surface is reached in the EMSI image at 2 s, while the front arrives at the same location in the RAM series at 3 s. An additional hint for proper alignment locally is the wavy structure of the CO front best seen at 1 s in the EMSI picture, which is nicely reproduced 1 s later in the following RAM image. The interpretation of the delay might be quite obvious: While the completely oxygen covered Pt(110) surface is still reconstructed in a 1×2 manner, the CO front lifts this reconstruction, but with a delay time originating from the small rate constant of the phase transition. These sequences offer an idea about the potential of using RAM and EMSI together, thereby achieving information about changing local coverages in the submonolayer region (EMSI) followed by structural information (RAM) due to changing coverages. Of course, these interpretations offer

4.2 IMAGING REACTION FRONTS IN HETEROGENEOUS CATALYSIS

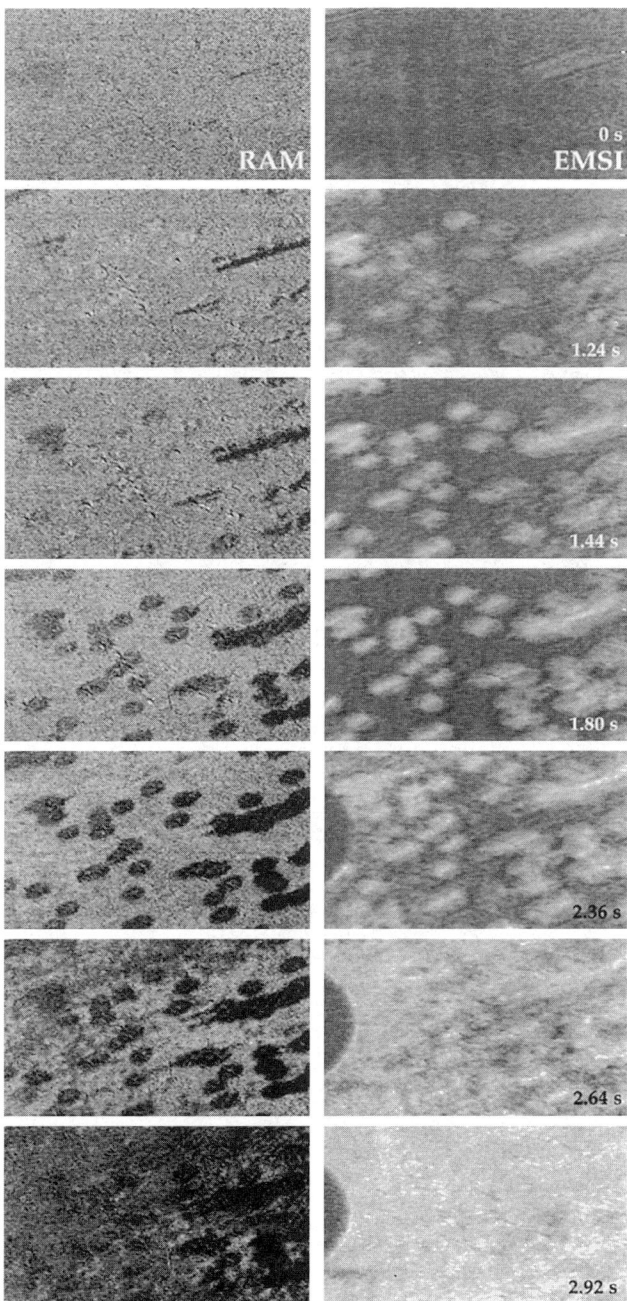

Figure 4.18. A series of images of O islands growth on a CO covered Pt(110) surface recorded simultaneously by RAM and EMSI. $p_{O_2} = 4 \times 10^{-4}$ mbar, T = 401 K, CO was closed 9 s before the first image from $p_{CO} = 2 \times 10^{-4}$ mbar, image size is 2×1.2 mm^2. Reproduced from Haas et al. [159].

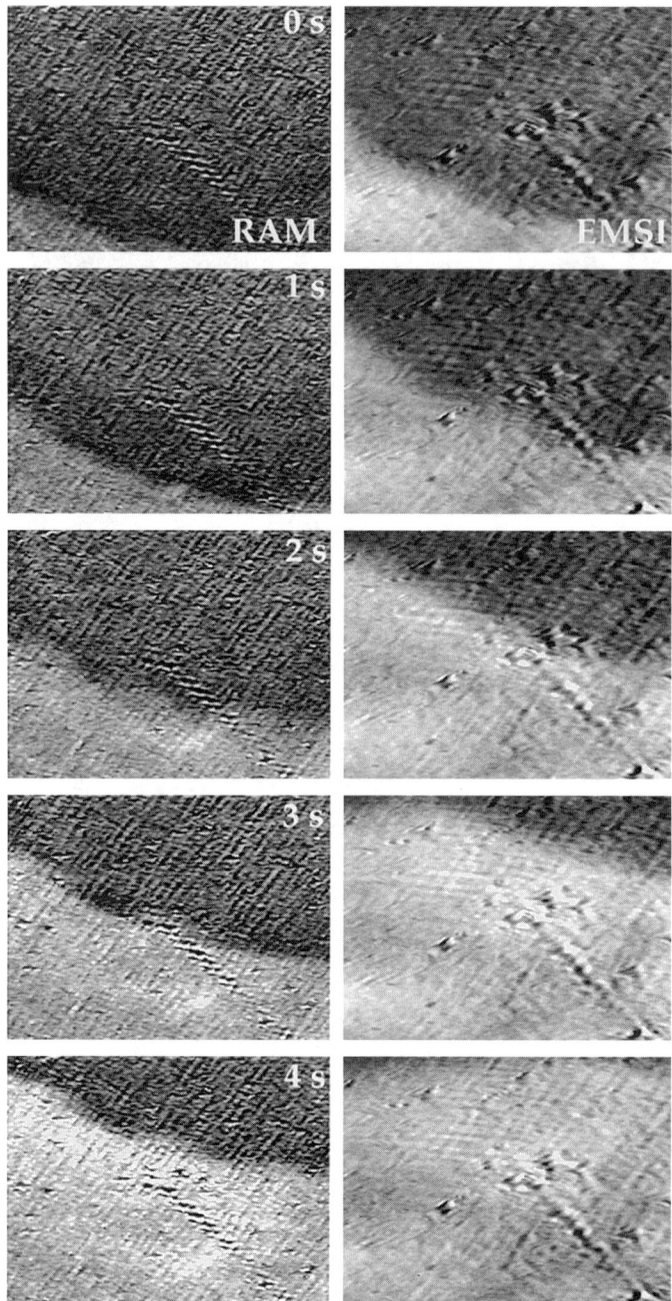

Figure 4.19. A single CO front observed on the same surface area simultaneously with RAM and EMSI at p_{O_2} = 1003 mbar and p_{CO} = 7 mbar, T = 378 K, imaged area is 1.85×1.54 mm^2.

4.2 IMAGING REACTION FRONTS IN HETEROGENEOUS CATALYSIS

a rather ideal and very optimistic view of the situation. As described earlier, both techniques are performed by rotating the analyzing polarizers to deepen the minimum of the signal, for instance of the oxygen covered surface. This "rotation" is of course arbitrary and a sort of "feeling" of the operator is needed to achieve the best contrast. It is not the polarizer position for true extinction, but a position at one or the other side close to zero, as it is illustrated in Figure 4.20 for RAM imaging three differently covered Pt(110) surfaces. The curve labeled CO + O/Pt(110) indicates a reaction taking place homogeneously on the surface. Depending on the setting of the analyzer in the positions a, b or c, the contrast for patterns will be different. A similar set of extinction curves exist for the EMSI setup. Since the choice for a setting is carried out for both methods independently, the contrast accomplished for RAM and EMSI may vary significantly. Admittedly, the procedure has to be standardized to guarantee an operator independent adjustment.

4.2.5 Infrared Imaging

Infrared (IR) imaging has been used for some time in the research of oscillating reactions under atmospheric pressures. A crude version using an IR sensitive detector was described in 1982 [12]. Improvements in the technique allowed

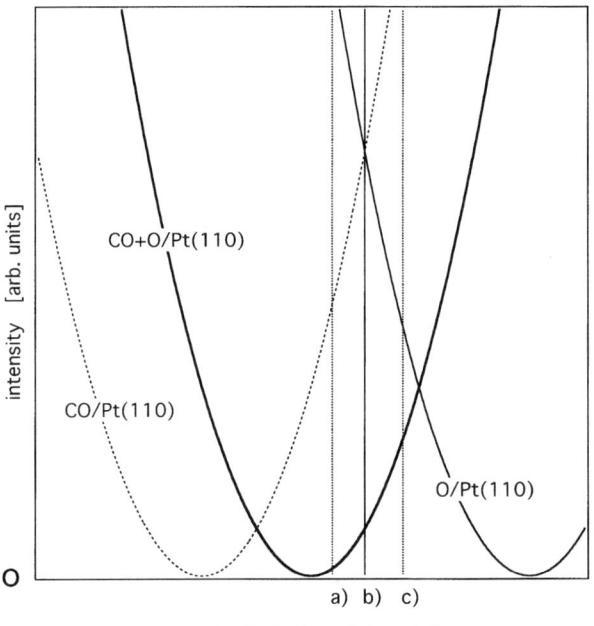

Figure 4.20. Schematic drawing for the extinction curves for RAM imaging different Pt surfaces, as indicated.

a lateral resolution of about 0.6 mm^2, a scale in which spatial temperature oscillations during hydrogen oxidation on a nickel foil [52] and a steadily rotating temperature pulse during hydrogen oxidation on a nickel ring [53] could be observed. Temperature resolution was just below 1 K, depending on the setting for range and level. Time resolution was about one image per second.

After the end of the Cold War in the early 1990s, a whole range of very sophisticated IR imaging systems, previously restricted for military use, became readily available. The camera system used in our laboratory was a Radiance PM-LAB version from Amber. The detector is an 256 × 256 InSb array, which is cooled by an integrated Stirling motor to 77 K. The sensitivity is specified to be $\Delta T = 0.025$ K at 300 K with a dynamic range of 12 bits. It has a standard video output with 25 full frames per second. Its spectral response lies between 3 and 5 μm. To image small objects, a 100 mm focal length Ge objective on an extension tube is used, allowing a lateral resolution of about 20 μm. To gain the full spectral range of IR radiation, a MgF_2 window was used at about 60 mm distance from the surface.

Even though this IR camera is a powerful tool, some intriguing problems can arise from the quite low IR emissivity of Pt samples. The (110) plane of Pt exhibits an emissivity value of just 0.02. Other surfaces in the vicinity at a lower temperature but with a higher emissivity may reflect IR light toward the sample or the camera, thereby reducing the quality of the desired image. Local changes in the emissivity may mislead the experimentalist to believe in a real temperature change. Sometimes a simple test can clear the situation: If there is suspicion that a contrast change during pattern formation is not originating from the temperature, only one of the reactants should be allowed to cover the sample. This signal change should be significantly smaller in comparison to the questionable reaction fronts previously observed.

In Plate 4.1 a sequence of EMSI and IR images shows CO oxidation at a partial pressure of O_2 where the reaction already generates substantial heat. To observe the pattern formation under stabilized conditions, the temperature of the crystal bulk was regulated via a fairly slow PID controller with a time constant of several seconds. The EMSI series on the left-hand side, shows an O front (yellow orange) moving toward the upper right corner between frames 0 and 0.96 s. The temperature front is located at the identical area on the surface, even showing the kink easily identified in the 0 s EMSI image. When the reaction front stops and jumps back between 0.96 and 1.44 s again the temperature front follows perfectly. The wavering motion of the reaction front between 1.44 and 2.40 s is clearly visible in the temperature image. Of course, since during the whole time about half of the imaged surface has been in the reactive state, dissipating the chemical energy into the bulk, the averaged temperature has increased by 2.5 K. Maximum temperature difference in the IR images for the dark red area of the frame at 0 s and the bright yellow part of the frame at 2.4 s is about 3.8 K, and a typical value for the temperature front is a ΔT of 1.1 K. While on the reactive surface, that is the O covered surface, no temperature gradient corresponding to the front movement can be detected. While on the

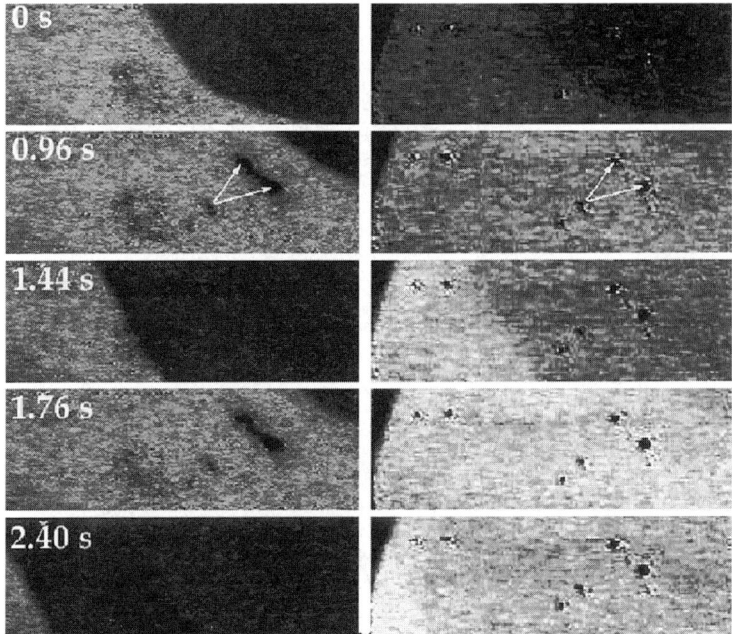

Plate 4.1. A single O front observed at a corresponding surface area simultaneously with EMSI and IR-imaging at p_{O_2} = 113 mbar and p_{CO} = 3 mbar, T = 453 K, imaged area for each frame is 3×1 mm^2; times are indicated in the EMSI frames (left-hand side), the same two defects on the surface are marked by arrows in the frames for 0.96 s. See color plate section.

nonreactive areas explicitly for the 2.40 s image a temperature gradient of about 0.6 K is found having a quite simple reason: it seems to depend on the velocity of the fronts varying between 4 and 20 mm/s. Only fairly slow fronts leave enough time to build a temperature gradient. In contrast to front velocities on Pt(100) where the oxygen fronts are always much faster. In contrast, either the CO or the O fronts could be the faster ones on the Pt(110) surface.

In Figure 4.21 a plot of the front location versus time is presented, from which the velocities can be easily calculated. This plot covers a longer time span than represented in Plate 4.1 to illustrate the variety of speeds. The series depicted in Plate 4.1 is indicated by two horizontal arrows in Figure 4.21. Clearly the front speeds of the last CO front is one of the fairly slow fronts with about 6 mm/s, while the other ones are in the 20 mm/s range. This is already at the time resolution limit for the video system: some fronts just jump across half of the entire imaged area between two frames.

A serious problem may arise from the temperature regulation since the PID controller tries, of course, to compensate the changes in temperature due to the reaction. On the other hand, one has to stabilize the temperature at a value, where pattern formation can occur and not only in a transient fashion.

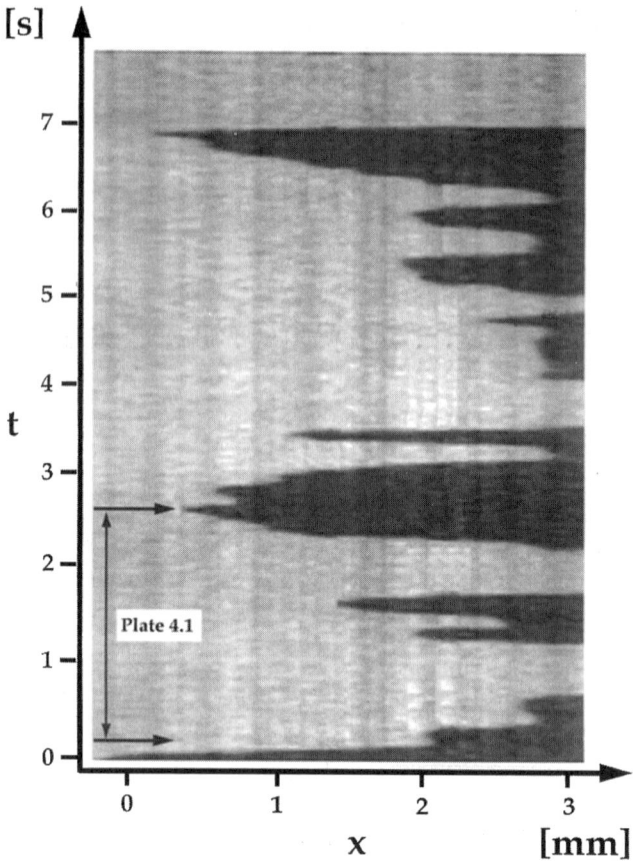

Figure 4.21. A 1d-space/time plot corresponding in part with Plate 4.1. To produce this plot ten horizontal lines across the middle of each frame of a 7 s long video sequence were integrated and mounted as a sequence. The parameters are given in Plate 4.1. The part of Plate 4.1 covered in this figure was marked with two arrows. Each line in the illustration represents a 40 ms time window localizing the fronts.

4.2.6 Summary

We have chosen the CO oxidation on Pt(110) as a model system for a heterogeneously catalyzed reaction, because it is one of the simplest reactions, but nevertheless, it is of great consequence to our environment. In the automobile's catalytic converter, or on similar catalysts in exhaust filters for power plants, this is one of the reactions that have to be performed with high reliability over many years. Therefore, it has been studied by numerous researchers in surface science. In addition to this quite practical importance, the CO oxidation on Pt surfaces exhibits under defined conditions a whole range of spatio-temporal patterns. Consequently, it has become the most widely studied nonlinear chemical reaction on a surface.

We began searching for spatio-temporal patterns using a simple approach, the SPM. It proved to be a good choice for a Pt(100) surface, which exhibited strong but slow oscillations. On this surface the coupling mechanism is dominated by diffusion of CO, and global coupling was not observed. All patterns discovered were irregular in appearance. The picture was governed by slow moving CO fronts and by oxygen fronts moving too fast to be imaged. The lateral resolution was appropriate for the length scale of the pattern.

The main goal in designing the PEEM was to significantly improve the time resolution, allowing the investigation of the much faster oscillations on the Pt(110) surface. On this surface at lower temperatures, the diffusion of CO controls the pattern formation as well, but nevertheless, a whole new range of patterns was found. Those patterns originating in microscopic steps of diffusion will give rise to mesoscopic patterns, such as spirals or target patterns. Investigations at elevated temperatures revealed an increase in the production rate and thereby introduced a new coupling mechanism, the gas phase coupling. It showed features like standing waves, which revealed a much faster time scale and a long-range order in space.

The next step was an increase of the partial pressures for the reactants to allow non-isothermal effects to come into play. Experimental methods using electrons for imaging purposes at higher pressures are not possible, therefore EMSI and RAM were introduced. Both methods can operate at any pressure using only light to achieve information about patterns on a surface. These methods permit the observation of reaction (i.e., concentration) fronts, where the dominant information transfer is ruled by temperature waves. Therefore, a commercial IR camera was used simultaneously with EMSI to image the fronts. Thereby, the region of interchange between reaction diffusion control and thermokinetic effects has been elucidated.

4.3 IMAGING IN ELECTROCHEMICAL ENVIRONMENTS

4.3.1 Introduction

In heterogeneous catalysis, systems are described by mass balance as well as by heat balance for the case of non-isothermal reactions. Correspondingly, patterns consist of different concentrations and possibly different temperatures. In an electrochemical system there is an additional balance equation that comes into play, the charge balance. This new feature introduces a new pattern-forming variable into the system, the electrical potential. Let us reflect how potential differences along the electrode can develop.

Apart from a few exceptions, dynamic instabilities in electrochemistry arise from an interplay of the current-potential characteristic and the external circuit, which is illustrated in Figure 4.22. Suppose that the current-potential curve possesses a region with a negative differential resistance. The negative differential resistance can occur in the stationary polarization curve, as shown in

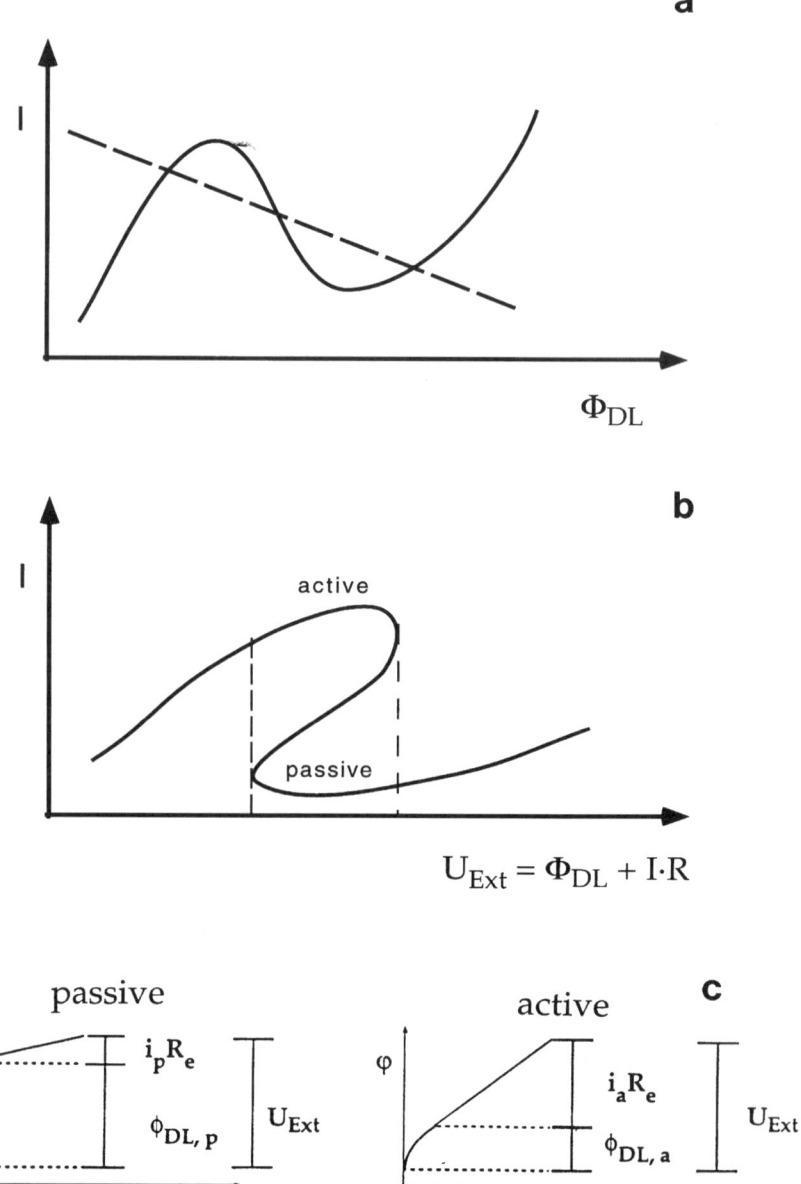

Figure 4.22. (a) Schematic of a current versus potential (ϕ_{DL}) curve with negative differential resistance and load line. (b) Resulting current versus externally applied voltage (U_{ext}) curve, exhibiting bistable behavior. (c) Potential as a function of the distance to the electrode z for the state with high current density (active) and low current density (passive) in the bistable regime for the same value of U_{ext}.

Figure 4.22a, or on a fast time scale in which case it is hidden in the stationary $dI/d\phi_{DL}$-curve. (For an extensive discussion of mechanisms that lead to these two types of current-potential curves and of their stability in different electrochemical operation modes see the recent reviews by Koper [8] and by Krischer [9].) The potential/current pair realized in a certain experiment (i.e., for a given external voltage in a potentiostatic experiment), depends on the ohmic resistance of the electrolyte. It is given by the intersection of the $dI/d\phi_{DL}$-curve and the "load line," which is defined by $I = (U_{ext} - \phi_{DL})/R$, where U_{ext} is the external voltage, ϕ_{DL} the potential drop across the double layer and R the electrolyte resistance. It is easy to see that three intersections (or stationary states) exist, if the slope of the load line ($|-1/R|$) is smaller than the slope of the current/potential curve in the region with the negative impedance, i.e., if $R > |d\phi_{DL}/dI|$. In this situation a measured current-voltage curve exhibits bistable behavior (Fig. 4.22b). Hence, there exists an interval of the external voltage U_{ext}, for which at one value of U_{ext} the system can be in one of two stable stationary states. These states are characterized by different potential drops across the double layer and, hence, different current densities as well as different voltage drops inside the electrolyte (Fig. 4.22c).

Consider now a situation in which part of the electrode is in one steady state and part of it in the other one. Because the potential drops across the double layer are different for the two cases, there is a potential gradient at the interface between the two states parallel to the electrode. This gradient causes ions to migrate, thereby lowering the electric field component parallel to the surface. Necessarily, in one of the two states, the potential will take on values, at which the autocatalytic process drives the system into the other state. Hence, in electrochemical systems it is the interplay of *migration* and reaction that causes the propagation of the interface, or, in other words, the formation of a reaction front. It also means that in electrochemical systems patterns are connected with a spatially inhomogeneous potential distribution at the electrode.

So far, we have discussed how bistable behavior can arise. Oscillations can emerge, if, in addition to a negative differential impedance in the current/potential curve and a considerable ohmic resistance of the electrolyte, there is a second variable that does not follow variations of the potential instantaneously. In general, this second variable is the concentration of a reacting species or an adsorbate. Again, if the oscillations are connected with spatial patterns, the patterns also have to show up in the potential drop across the double layer or the ohmic potential drop inside the electrolyte.

Our discussion, so far, has been on potentiostatic conditions. Under galvanostatic conditions, the potential drop across the double layer again plays the role of the autocatalytic species [8,9,89] and hence also in this case, any spatial pattern is connected with a spatially inhomogeneous potential distribution. These potential variations are detected with both techniques for imaging patterns at electrode surfaces as in the following discussion. With potential probes different $I * R$ drops in the electrolyte are registered; surface plasmon microscopy is based on the dependence of the surface plasmon resonance on the potential drop across the double layer.

4.3.2 Potential Probes

The oldest, though still important methods for the study of spatio-temporal patterns in electrochemical systems are those techniques that probe the local potential inside the electrolyte close to the working electrode with a small indicator electrode. Scanning indicator or reference electrode techniques are often used in other contexts in electrochemistry, especially to study corrosion. In this chapter we do not give a general overview on potential microprobe techniques, but only emphasize those aspects of its application that have to be considered when studying reaction fronts. These aspects are, on the one hand, the fast time scales that typically arise in electrochemical pattern formation, and, on the other hand, the sensitivity of the dynamics on the transport of the reactants. The latter point implies first, that much care has to be taken in order to avoid any disturbance of the mass transport by the potential probe, and second that the setup should allow the fine-tuning of the transport in order to enable the study of different dynamic regimes.

Both points together have restricted the use of potential probes for the study of electrochemical pattern formation to one-dimensional electrode geometries. Most of the studies have been carried out with linear wires [38,39,41,90–92]. For an electrochemical experiment, the wires had a considerable length ranging from 10 cm to 1 m. In these cases, the spatial resolution was achieved with several stationary indicator electrodes. The number of the potential probes varied between two and sixteen, yielding a poor spatial resolution of one to several cm. In contrast, the temporal resolution is only restricted by the response time of the probes, which can be easily brought into the μs regime, as discussed in the following paragraphs.

In more recent studies [93] rotating ring electrodes were used in connection with a stationary potential probe (Fig. 4.23). This setup has the advantage of a

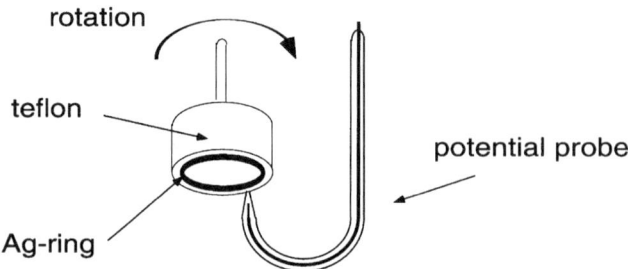

Figure 4.23. Setup for measuring spatio-temporal pattern formation on a ring electrode with a stationary potential probe.

4.3 IMAGING IN ELECTROCHEMICAL ENVIRONMENTS

well-defined transport of the reacting species that can be optimally controlled (as it is the case for any rotating electrode). The spatial resolution is simply achieved by measuring as many points as desired during one rotation of the electrode. Hence, in this arrangement the spatial resolution is restricted by the diameter of the probe, which can be as small as a few μm. However, the price for this greatly improved spatial resolution is a longer response time of the potential probe. It has to be checked, whether the response time is compatible with the desired recording frequency and, of course, with the time scale of the dynamics being studied. The temporal resolution at a certain location along the ring is obviously given by the rotation rate of the electrode. As the mass transport, and hence the rotation rate, crucially determines the dynamic behavior, the temporal resolution cannot be chosen independently from the dynamic regime. It reduces the applicability of this arrangement in the oscillatory regime, which occurs typically at lower rotation rates than the bistable regime. As a rule of thumb, one should measure at least ten points during one oscillatory cycle, which means for the latter setup the rotation frequency should be ten times higher than the oscillation frequency. However, often the two frequencies lie close to each other. In this case, one has to use either several potential probes or a completely different technique such as surface plasmon microscopy, which is discussed in the next section.

The potential probes employed so far consisted all of glass tubes that were pulled out to a capillary at the tip and contained a reference electrode inside the tube (Fig. 4.24). As previously mentioned, the diameter of the tip restricts the

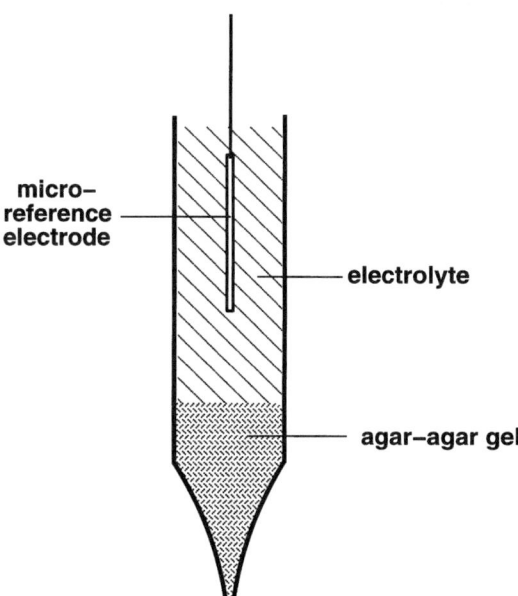

Figure 4.24. Potential probe suitable for measuring fast propagating potential patterns.

spatial resolution in the second setup. With the patch clamp technique [94,95] it is possible to produce capillaries with openings as small as 0.1 μm in diameter. Idealized, the capillary can be viewed as a parallel connection of a capacitor and a resistor, where the capacitor symbolizes the glass and the resistor the solution inside the capillary [96]. Hence, the response time of the probe τ can be roughly estimated by $\tau = R_{probe} \, C_{probe}$. For a fast response time, a fast broadening of the thin tip is necessary as well as a high conductivity of the solution inside the capillary, which can be achieved by filling the tip with agar-agar gel saturated with a salt of high solubility such as NaCl or Na_2SO_4. The agar-agar gel prevents the electrolyte from contamination with the solution inside the tip. In this way, response times of 10 μs were achieved for a 200 μm probe. In comparison, a 1 mM electrolyte yields 0.3 s response time [96].

Examples of different electrochemical patterns measured with potential probes are displayed in Figures 4.25, 4.26, and 4.27. Figure 4.25 comprises some of the data obtained by the groups of Bonhoeffer and of Franck [90]. The first three examples show *fronts*, i.e., transitions from a passivated to an active state in the bistable regime during the electrodissolution of a Fe, Au, and Zn wire, respectively. The last two curves display examples of *pulses* in an *excitable* medium, again from metal dissolution reactions. Excitable means that a small, local perturbation of a homogeneous, stable stationary state induces a large excursion of the variables, before they return to the steady state again. In a spatially extended system, the disturbance propagates in space, which results in the pulse-like structure displayed in Figures 4.25d and e. In all examples of Figure 4.25, just two stationary electrodes were used, and from the time lag between the transitions at the two probes spatio-temporal properties such as the velocity of the fronts or pulses were extracted. In all five examples the readings of the two probes seem to be just time shifted versions of each other. This indicates that the structures propagate with a constant shape and a constant velocity, and hence represent traveling waves. It is striking that typical propagation velocities and length scales are considerably larger than in catalytic systems. The velocity of the traveling waves ranges between 10 cm/s and about 1 m/s. The widths of the interface between the active and passive state in Figures 4.25a, b, and c vary between 10 and 20 cm. The widths of the pulses in Figures 4.25d and e is more than twice as large.

Recently it was found that in electrochemical systems the velocity of fronts is often not constant but increases with time [93,97]. An example of such an accelerated front, measured with a rotating ring electrode and a stationary potential probe, is shown in Figure 4.26. It was observed in the bistable regime of the reduction of peroxidisulfate and constitutes the first example of an electrochemical wave that is not linked to metal dissolution of the working electrode. The average velocity of the fronts in this system was also found to range between a few cm/s and m/s [93]. The main parameter that determines the velocity is the conductivity of the electrolyte; the higher the conductivity the faster are the waves. This dependence was found earlier for metal dissolution reactions by Bonhoeffer and Renneberg [38] and by Franck [39] and indicates that the main

4.3 IMAGING IN ELECTROCHEMICAL ENVIRONMENTS

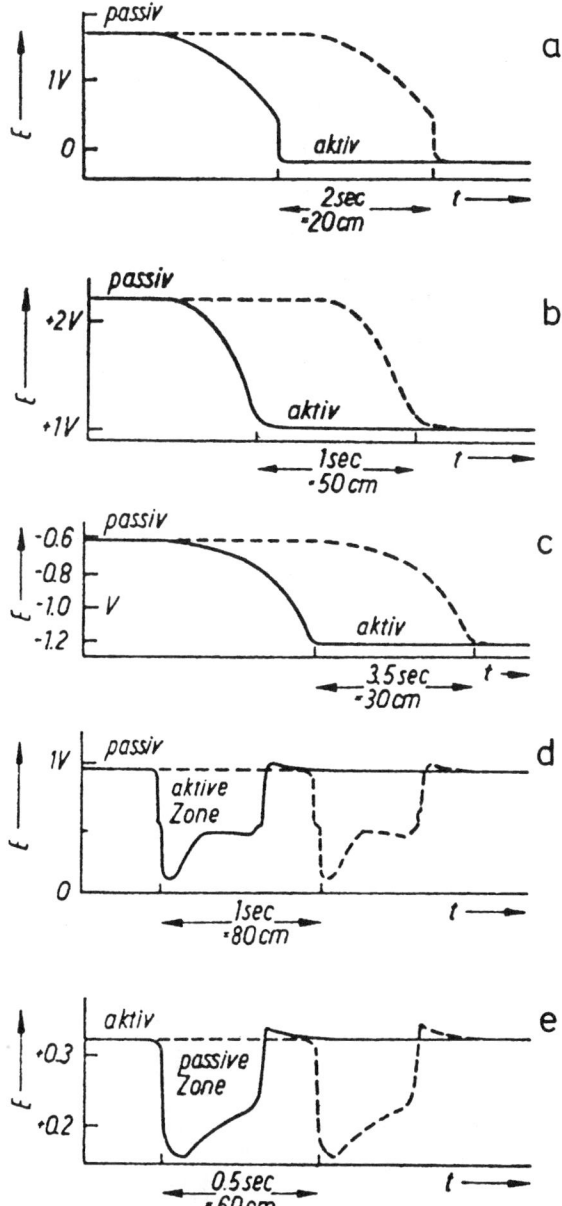

Figure 4.25. Potential time curves measured at two different positions along a metal wire during (a) Fe dissolution in 1 N H_2SO_4; (b) Au corrosion in 1 N HCl/2N NaCl; (c) Zn dissolution in 4 N NaOH; (d) activity wave of Fe in 12 N HNO_3 with successive repassivation; (e) passivation wave during CO dissolution in 1.3 N CrO_3 + 1 N HCl with successive reactivation. The distance between the potential probes corresponds to the distance given under each curve. Reproduced by permission of VCH [90].

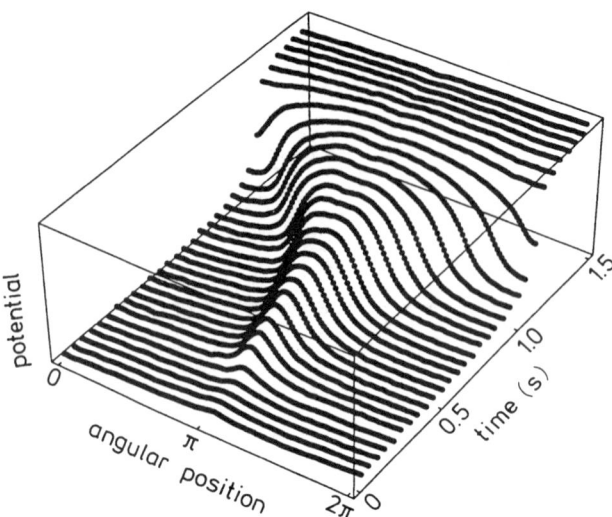

Figure 4.26. Spatio-temporal plot of the local potential during an accelerated transition from the passive (low current density) to the active (high current density) state in the bistable regime of the $S_2O_2^{2-}$ reduction. Reproduced from Flätgen and Krischer [93].

transport mechanism is migration. A recently developed spatio-temporal model that takes only migration into account is capable of reproducing accelerated and constantly propagating fronts as well as the dependence on the conductivity and the order of magnitude of the front velocity [98].

Figure 4.27 displays an example of pattern formation in the oscillatory regime of Ni dissolution under galvanostatic conditions. This pattern was mea-

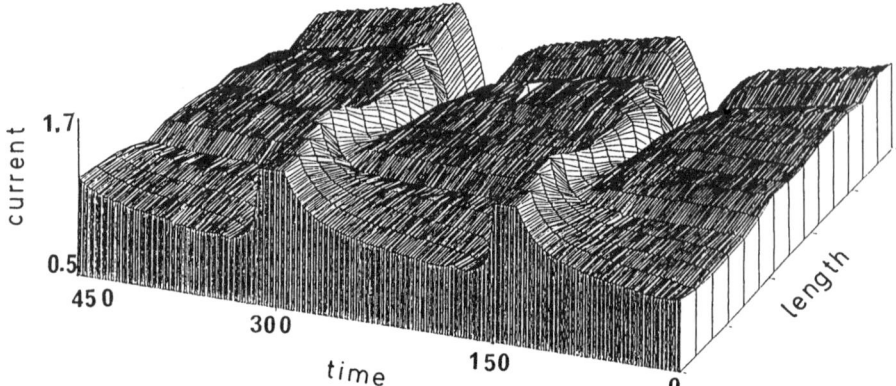

Figure 4.27. Local current distribution of antiphase oscillations during the galvanostatic dissolution of Ni. Reprinted with permission from Nature [91]. Copyright (1988) Macmillan Magazines Ltd.

sured with 16 stationary potential probes. It represents the first example of an antiphase oscillation in a chemically reacting system [91]. (The name *antiphase oscillation* refers to the phase shift of approximately 180° between the local oscillation at the two ends of the wire.)

Up to now, most of what is known about spatial pattern formation in electrochemical systems was measured by means of potential probes. Their advantages are obvious: they are easy to use, inexpensive, and can yield quantitative results. We are convinced that in the future, they will also reveal important basic properties that govern pattern formation in electrochemical systems. At the moment, the understanding of when and how spatial structures are formed is still in its infancy. The most severe drawback of potential probes, in our opinion, is that the investigation of two-dimensional patterns is not possible, due to the shielding problems as well as the fast time scales involved in the dynamics in electrochemical systems. At the present time there are only two techniques available, as mentioned in the introduction, that yield two-dimensional pictures: visible light microscopy and surface plasmon microscopy. The first one is restricted to systems that undergo severe (at least nanometers thick) changes of the surface and has been applied only for electrodissolution reactions. In contrast, surface plasmon microscopy is even sensitive to small variations of the electrode potential or compositional changes within one monolayer. This approach is discussed in much detail in the next section.

4.3.3 Surface Plasmon Microscopy at the Ag-Film Electrode

4.3.3.1 Surface Plasmons in Spectroscopy and Electrochemistry The reflection of light by metal surfaces has been a well-known phenomenon for centuries if not for millennia. An incident electromagnetic (EM) wave forces the metal electrons to oscillate, creating thereby an opposing EM field. The latter annihilates the incident radiation but produces (at least partly) new radiation in the reflection direction. Metal electron oscillations of this type belong to a specific class of electromagnetic waves, the surface plasmons (SPs). They play a crucial role in metal optics, either in the form of virtual or real excited SPs [99]. The latter case, possible under suitable experimental conditions, results in an enlarged EM field at the surface due to SPs having a longer lifetime compared to the case of virtual SP excitation.

Metals such as Ag, Cu, and Au possess an electronic configuration close to a "free electron system." Hence, they support the propagation of (comparatively undamped) surface plasmons along the surface. Their excitation can occur by electrons or photons. The enhancement of the associated EM field reacts sensitively to interfacial changes, which makes the use of SP spectroscopy particularly attractive for electrochemical studies. Therefore, the study and use of surface plasmons has been a hot topic in science throughout the last decades. A few examples shall highlight the importance of SP for metal optics and surface studies.

In 1968 Otto discovered a way to excite surface plasmons by incident light:

a prism/gap/metal configuration is used to pass light through a prism towards the metal employing the so-called attenuated total reflection (ATR) condition [100–102]. At an angle of incidence above a critical angle the light is totally reflected at the prism/gap interface. Yet, there is an evanescent wave that can interact with the metal electrons, provided the gap is thin enough. There exists a specific angle of incidence, at which surface plasmons can be excited. This configuration is denoted as the "Otto-configuration." Subsequently, also in 1968, Kretschmann reported the excitation of SP using a prism/metal film/dielectric configuration, now well known as the Kretschmann configuration [103,104] (see Figure 4.28). Though the former method can be applied to crystalline samples in contact with an electrolyte, it is hampered by the need to adjust the gap width usually between 500 and 1000 nm precisely enough and to work in a thin layer configuration. In the Kretschmann configuration this problem is avoided in an elegant way. However, one relies on the ability to produce thin metal films with the desired thickness and homogeneity. Today, the Kretschmann configuration is more commonly used, because the glass/thin metal film/electrolyte configuration is more versatile.

Soon after the discovery of Surface Enhanced Raman Scattering (SERS) [105,106] it turned out that the "giant" enhancement in SERS has two sources, the so-called "chemical" and "electromagnetic" enhancements. The latter is based on the excitation of (localized) SPs provided by the surface roughness leading to an increase of the electromagnetic field at the metal/dielectric interface compared to the incident electromagnetic field; its gain is denoted by the term g. In the electromagnetic enhancement of the Raman scattering, there is an enhanced coupling of the incident photon field to the surface plas-

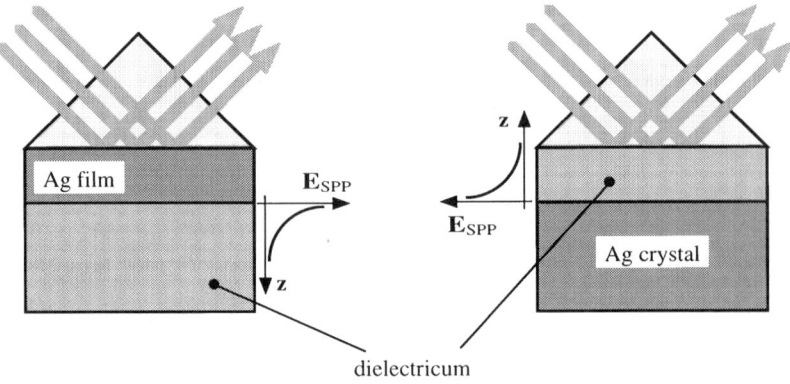

Kretschmann configuration   **Otto configuration**

Figure 4.28. Otto (prism/dielectricum/metal) and Kretschmann (prism/metal-film/dielectricum) configuration for excitation of SPs.

4.3 IMAGING IN ELECTROCHEMICAL ENVIRONMENTS

mon field and of the scattered surface plasmon field to the field of scattered photons with essentially the same gain for both processes. Thus, the differential Raman cross section, $(d\sigma/d\Omega)$, is enhanced roughly by the 4th power of g: $(d\sigma/d\Omega)_{EM} \propto g^4 (d\sigma/d\Omega)_{NRS}$; where the index NRS denotes the normal (i.e., not surface enhanced) Raman scattering and the index EM the so-called "electromagnetic enhancement." [107] Consequently, the direct excitation of SP under ATR condition has been used to test this model and to enlarge Raman scattering of adsorbates at smooth surfaces [108–111]. As a parallel approach to the Kretschmann configuration metal gratings were used to magnify the Raman scattering [112–114].

Another application of SP is the strongly enhanced photoelectron emission from rough films by the excitation of localized surface plasmon modes reported by Douketis [115]. A somewhat similar approach is the enhanced photoemission in scanning tunneling microscopy at silver films as observed by Gimzewski et al [116,117].

An increase of the electromagnetic field is particularly attractive in the field of nonlinear optics, such as second harmonic generation (SHG) and sum frequency generation (SFG). In fact, as early as 1974, enhanced harmonic radiation was observed [114,118–129]. The most advanced nonlinear approach in this context might be seen in the enhanced surface sum frequency generation [121,130].

Upon the invention of the STM, modified approaches were developed, employing the excitation of SP by one or another way: examples are the near field optical scanning microscopy [131,132], the photon scanning tunneling microscope [133,134], the scanning tunneling optical microscopy [135], and the electrooptical waveguide spectroscopy and microscopy [136].

Finally, surface plasmons were directly used for optical imaging. Knobloch et al. employed a combination of spectroscopy and imaging, the Raman imaging with surface plasmons [137,138]. Surface plasmon polariton fields and interferometry were used by Rothenhaeusler et al. to image the structure of thin organic coatings at a silver film [139,140]. Yeatman applied the surface plasmon microscopy to image thin coatings and to evaluate the spatial resolution of this technique [141,142].

In the next section we shall address first the type of excitations the surface plasmons represent and their excitation conditions. We also discuss how and to what extent a change in the electronic configuration of the interface induced by the electrode potential can affect the SP excitation. Based on this introduction to surface plasmons, we illustrate the use of SP microscopy to monitor pattern formation. The example employed here is the reduction of peroxidisulfate at the silver film electrode, which exhibits bistable, so-called complex bistable, and oscillatory kinetics. We shall see that these three regimes are characterized by distinct fast and slow(er) processes associated with the propagation of reaction fronts along the electrode surface.

4.3.3.2 Surface Plasmons For both of the earlier mentioned Otto and Kretschmann configurations, the secret to success is the use of ATR conditions. Photons traveling through a glass prism have a lower speed than photons in vacuum, $c \rightarrow c/n_{\text{prism}}$, where n_{prism} is the refractive index of glass, which depends weakly on the energy of light, $\hbar\omega$. The wave vector of the photons traveling through vacuum, k, is proportional to $\hbar\omega$: $k = \hbar\omega/c$. If we substitute c by the quantity denoted, we obtain the length of the k-vector of photons traveling through a dielectric: $k_{\text{prism}} = n_{\text{prism}} k$. That means these photons have a momentum enlarged by $n_{\text{prism}} = \sqrt{\varepsilon_{\text{prism}}}$, where $\varepsilon_{\text{prism}}$ is the dielectric function of the prism.

Electromagnetic waves associated with specific modes of collective electron oscillations have dispersion relationships different from free propagating EM waves, as illustrated in Figure 4.29. It shows the dispersion of an electromagnetic wave in various environments (i.e., the photon energy, $\hbar\omega$, versus the momentum, k): There are four dispersion curves given: $k_{\text{VP}}^{\text{Me}}(\omega)$ for volume plasmons, $k_{\text{SP}}^{\text{Me}}(\omega)$ for surface plasmons, and $k_{\text{photon}}^{\text{H}_2\text{O}}(\omega)$ and $k_{\text{photon}}^{\text{SFL56}}(\omega)$ for photons traveling through the aqueous electrolyte or through the prism, both under gracing incidence (it is then for p-polarization, that the component of incident electric field strength parallel to the surface is at maximum $E_x \cong |\mathbf{E}|$). Note

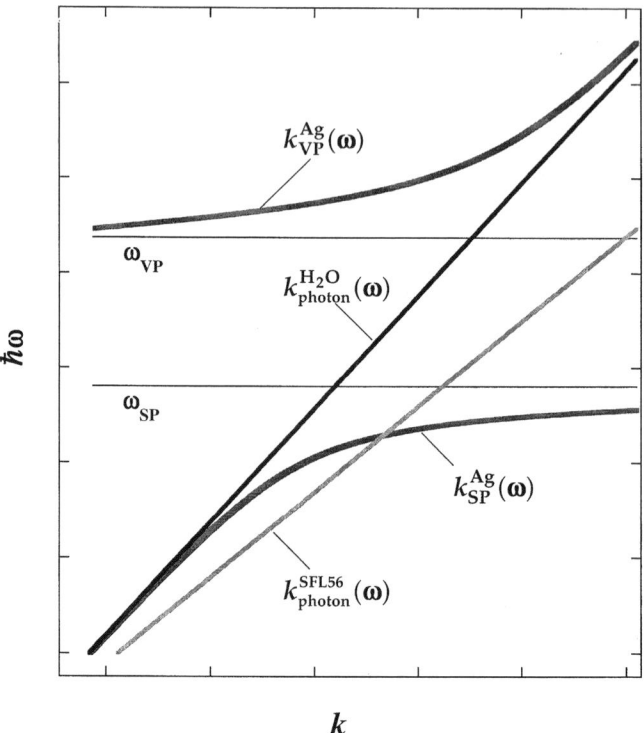

Figure 4.29. Scheme of the dispersion of photons, volume plasmons, and surface plasmon polaritons.

4.3 IMAGING IN ELECTROCHEMICAL ENVIRONMENTS

that $k(\omega) = |\mathbf{k}(\omega)|$. In the following, we omit the multiple indexing for the sake of clarity. As can be seen in Figure 4.29, there is, in general, no match of the momenta of photons traveling through water and the momenta of either volume plasmons or SPs. Volume plasmons have lower momenta, but SPs have larger momenta than photons in the electrolyte at the same energy: $k_{H_2O}(\omega) > k_{vp}(\omega)$ and $k_{H_2O}(\omega) > k_{SP}(\omega)$. For volume plasmons and SPs, the phase matching can be established with the wave vector component of photons, $k_x = |\mathbf{k}|\sin\alpha$, which runs parallel to $\mathbf{k}_{SP}(\omega)$. For this reason SP excitation can only occur using incident p-polarization. Thus, for volume plasmons, phase matching conditions can be achieved for all energies above $\hbar\omega_{vp}$, such that $k_{H_2O,x}(\omega) = n_{H_2O}\, k(\omega)\sin\alpha = k_{vp}(\omega)$. An excitation of SP can occur only at energies below the limiting energy, $\hbar\omega_{SP}$, but it appears to be forbidden because of the mismatch of the momenta: those of photons are generally smaller than those of SPs: $k_{photon}^{H_2O}(\omega) < k_{SP}^{Ag}(\omega)$. As becomes clear from the corresponding dispersion curve in Figure 4.29, photons can be tuned into the phase-matching condition when passing them through a dielectric with sufficiently large refractive index (such as glass) [100–104,143]. In this case, there is an angle of incidence, α, at which the photon wave vector component parallel to the surface equals the wave vector of surface plasmons:

$$k_{prism,x}(\omega) = n_{prism}k(\omega)\sin\alpha = k_{SP}(\omega) \quad (1)$$

For a two-layer system the dispersion of surface plasmons, excited at metal/dielectric phase boundary (as dielectric we consider either the adjacent electrolyte, the gas phase, or the vacuum), is given by the following relationship:

$$\tilde{k}_{SP}(\omega) = \frac{\omega}{c}\left(\frac{\varepsilon_d \varepsilon_M}{\sqrt{\varepsilon_d + \varepsilon_M}}\right) \quad (2)$$

where $\tilde{k}_{SP}(\omega) = k_{SP}(\omega) - i\kappa_{SP}(\omega)$ is the complex momentum vector of the SP wave and $k_{SP}(\omega)$ and $\kappa_{SP}(\omega)$ are its real and imaginary parts, respectively; ε_d and ε_M are the dielectric functions of the dielectric and of the metal film, respectively. Note, that the (complex) dielectric functions are functions of the frequency ω. As is well known, the following relationship takes place between the dielectric function and the refractive index: $\varepsilon = (n - ik)^2$ where n and k denote the real and imaginary parts of ε. For metals, $n_M(\omega)$ and $k_M(\omega)$ vary distinctly with frequency. For some metals such as silver, there exist a frequency region, where $n_M(\omega)$ is rather small, but $k_M(\omega)$ is growing substantially with decreasing ω. Then $\mathrm{Re}(\varepsilon_M(\omega)) = (n(\omega))^2 - (k(\omega))^2$ varies nearly monotonically from positive to (large) negative values with decreasing ω, whereas $\mathrm{Im}(\varepsilon_M(\omega)) = -2in(\omega)k(\omega)$ remains rather small. Due to this property of $\varepsilon_M(\omega)$, there is in general—according to eq. (2)—one frequency, ω_{SP}, for which the real part of the denominator in eq. (2), $\text{Re}(\epsilon_d(\omega_{SP}) + \epsilon_M(\omega_{SP})) = 0$ and, therefore, $k_{SP}(\omega_{SP}) \to$ maximum. This frequency is the upper limit of the surface plasmon frequencies; that is, the frequency range of the surface plasmon branch is $0 < \omega < \omega_{SP}$, where ω_{SP} is related to the frequency of volume plasmons by $\omega_{SP} = \omega_{VP} (1 + \epsilon_d)^{-1/2}$. Since $\tilde{k}_{SP}(\omega)$ is complex, a damping of the SP wave restricts its propagation along the surface to several μm in the visible frequency regime.

As already mentioned, the excitation of SP by light can be achieved using the Kretschmann configuration, which is a three-layer system. The corresponding analytical expression for $\tilde{k}_{SP}(\omega)$ is rather complicated and is not given here. Nevertheless, eq. (2) represents a fairly good approximation if we use ε_d for water and ε_M for silver. Together with eq. (1) it describes the excitation of SPs in a qualitative way. Numerical calculations for a three-layer system show that the SP excitation occurs in a narrow interval of the angle of incidence, where the reflectivity of the prism/silver film/electrolyte system exhibits a minimum, indicating that both the energies and momenta of the incident photons match those of the SPs. The deeper the minimum, the more incident photons are converted into SPs. Since the SP wave is essentially an evanescent wave, the EM field is centered at the metal surface with a field strength about tenfold higher than that of the incident wave. Consequently, the EM intensity at the surface is about 100 times larger than that of the incident wave, which means that at the surface all *one-photon* processes have a 100 times enhanced cross section [144]. Electrochemical processes, which vary the electronic distribution of interfacial states, can alter the height of the minimum, the resonance angle and the width of the dip. This influence will be illustrated here in some detail.

A three-phase model describes the surface plasmon excitation by photons in general terms. In order to take into account the influence of the applied potential and adsorption (which can alter the electronic and geometrical structures of the surface) on the SP excitation condition, one has to extend this three-layer model, or one has to use a completely different approach. Since a sufficient quantum-mechanical modeling of the SP excitation is still missing, we proceed by extending the three-phase model to a five-phase model. One should note that this approach, in fact, will give answers on how the excitation conditions are changed due to the presence of additional two layers with distinct optical properties. Its basic limitation, however, is the unsolved question of how to correlate the influence of potential and adsorption on the electronic and geometrical structures of the surface with specific optical properties of the intermediate layers.

Let us denote these layers by indices 0 to 4 (see Fig. 4.30): phase 0 denotes the prism, phase 1 is the metal film usually with the thickness set to $d_1 = 505$ Å, phase 2 is the (possibly optically different) metal adlayer, phase 3 is representative for the electric double layer, and phase 4 denotes the bulk electrolyte. In cases where the adlayer has the same refractive index as the metal film, we do not specify phase 2 explicitly, and analogously we proceed for the phase 3 (electric double layer). Note that in our calculations (for a three-layer configuration) only a thickness of 505 Å yields a total extinction of the reflectivity.

4.3 IMAGING IN ELECTROCHEMICAL ENVIRONMENTS

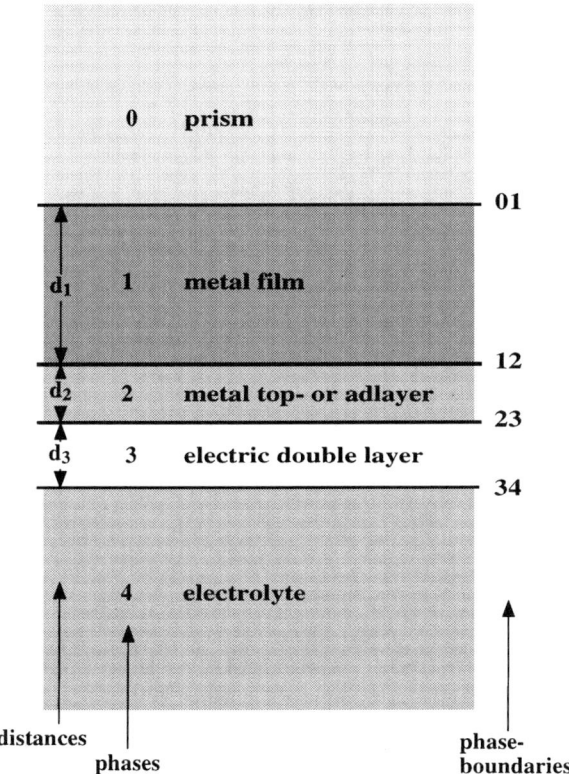

Figure 4.30. Scheme of a five-layer system to represent the prism/metal-film/electrolyte configuration.

Nevertheless, a variation of the thickness by a few Å does not change the presented results; see also the following discussion of Figure 4.32.

In the absence of a silver film, a high reflectivity at the prism base is given only for angles of incidence larger than the critical angle of total reflection. In the presence of a silver film, a film thickness of about 500 Å is sufficient to obtain a high reflectivity at nearly all angles of incidence. This is shown in Figure 4.31, reproducing the calculated reflectivity for a prism/silver film/air of prism/metal film/electrolyte configuration for incident s- and p-polarization. Figure 4.31 depicts three situations: the left curve(s) holds for a prism/metal film/air configuration ($n_0 = 1.766$; $n_2 = n_3 = n_4 = 1$; $\lambda_L = 632.8$ nm). For s-polarization the reflectivity is for all angles of incidence close to 100% (dotted curve). For p-polarization (solid line) the phase matching condition occurs at about 35.25°. The calculated reflectivity for p-polarized light drops to a low value, for an Ag film thickness of 505 Å down to about 1%; it indicates the excitation of SPs. The SP wave propagates along the surface for some distance decaying to electron hole pairs and heat [145]; it shows an average propagation

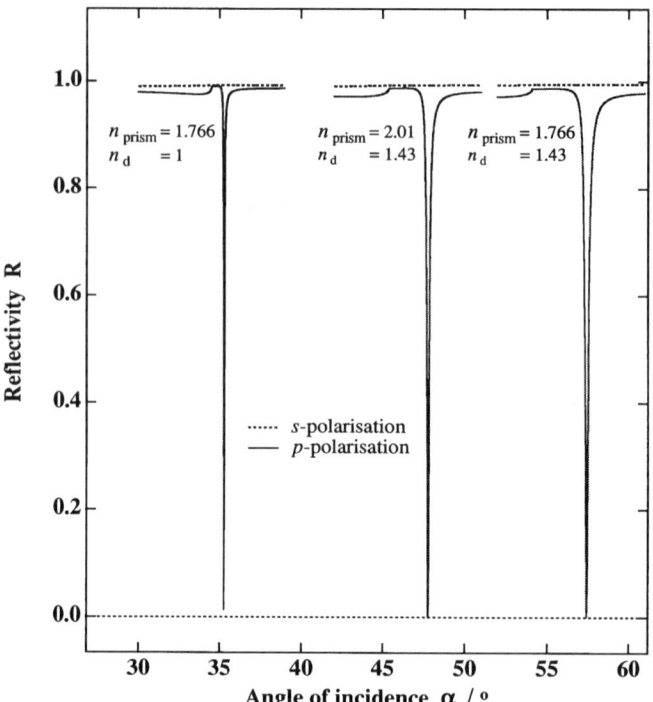

Figure 4.31. Calculated reflectivity of a prism/Ag film/dielectricum (Kretschmann) configuration. n_{prism} = 1.766 (sapphire) or 2.01 (LASF35); $d_{Ag\text{-film}}$ = 505 Å; n_{Ag} = (0.04, 5.2); n_d = 1 (vacuum or air) or 1.43 (electrolyte). Solid lines: p-polarization; dotted lines: s-polarization.

length of about 22 μm at the laser wavelength of 632.8 nm [146]. Of course, the phase matching condition depends sensitively on the refractive indices of both dielectrics, the glass and the electrolyte. If the glass prism covered with the silver film is in contact with an aqueous electrolyte, there is a very different phase matching condition leading to the minimum reflectivity for p-polarization around α = 57.4° and the reflectivity dropping down to about 0%. This is shown in Figure 4.31, right curve for $n_2 = n_3 = n_4$ = 1.43, but otherwise identical parameters. The middle curves in Figure 4.31 show the reflectivity for the case of a prism having a refractive index of n_0 = 2.01 (LASF35 Schott, Mainz, Germany) and with the other parameters being identical with the previous case.

Let us further illustrate the sensitivity of the SP excitation to the (local) environment by additional examples. The influence of the Ag film thickness is shown in Figure 4.32 for d_1 = 400, 505, and 600 Å. We mainly notice a change in the width of the reflectivity minimum and a slight shift of its angular position with rising thickness. The thickness of 505 Å turned out to be the absolute minimum of the reflectivity with the other parameters given.

4.3 IMAGING IN ELECTROCHEMICAL ENVIRONMENTS

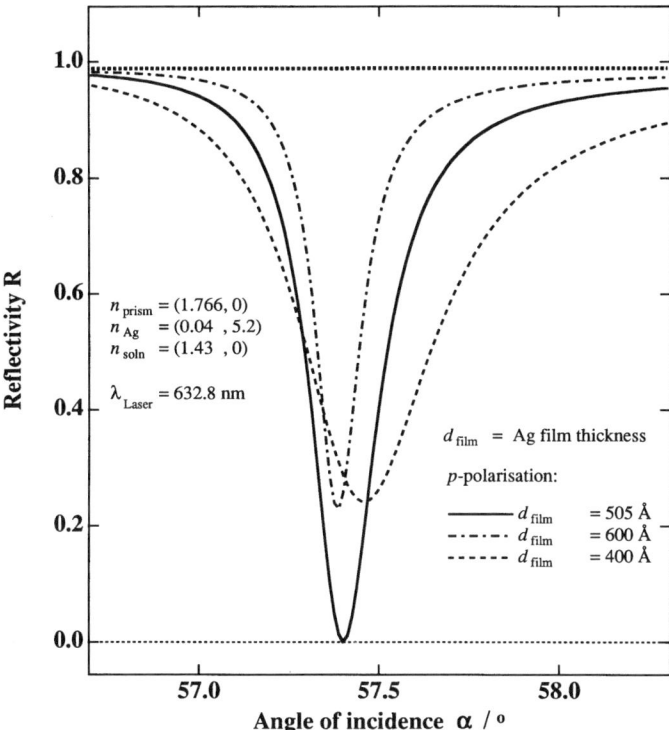

Figure 4.32. Metal film thickness dependence of SP excitation. $n_{\text{prism}} = 1.766$ (sapphire); $n_{\text{Ag}} = (0.04, 5.2)$; $n_d = 1.43$; $d_{\text{Ag-film}} = 400$, 505, or 600 Å; $\lambda_L = 632.8$ nm.

We are interested in how the state of the metal surface and its vicinity alter the SP excitation. In an electrochemical environment the state of the metal electrode surface is controlled by the applied potential. Via the interfacial free energy the applied potential determines the charge density of the metal and, thus, the amount of adsorption including specific adsorption. In turn, adsorption of ions and molecules can also change the charge density at the interfaces. Adsorption, in particular specific adsorption, results in orbital mixing—at least to some extent. In other words, the change in the charge density and adsorption affects the electronic configuration of the metal surface. Consequently, its optical properties are different from those of the bulk material, in particular they depend on electrode potential and adsorption [147,148].

We distinguish the influence of the optical properties of the interface on the SP excitation between two main categories: the first category refers to the adjacent electrolyte. There are electrochemical processes such as the reduction of ions (here $S_2O_8^{2-} + e^- \rightarrow SO_4^- \cdot + SO_4^{2-}$) and the accumulation of ions in the diffuse double layer, which may change the refractive index of the electrolyte in this region. The effective thickness of the adjacent electrolyte layer sampled by the SP is about $d_{4'} \leq 1200$ Å. Note that this adjacent electrolyte layer is denoted by the subscript 4'; its location is between d_3 and d_4, but it is not shown in Figure 4.29 because we refer to it only in this paragraph. For the case where this adjacent layer has a refractive index different from the bulk electrolyte, there is a shift of the resonance angle, the magnitude of which is dependent on the thickness of this layer. We define an effective thickness to be the thickness for which about 80% of maximal angle shift is achieved. For the parameters $n_0 = 1.777$, $n_1 = (0.04, 5.2)$, $d_2 = d_3 = 0$ (!), $n_4 = 1.43$ and $\lambda = 632.8$ nm, a variation of $n_{4'} = 1.430 \rightarrow 1.429$ with $d_{4'} \gg 1200$ Å leads to a shift of the SP resonance angle from $57.400°$ to $57.344°$. Considering, however, the very low ion concentration in the electrolyte, the change of the refractive index, $n_{4'} \rightarrow 1.429$, is certainly an upper limit. In reality, it will be much smaller and, therefore, an ion accumulation in the diffuse layer cannot cause a shift of the SP resonance angle on the order of $0.1°$ as seen in the experiments. Hence, subsequently we will not distinguish between the phases 4' and 4.

Typical experimental reflectivity curves are displayed in Figure 4.33 for two different values of the potential; they are clearly shifted relative to each other.

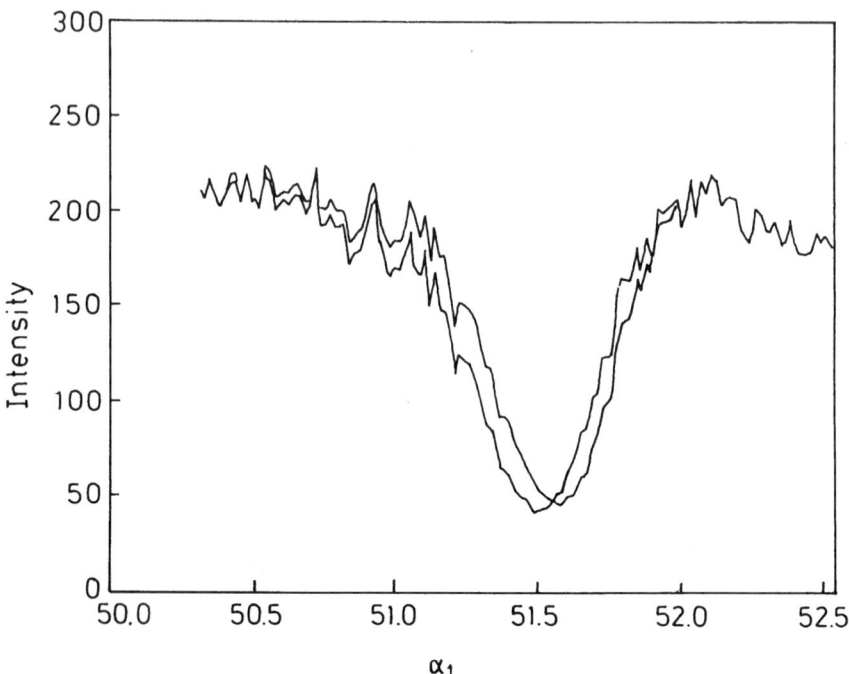

Figure 4.33. Reflectivity for incident light ($\lambda_L = 632.8$ nm) of the prism thin Ag film/electrolyte system versus angle of incidence measured for two potentials. Reproduced from Flätgen et al.[160]

Obviously, the SP resonance condition depends on the potential, a phenomenon known for a long time [47,148,149]. This potential dependence of the reflected intensity provides sufficient contrast for its use in imaging.

The second category concerns the interfacial regime, i.e., the top metal layer (phase 2) and the electric double layer (phase 3). The former is denoted also as metal adlayer to cover the case of deposition of distinct metal deposits or the case of a more complex metal adsorbate structure such as oxide atoms in the subsurface position. Its thickness (d_2) ranges from about 1 to a few Å. The adjacent electric double layer can contain specific adsorbed ions at the location of the inner Helmholtz plane (IHP). Specific adsorption means that there is some sort of covalent bonding of these ions to the metal atoms.

The new states formed here by covalent bonding can give rise to optical transitions. Consequently, the refractive index of the electric double layer may become complex to account for the absorptivity. In turn, the orbital mixing between adsorbate and electronic states of the metal surface may change the absorptivity of the metal adlayer and, thus, a substantial variation of k_2 is expected (note $\hat{n} = n - ik$). Figure 4.34 illustrates how such an effect alters the SP

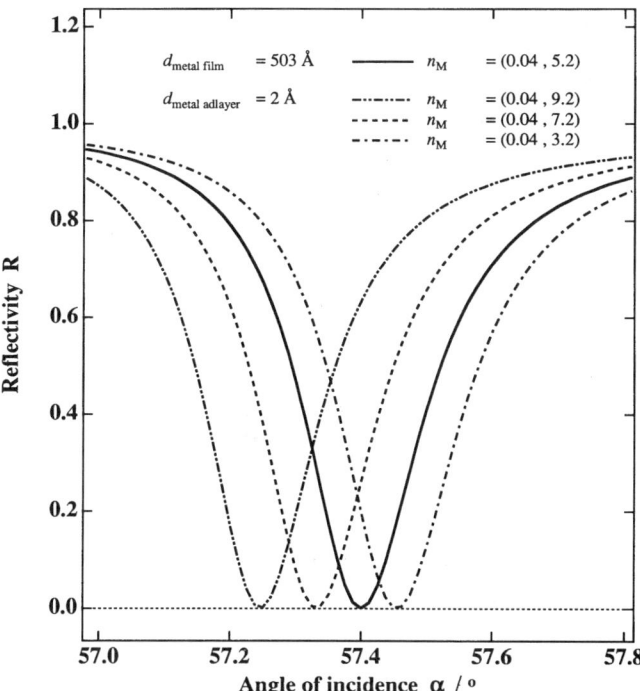

Figure 4.34. Influence of metal adlayer on SP excitation. $n_{\text{prism}} = 1.766$; metal film: $d_{\text{Ag film}} = 503$ Å, $n_{\text{Ag film}} = (0.04, 5.2)$; metal adlayer: $d_{\text{adlayer}} = 2$ Å, $n_{\text{adlayer}} = (0.04, 3.2), (0.04, 5.2), (0.04, 7.2), (0.04, 9.2)$; electrolyte: $n_{\text{electrolyte}} = 1.43$; $\lambda_L = 632.8$ nm.

resonance condition. For a metal adlayer of 2 Å thickness and k_2 ranging from 3.2 to 9.2 a substantial angular shift occurs of about 0.2°. However, the reflectivity minimum is unaffected as well as the width of the reflectivity dip.

Somewhat different effects on the SP resonance are seen when the refractive index of the electric double layer, n_3, is varied with $d_3 = 3$ Å: raising Im(n_3), i.e., raising $n_3 = (1.43, 0)$ to $(1.43, 2)$, increases the resonance angle by about 0.1°, but it also alters the reflectivity minimum and the width of the reflectivity dip (Fig. 4.35). Contrary to this, slight changes in Re(n_3) (not shown) have only a minor impact on the SP resonance angle, width and reflectivity minimum.

Under the experimental conditions reviewed in the present article, a substantial shift in the SP resonance angle (i.e. a shift by about 0.1°) is observed (see Fig. 4.33). This result can be explained only if a significant change in the absorptivity of the "metal adlayer" or of the electric double layer is induced

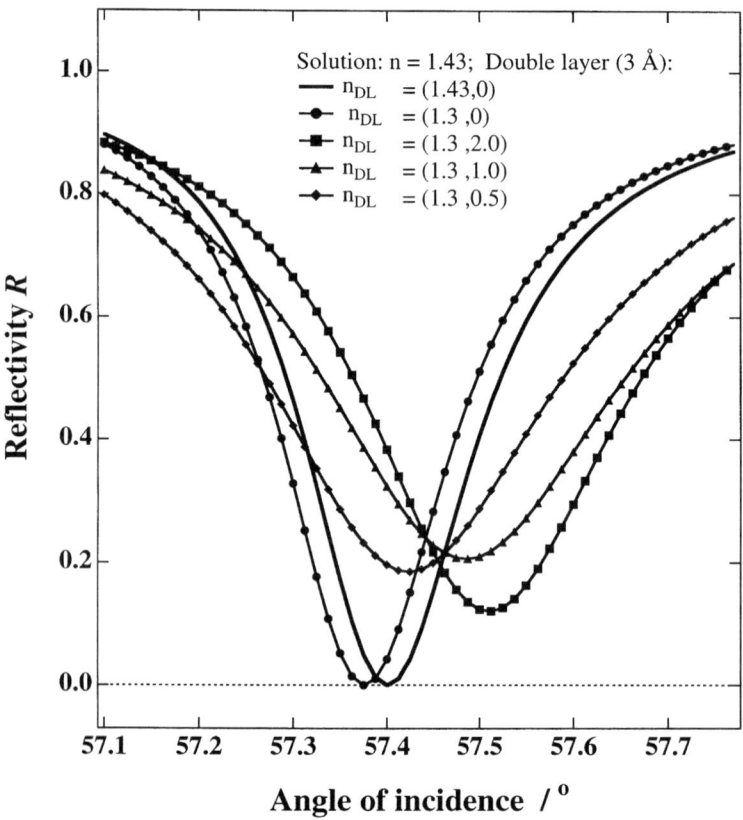

Figure 4.35. Influence of dielectric double layer properties on SP excitation: $n_{prism} = 1.766$; metal film: $d_{Ag\,film} = 505$ Å, $n_{Ag\,film} = (0.04, 5.2)$; double layer; $d_{DL} = 3$ Å, $n_{DL} = (1.43, 0), (1.43, 0.5), (1.43, 1), (1.43, 2)$; electrolyte: $n_{electrolyte} = 1.43$; $\lambda_L = 632.8$ nm.

by the electrode potential and/or adsorption of ions. In other words, the latter two effects cause a substantial change of the electronic configuration of the interface, sufficient to move the resonance angle by about 0.1°.

4.3.3.3 SP Microscopy at Ag Film Electrodes In SP microscopy the excitation of SPs is exploited to obtain spatially resolved information on the properties of the electrode/electrolyte interface. This is achieved by broadening the p-polarized laser beam and illuminating the major part of the Ag film in a Kretschmann configuration at a special angle of incidence. The illuminated part of the electrode is imaged onto a screen. If the properties of the interface differ at different locations, so do the resonance conditions for SP excitation. Hence, at the angle of incidence used, only parts of the electrode surface are in resonance (i.e., their reflectivity is at minimum). The other parts of the electrode show higher reflectivities. Consequently, on the screen an intensity image is obtained that maps different optical properties of the interface. If the latter are associated, for instance, with different coverages, then the intensity image can be identified with a concentration pattern. If the interfacial optical properties are due to different double layer potentials, the image renders a potential pattern.

An experimental setup of an SP microscope is shown in Figure 4.36. A Kretschmann configuration is used. On the prism base (LASF35 or LASF9, Schott, Mainz, Germany), a thin silver film is evaporated serving as a working electrode; it has a thickness of about 50 nm. An alternative experimental setup consists of a little sapphire plate with the evaporated silver film on its lower side. Its upper side is brought into optical contact with a sapphire prism or a glass prism of nearly identical refractive index (SFL56, Schott, Mainz, Germany). In these cases, CH_2I_2 serves as index-matching liquid to establish the optical contact between the sapphire plate and the prism. The prism (eventually together with the sapphire plate) is mounted to an electrochemical cell such that only the base of the prism with the silver film is in contact with the electrolyte. A defined convection is achieved with an impinging jet that is placed close to the freely accessible working electrode.

The working electrode is illuminated from behind through the prism using a p-polarized He-Ne laser beam. The shown arrangement of mirror and lenses (before the laser beam enters the prism) serves to increase the beam diameter. The whole cell can be rotated around its vertical axis passing through the center of the reflection spot. By rotation the reflectivity minimum can be found, and the angle of incidence is usually fixed close to this minimum. After reflection at the working electrode, the laser beam, showing some dispersion in k-space due to locally different reflectivity, exits the prism. This radiation is collected by a lens and is imaged onto a CCD.

A Dalsa CCD with a full-frame transfer architecture can be used to store 800 frames per second with 128×128 pixels. This high temporal resolution is often necessary in order to resolve the fast dynamics characteristic for oscillatory electrochemical reactions. The camera can be connected to an image processing board: in general it will be necessary to further process the images in order to

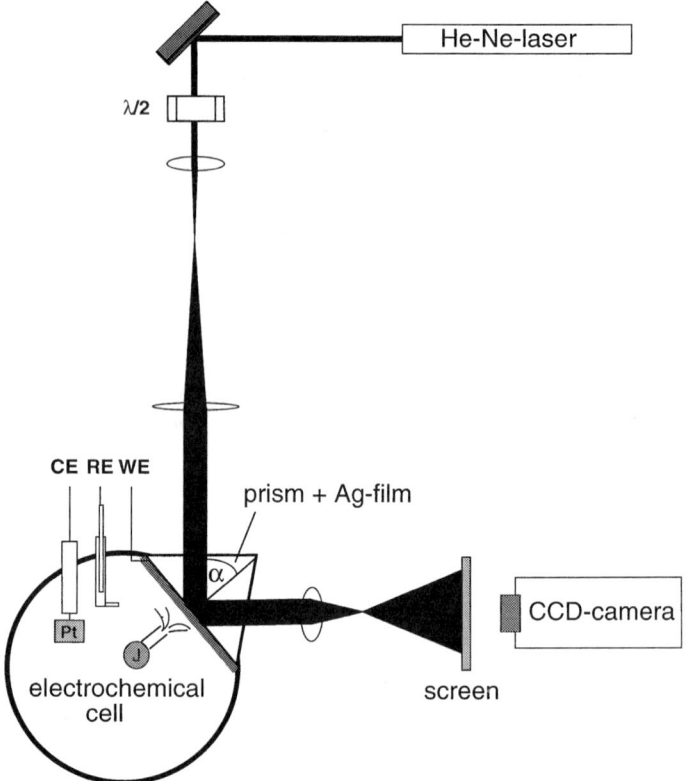

Figure 4.36. Experimental setup of SPP microscopy.

eliminate intensity difference due to the Gaussian intensity profile of the laser beam or inhomogeneities in the film thickness.

The drawback of this optical technique is that it is restricted to a few metals such as silver, gold, or copper, to a thin film geometry, and to the danger of film dissolution during electrochemical processes. So far, however, it represents the only relatively fast, nonperturbative imaging method for electrochemical double layer processes.

Let us now illustrate the use of SP microscopy for the study of patterns in electrochemical systems exhibiting dynamic instabilities. As explained previously, in general the two stable states in the bistable regime of an electrochemical reaction are associated with different potential drops across the double layer and, hence, also with different resonance conditions for SP excitations.

Figure 4.37a shows a current-voltage characteristic of the reduction of peroxidisulfate under conditions of bistable behavior. Simultaneously, the current and the average intensity of the reflected beam at a constant angle of incidence was recorded, shown in Figure 4.37b. Evidently, the bistable behavior in the two curves corresponds to each other. The sudden jump from a low-current

4.3 IMAGING IN ELECTROCHEMICAL ENVIRONMENTS

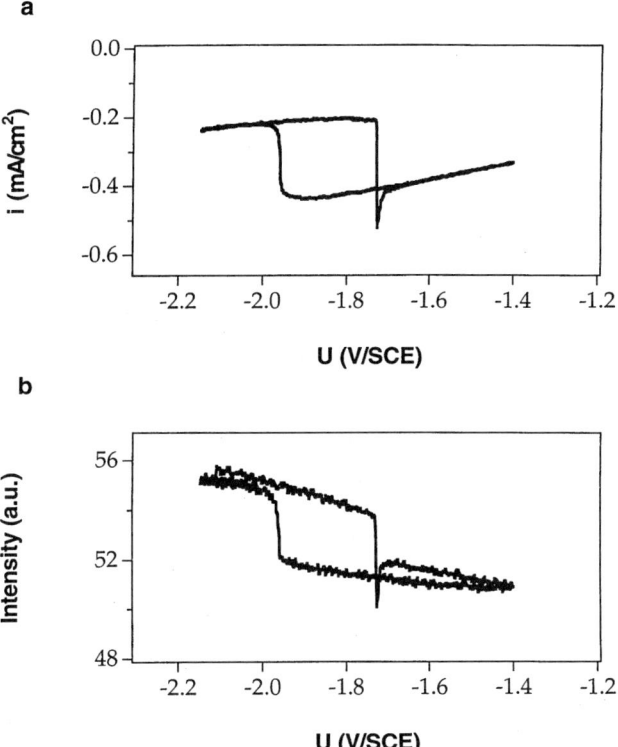

Figure 4.37. (a) Current-voltage and (b) intensity-voltage characteristics of the peroxidisulfate reduction. An impinging jet was used to establish a mass transfer of the electrolyte (2 mM $Na_2S_2O_8$) to the electrode surface analogous to the effect of rotating the electrode. Reproduced from Flätgen et al.[46]

density state to a high-current density state, and vice versa, is accompanied by corresponding jumps in reflectivity. Careful comparison of how current and intensity change with the applied voltage indicates subtle differences among the optical and electrochemical measurements. Contrary to the current density characteristic outside the bistable region, the reflectivity shows a hysteresis in the positive and negative potential scans. Obviously, the optical measurements reveal a somewhat different state of the surface in the two scan directions, which is undetectable by the current density measurement alone. This indicates that the chemisorbed species in the two scan directions are different in number and possibly also in nature.

Because of the low electrolyte concentrations, there is a noticeable solution resistance. Thus, the potential drop across the interface is not equal to the externally applied potential: $\phi_{DL} \neq U_{ext}$. The former, however, controls the current density and resonance condition for SP excitation. It is, therefore, illustrative to correct the curves against the $I * R$ drop in the electrolyte; the result is shown

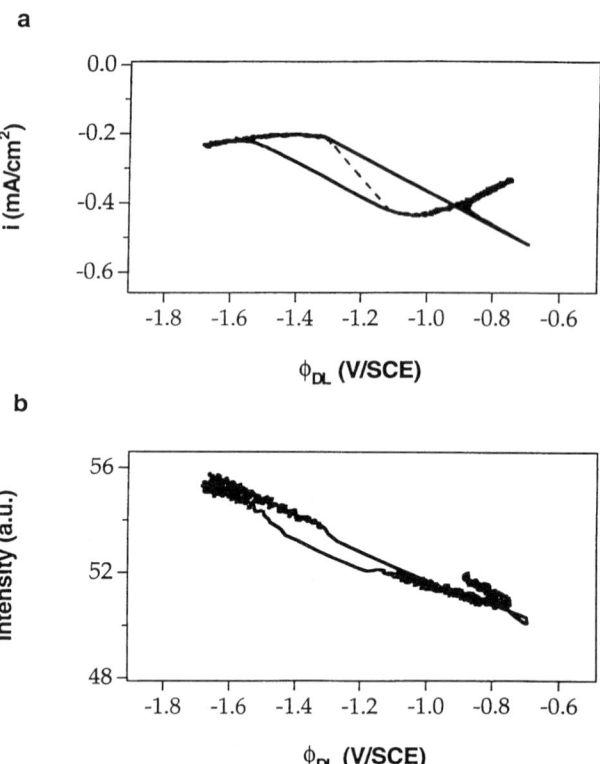

Figure 4.38. (a) Current-voltage and (b) intensity-voltage curves of Figure 4.37 versus φ_{DL} obtained after *IR* correction; R = 3.6 kΩ. Reproduced from Flätgen et al.[160]

in Figure 4.38. Evidently, there is a region of negative resistance in the corrected current density curve (i.e., $dj/d\phi_{DL} < 0$), and the unstable steady state, not detectable in the experiment, is indicated by the dashed line. Contrary to the current density in Figure 4.38a, the reflected intensity depends, to a first approximation, linearly on the potential ϕ_{DL}.

This potential dependence of the local reflectivity is the basis of the SP microscopic images shown subsequently for dynamically different regimes in the bistable and the oscillatory region. The first appears to be the most simple one, because it involves a monotonic transition from one steady state to another one. A typical example is shown in Plate 4.2, with Plate 4.2 depicting the transient of the current density from the low-current to the high-current density state. (See color plate section.) Plate 4.2b shows a series of images recorded at the indicated delay times. At 32 ms the current density starts to rise monotonically and reaches the active state at about 73 ms. The sequence of images recorded during the transition shows the following: as long as there is no rise in the current density, the images show the electrode in its passive state. As the current

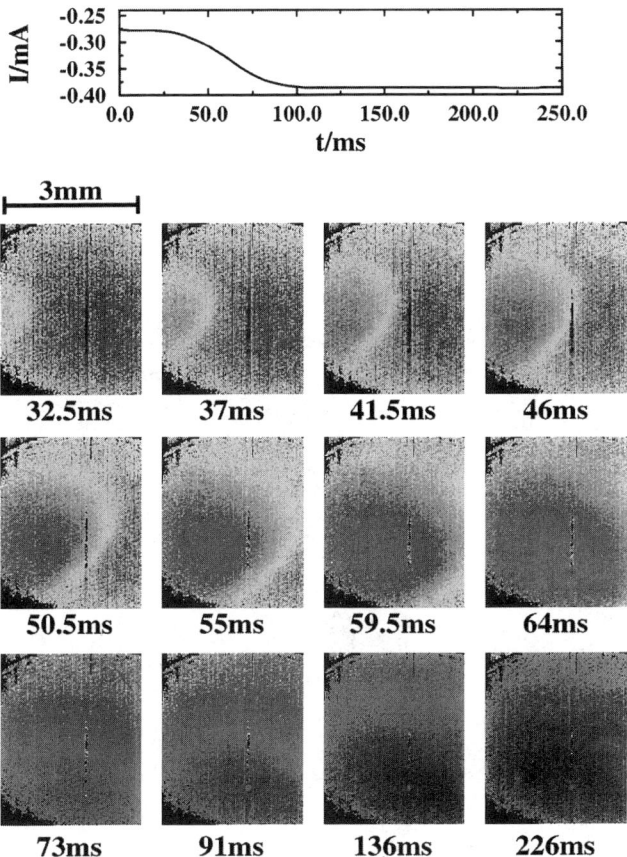

Plate 4.2 (a) Time trace of the global current during a transition in the bistable regime. (b) SPP microscope images of the electrode during this transition. Reproduced from Flätgen et al.[46] See color plate section.

density begins to increase, a small nucleus is appearing at the lower left rim of the electrode disk. The next seven images show that this nucleus is growing and spreading over the surface, until the surface appears to be homogeneous in the active state. The final four images show further, but slow and weak changes in the images at times, where the current density is already in its steady state. It should be emphasized that repeated experiments revealed the same series of spatial developments (i.e., the creation of a small nucleus at the rim of the electrode, its spreading over the surface, and the final slow changes); only the actual location of the nucleus along the rim differed from experiment to experiment.

When reducing the convection, the system is still in the bistable regime, but the transients are more complex. They overshoot the stationary state and successively relax toward it from high current densities. We refer to this behavior as *complex bistable*. Plate 4.3 illustrates this behavior, again with a current density

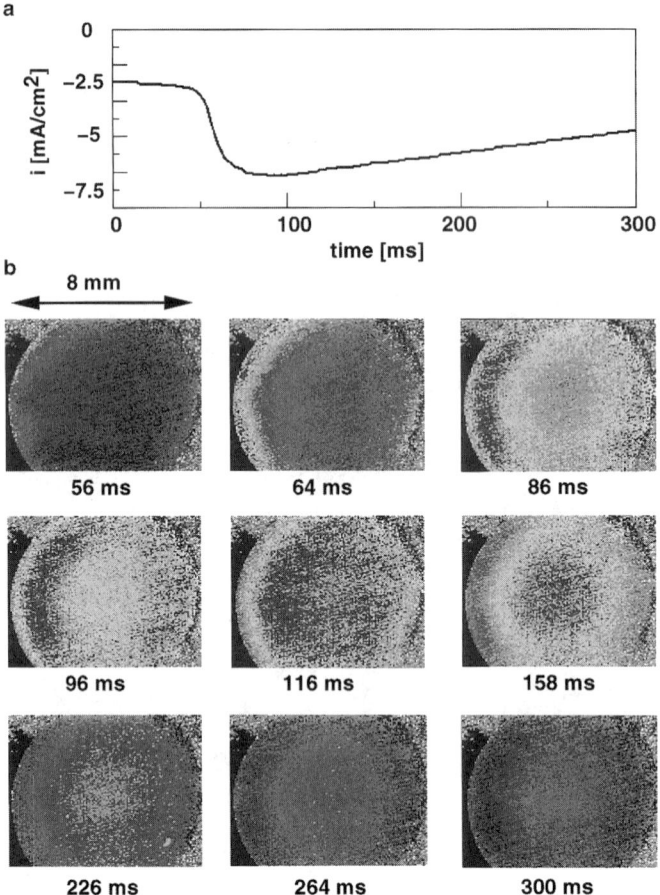

Plate 4.3 (a) Time trace of the global current during a transition in the bistable regime at parameter values closer to the oscillatory regime than in Plate 4.2. Note that the relaxation to the steady state is not completely shown. (b) SPP microscope images during this transition. (The color is reversed with respect to Plate 4.2 as the angle of incidence lay on the other side of the resonance minimum.) Reproduced from Flätgen et al.[160] See color plate section.

transient in Plate 4.3a and a corresponding series of SP images in Plate 4.3b. After 54 ms the current density starts to increase towards the active state; it reaches its maximum at 86 ms, thereby overshooting the level of the stationary state; then it relaxes back to this level within several hundred milliseconds (this is not shown completely). The first three images, taken during the rising part of the current density, show that actually two waves (corresponding to the state of high reflectivity) are propagating across the surface, one originating at the left edge, the other one at the right edge a little later. At $t = 116$ ms (about 30 ms after the current maximum), the electrode appears to be in a fairly homoge-

4.3 IMAGING IN ELECTROCHEMICAL ENVIRONMENTS

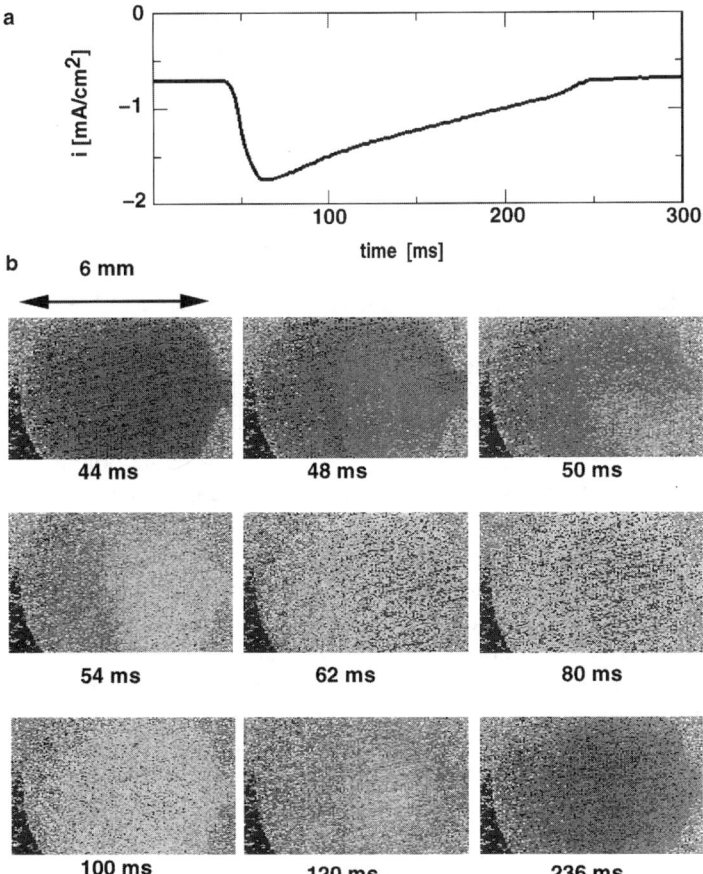

Plate 4.4. Time trace of the global current during an oscillation of 2 Hz frequency. (b) SPP microscope images during the oscillation. Reproduced from Flätgen et al.[160] See color plate section.

neous state. A nearly radial wave is then created along the rim of the electrode, which propagates toward its center, transferring the electrode to a state with a reflectivity only slightly different from the passive state.

Oscillations can be achieved by further reducing the convection. The current density curves resemble that of the previous example. An oscillation is characterized by a fast increase of the signal, a slower relaxation back to its initial value, and a stationary phase. In the example presented here the oscillation frequency is about 2 Hz. The first SP image in Plate 4.4b shows the electrode in the quasi-stationary, low current state. Simultaneous with the rise of the current, a nucleus possessing a high current density appears in the lower right corner. Its growth and propagation are seen in the next four images. The last of them displays a nearly homogeneously distributed state with maximum reflectivity achieved at $t = 80$ ms, whereas the current density reached its maximum already around $t \sim 60$ ms. The following three selected images (note that every 2 ms an image was recorded) display the rather slow transition at the surface back to its passive state. It is likely that this transition is again accompanied by propagating wave fronts. However, the high noise level makes a clear assignment difficult.

4.3.3.4 Discussion Employing SP microscopy during the reduction of $S_2O_8^{2-}$ at a silver electrode, we were able to monitor the transitions in the bistable regime and the oscillatory region. Both of these dynamic regimes exhibited waves propagating across the electrode surface. For a deeper understanding of the dynamics involved, one has to evaluate how the electrode potential influences the SP resonance. The linear dependence of the reflectivity on the electrode potential (Fig. 4.38) may indicate that the reflectivity response under SP excitation is a linear function of the potential, e.g., the reflectivity probes the potential. However, experiments and calculations reported by different groups point to a more complicated situation [47,48,148–152]. Most important is the asymmetry in the SP resonance versus potential curves: at potentials negative to a characteristic (threshold) potential, U_{th}, the SP resonance does not exhibit any dependence on the potential, whereas at potentials positive to U_{th} a significant shift of the position of the SP excitation is noted. Calculations have shown that a change in the electron density of the metal alters the SP resonance (e.g., the size of the k_{SP} vector) in a minor way. This situation may change, when we consider specific adsorption, where both the enlarged static electric field across the interface and the orbital mixing between metal and adsorbate states affect the k_{SP} vector.

These effects can be modeled in the framework of our multiphase calculations by introducing a thin metal adlayer with considerably different optical properties than the rest of the silver film. Associating a change of the electron density with a change in $\text{Re}(n)$ we observe a minor influence of the SP resonance during the variation of n; contrarily, a change in $\text{Im}(n)$ significantly altered the SP resonance, that is, it can be associated with a change of the electronic distribution of the aforementioned adlayer. We expect such an effect for the case of specific adsorption, here of SO_4^{2-}, one of the products of $S_2O_8^{2-}$ reduction. As far as electrode charging and orbital mixing follows the changes in surface coverage of the adsorbate, we observe potential waves, or coverage waves, spreading over the surface.

Note in this context that any change in the electronic structure of the interface (due to adsorption, reconstruction, restructuring, oxidation, etc.) alters the SP resonance condition. Thus, these distinct processes can be monitored and separated using SP microscopy, in particular if they occur on different time scales. In our case, the change in potential and adsorption (of SO_4^{2-}) essentially occurs on the same time scale; therefore, we observe coverage waves (shown in Plate 4.2 by the first five images) on a time scale of about 60 ms. Slower processes (spanning about 150 ms) also occur as, for example, the last four images of Plate 4.2 show. They are not connected with changes in the current

4.3 IMAGING IN ELECTROCHEMICAL ENVIRONMENTS

densities, and therefore cannot be associated with changes of ϕ_{DL}. Because, in our experience, restructuring processes occur on a time scale of minutes up to hours, the last four images of Plate 4.2 show a change in the state of the surface, possibly due to the adsorption of a reaction intermediate of the reduction of the $S_2O_8^{2-}$.

Changes in the SP intensity on a different time scale are also evident in Figure 4.37, Plate 4.3, and Plate 4.4. Figure 4.37b exhibits a hysteresis in the active and passive branches, which is absent in the current density given in Figure 4.37a. Both Plates 4.3 and 4.4 show a substantial temporal displacement of intensity and current maxima. This indicates that what we call the *state of the surface* is not only determined by the (relative fast) changes in potential and SO_4^{2-} coverage but also by slower processes following reaction and adsorption.

The preceding interpretation of the images has a tentative character, as our knowledge about the nature of the interfacial changes is rather poor, in particular for changes occurring on the longer time scale. Yet, we can achieve conclusions on the characteristic times and lengths of patterns in electrochemical systems.

Let us consider the transition in the bistable regime shown in Plate 4.2. The propagation of the reaction front occurs with a relatively high speed of about 10 cm/s. In other investigations we noted the rise of the propagation velocity with the conductivity of the electrolyte; at a concentration of 10 mM of SO_4^{2-}, velocities on the order of meters per second were observed [93]. Contrary to that, velocities in systems such as the CO oxidation at Pt single crystals under low pressure are smaller by a factor of 10^3. Next, the width of the front is around 0.1 mm, which again is more than one order of magnitude larger than in the case of CO oxidation under low pressure.

Several distinct parameters control pattern formation and its kinetics such as potential, ion concentration, and the rate of mass transfer; the latter depends, obviously, on the former parameters, but also on other experimental conditions such as convection. Because the mass transfer rate acts on several characteristic times of the nonlinear dynamics, its variation changes the behavior of the system. At sufficiently high mass transfer rates, the system is in a *bistable regime* (Plate 4.2); lowering the mass transfer rates, the dynamics becomes *complex bistable* (Plate 4.3); and last but not least, for even lower mass transfer rates one enters the *oscillatory regime* (Plate 4.4). A more detailed discussion of the dynamical behavior of nonlinear electrochemical systems is beyond the scope of the present paper; we would like to refer the reader to the recent reviews [8,9].

4.3.3.5 Final Remarks Optical methods open a new avenue for imaging local differences in dynamic processes at the electrode/electrolyte interfaces, providing, therefore, new insight into electrochemical systems exhibiting nonlinear behavior.

Surface plasmon microscopy, applied to electrochemical systems, has several important advantages: (i) It allows the imaging of the entire surface; (ii) it provides a high temporal resolution (about 1 ms); (iii) it is a nonperturbing technique; and (iv) the electrode is freely accessible. This method is, therefore, suitable to achieve a better understanding of the processes leading to spatio-temporal patterns in electrochemical systems. It can be used to study completely different questions such as nucleation/growth phenomena or phase transitions provided the associated states of the surface (due to different coverages, for instance) have different optical properties.

Other optical techniques, when used to monitor pattern formation and reaction fronts at electrodes, essentially have the same advantages as listed. Here we would like to mention electroreflectance, ellipsometry, and second harmonic generation (SHG). These three methods can be applied at all interfaces (in contrast to SP microscopy which is basically restricted to thin silver or gold films). The use of SHG is particularly interesting because of two additional advantages: (i) The optical signal stems exclusively from the interfacial region; (ii) it shows a remarkable dependence on electrode potential as well as on adsorption and distinct adsorption geometries. Currently, we are setting up a corresponding experiment to employ this nonlinear optical process in the form of SHG microscopy [153].

4.4 CONCLUSIONS AND FUTURE DIRECTIONS

In this chapter we have focused our interest mainly on the multiple experimental methods used to image reaction fronts at surfaces and interfaces. All the discussed techniques have been developed only during the last several years but, nevertheless, proved to enhance the knowledge about nonlinear chemical reactions on surfaces. They open new avenues of basic understanding to improve such important issues as selectivity in a reaction and increasing the turnover rate, while avoiding runaway situations. The optical methods especially are most promising in closing the so-called "pressure gap." Most techniques capable of unraveling local inhomogeneities in surface science cannot be applied at high pressure, not to mention the liquid phase in electrochemistry. Even though their spatial resolution is normally diffraction limited, they inherently possess the advantage of real-time imaging and therefore have a great potential of observing fast dynamical processes at surfaces and interfaces.

We are convinced that in the future these techniques will be applied for many different questions in surface science and electrochemistry beyond the studies of nonlinear phenomena.

ACKNOWLEDGMENTS

The authors thank G. Ertl for his continuing, stimulating interest and generous support of their work on imaging of reaction fronts during the past years. The authors are indebted to R. Colen and G. Haas for fruitful collaboration in these studies, and to M. Sheintuch for providing the original of Figure 4.27.

LIST OF ACRONYMS

ATR	attenuated total reflection
DL	double layer
CCD	charge coupled device
CSTR	continuous stirred tank reactor
EMSI	ellipso-microscope for surface imaging
ML	monolayer
EM	electromagnetic
IR	infrared
$I * R$	potential drop due to ohmic resistor: $U = IR$
IHP	inner Helmholtz plane
M	metal
NRS	normal Raman scattering
PEEM	photoemission electron microscope
RAM	reflection anisotropy microscope
pzc	point of zero charge
SCE	saturated calomel electrode
SERS	surface enhanced Raman scattering
SFG	sum frequency generation
SHG	second harmonic generation
SP	surface plasmon
SPM	scanning photoemission microscope
STM	scanning tunneling microscopy
U_{ext}	(externally measured) electrode potential
U_{th}	threshold potential
vp	volume plasmons
c	speed of light/concentration of ions
$d_{Ag\,film}$	thickness of the silver film
g	enhancement factor for SERS
$\hbar$	Planck's constant/2π
k	length of the momentum vector of light in vacuum $k = |k|$; imaginary part of the dielectric constant ε
k_{prism}	length of the momentum vector of light in the glass prism ($k_{prism} = |\mathbf{k}_{prism}|$)
$k_{prism,x}$	x-component of the momentum vector of light in the glass prism ($\parallel$ to surface)
k_{vp}	length of the momentum vector of volume plasmons of the metal
k_{SP}	length of the momentum vector of surface plasmons
n_j	refractive index of the medium j (j = prism, M, adlayer, electrolyte, gap, DL)
α	angle of incidence
ε_j	dielectric function (j = prism, d, M: refers to prism, dielectricum, metal, respectively)

λ_L wavelength of the laser
ϕ_{DL} potential drop in the double layer
$(d\sigma/d\Omega)$ differential Raman cross section
ω, ω_j frequency of electromagnetic wave (j = vp, SPP: indicates volume plasmons or surface plasmon polaritons, respectively)

REFERENCES

1. A. S. Mikhailov, *Foundations of Synergetics I, 2nd ed.* (Springer, Berlin, 1994).
2. *Chemical Waves and Patterns*, eds. R. Kapral and K. Showalter (Kluwer, Dordrecht, 1995).
3. H. Haken, *Synergetics. An Introduction* (Springer, Berlin, 1978).
4. H. Haken, *Advanced Synergetics* (Springer, Berlin, 1987).
5. F. Schüth, B. E. Henry and L. D. Schmidt, *Adv. Catalysis*, **39** (1993) 51.
6. R. Imbihl and G. Ertl, *Chem. Rev.*, **95** (1995) 697.
7. M. Eiswirth and G. Ertl, in ref. 2, p. 447.
8. M. T. M. Koper, *Adv. Chem. Phys.*, **95** (1996) 161.
9. K. Krischer, in *Modern Aspects of Electrochemistry*, **32**, Eds. J. O. M. Bockris, B. E. Conway and R. E. White (Plenum Press, New York, in press).
10. V. V. Barelko and Y. E. Volodin, *Kinetika i Kataliz*, **17** (1976) 112.
11. V. V. Barelko, I. I. Kurochka, A. G. Merzhanov and K. G. Shkadinskii, *Chem. Eng. Sci.*, **33** (1978) 805.
12. S. A. Zhukov and V. V. Barelko, *Sov. J. Chem. Phys.*, **4** (1982) 883.
13. P. Pawlicki and R. A. Schmitz, *Chem. Eng. Prog.*, **83** (1987) 40.
14. G. A. Cordonnier and L. D. Schmidt, *Chem. Eng. Sci.*, **44** (1989) 1983.
15. L. L. Lobban, G. Philippou and D. Luss, *J. Chem. Phys.*, **93** (1989) 733.
16. J. C. Kellow and E. E. Wolf, *Chem. Eng. Sci.*, **45** (1990) 2597.
17. C. C. Chen, E. E. Wolf and H. C. Chang, *J. Phys. Chem.*, **97** (1993) 1055.
18. M. P. Cox, G. Ertl and R. Imbihl, *Phys. Rev. Lett.*, **54** (1985) 1725.
19. H. H. Rotermund, G. Ertl and W. Sesselmann, *Surf. Sci.*, **217** (1989) L383.
20. H. H. Rotermund, S. Jakubith, S. Kubala, A. von Oertzen and G. Ertl, *J. Electron. Spectrosc. Relat. Phenom.*, **52** (1990) 811.
21. H. H. Rotermund, S. Jakubith, A. von Oertzen and G. Ertl, *J. Chem. Phys.*, **91** (1989) 4942.
22. H. H. Rotermund, W. Engel, M. Kordesch and G. Ertl, *Nature*, **343** (1990) 355.
23. W. Engel, M. E. Kordesch, H. H. Rotermund, S. Kubala and A. von Oertzen, *Ultramicroscopy*, **36** (1991) 148.
24. W. Engel, Symposium A1 der Deutschen Gesellschaft für Elektronenmikroskopie, Leipzig, DGE Conference, Sept. (1995).
25. W. Telieps and E. Bauer, *Ultramicroscopy*, **17** (1985) 57.
26. B. Rausenberger, W. Swiech, C. S. Rastomjee, M. Mundschau, W. Engel, E. Zeitler and A. M. Bradshaw, *Chem. Phys. Lett.*, **215** (1993) 109.

27. B. Rausenberger, W. Swiech, W. Engel, A. M. Bradshaw and E. Zeitler, *Surf. Sci.*, **278/288** (1993) 235.
28. K. C. Rose, B. Berton, R. Imbihl, W. Engel and A. M. Bradshaw, *Phys. Rev. Lett.*, **79** (1997) 3427.
29. S. Swiech, B. Rausenberger, W. Engel, A. M. Bradshaw and E. Zeitler, *Surf. Sci.*, **294** (1993) 297.
30. K. C. Rose, D. Battogtokh, A. Mikhailov, R. Imbihl, W. Engel and A. M. Bradshaw, *Phys. Rev. Lett.*, **76** (1996) 3582.
31. Y. Uchida, G. Lehmpfuhl and J. Jäger, *Ultramicroscopy*, **15** (1984) 119.
32. Y. Uchida, G. Lehmpfuhl and R. Imbihl, *Surf. Sci.* **234** (1990) 27.
33. H. H. Rotermund, G. Haas, R. U. Franz, R. M. Tromp and G. Ertl, *Science*, **270** (1995) 608.
34. H. H. Rotermund, G. Haas, R. U. Franz, R. M. Tromp and G. Ertl, *Appl. Phys. A*, **61** (1995) 569.
35. M. F. H. van Tol, A. Gielbert and B. E. Niewenhuys, *Catal. Lett.*, **16** (1992) 297.
36. V. Gorodetskii, J. H. Block, W. Drachsel and M. Ehsasi, *Appl. Surf. Sci.*, **67** (1993) 198.
37. V. Gorodetskii, J. Lauterbach, H. H. Rotermund, J. H. Block and G. Ertl, *Nature*, **370** (1994) 276.
38. K. F. Bonhoeffer and W. Renneberg, *Z. Phys.*, **118** (1941) 389.
39. U. F. Franck, *Z. Elektrochem.*, **55** (1951) 154.
40. K. F. Bonhoeffer and G. Vollheim, *Z. Naturforsch.*, **8b** (1953) 406.
41. R. Suzuki, *Adv. Biophys.*, **9** (1976) 115.
42. A. Pigeau and H. B. Kirkpatrick, *Corrosion*, **25** (1969) 209.
43. J. L. Hudson, J. Tabora, K. Krischer and I. G. Kevrekidis, *Phys. Lett. A*, **179** (1993) 355.
44. R. D. Otterstedt, N. I. Jaeger and P. J. Plath, *Int. J. Bifurcation and Chaos*, **4** (1994) 1265.
45. C. Wang, S. Chen and X. Yu, *Electrochim. Acta*, **39** (1994) 577.
46. G. Flätgen, K. Krischer, B. Pettinger, K. Doblhofer, H. Junkes and G. Ertl, *Science*, **269** (1995) 668.
47. D. M. Kolb, in *Surface Polaritons: Electromagnetic Waves at Surfaces and Interfaces*, eds. V. M. Agranovich and D. L. Mills (North-Holland, Amsterdam, 1982) p. 299.
48. F. Chao, M. Costa and A. Tadjeddine, *J. Electroanal. Chem.*, **329** (1992) 313.
49. M. Eiswirth, A. Freund and J. Ross, *Adv. in Chem. Phys.*, **90** (1991) 127.
50. Y. E. Volodin, V. N. Zvyagin, A. N. Ivanova and V. V. Barelko, *Adv. in Chem. Phys.*, **77** (1990) 551.
51. Y. E. Volodin, V. V. Barelko and A. G. Merzhanov, *Sov. J. Chem. Phys.*, **5** (1982) 1146.
52. L. Lobban and D. Luss, *J. Chem. Phys.*, **93** (1989) 6530.
53. S. L. Lane and D. Luss, *Phys. Rev. Lett.*, **70** (1993) 830.
54. C. T. Rettner and D. J. Auerbach, *Science*, **263** (1994) 365.

55. M. Feinberg and D. Terman, *Arch. Rational Mech. Anal.*, **116** (1991) 35.
56. M. Eiswirth, J. Bürger, P. Strasser and G. Ertl, *J. Phys. Chem.*, **100** (1996) 19118.
57. B. C. Sales, J. E. Turner and M. B. Maple, *Surf. Sci.*, **114** (1982) 381.
58. N. Hartmann, R. Imbihl and W. Vogel, *Catal. Lett.*, **28** (1994) 373.
59. S. Ladas, R. Imbihl and G. Ertl, *Surf. Sci.*, **219** (1989) 88.
60. M. R. Bassett and R. Imbihl, *J. Chem. Phys.*, **93** (1990) 811.
61. R. Imbihl, M. P. Cox, G. Ertl, H. Müller and W. Brenig, *J. Chem. Phys.*, **83** (1985) 1578.
62. K. Krischer, M. Eiswirth and G. Ertl, *J. Chem. Phys.*, **96** (1992) 9161.
63. T. Engel and G. Ertl, *Adv. in Catal.*, **28** (1979) 1.
64. E. Wicke, *Chem. Ing. Techn.*, **46** (1974) 365.
65. G. Ertl, P. R. Norton and J. Rüstig, *Phys. Rev. Lett.*, **49** (1982) 177.
66. G. Ertl, in *Catalysis, Science and Technology*, eds. J. R. Anderson and M. Boudart, 4 (Springer, Heidelberg, 1983) p. 257ff.
67. G. Ertl, *Surf. Sci.*, **152/153** (1985) 328.
68. T. Gritsch, D. Coulman, R. J. Behm and G. Ertl, *Phys. Rev. Lett.*, **63** (1989) 1086.
69. H. H. Rotermund, J. Lauterbach and G. Haas, *Appl. Phys. A*, **57** (1993) 507.
70. J. Lauterbach, K. Asakura and H. H. Rotermund, *Surf. Sci.*, **313** (1994) 52.
71. A. von Oertzen, A. Mikhailov, H. H. Rotermund and G. Ertl, *Surf. Sci.*, **350** (1996) 259.
72. H. H. Rotermund, *Physica Scripta T*, **48** (1993) 549.
73. R. H. Fowler, *Phys. Rev.*, **38** (1931) 45.
74. M. Bär, S. Nettesheim, H. H. Rotermund, M. Eiswirth and G. Ertl, *Phys. Rev. Lett.*, **74** (1995) 1246.
75. S. Nettesheim, A. von Oertzen, H. H. Rotermund and G. Ertl, *J. Chem. Phys.*, **98** (1993) 9977.
76. H. H. Rotermund, S. Jakubith, A. von Oertzen and G. Ertl, *Phys. Rev. Lett.*, **66** (1991) 3083.
77. M. Bär, M. Eiswirth, H. H. Rotermund and G. Ertl, *Phys. Rev. Lett.*, **69** (1992) 945.
78. F. Mertens and R. Imbihl, *Chem. Phys. Lett.*, **242** (1995) 221.
79. G. Veser, F. Esch and R. Imbihl, *Catal. Lett.*, **13** (1992) 371.
80. F. Mertens and R. Imbihl, *Nature*, **370** (1994) 124.
81. F. Mertens and R. Imbihl, *Surf. Sci.*, **347** (1996) 355.
82. A. von Oertzen, H. H. Rotermund and S. Nettesheim, *Surf. Sci.*, **311** (1994) 322.
83. M. D. Graham, M. Bär, I. G. Kevrekidis, K. Asakura, J. Lauterbach, H. H. Rotermund and G. Ertl, *Phys. Rev. E*, **52** (1995) 76.
84. R. F. Cohn, J. W. Wagner and J. Kruger, *Applied Optics*, **27** (1988) 4664.
85. R. Reiter, H. Motschmann, H. Orendi, A. Nemetz and W. Knoll, *Langmuir*, **8** (1992) 1784.
86. R. M. A. Azzam and N. M. Bashara, *Ellipsometry and Polarized Light* (North-Holland, Amsterdam, 1977).
87. R. U. Franz, Doctoral Thesis (Technische Universität Berlin, 1996).

88. J. Lauterbach and H. H. Rotermund, *Surf. Sci.*, **311** (1994) 231.
89. W. Wolf, K. Krischer, M. Lübke, M. Eiswirth and G. Ertl, *J. Electroanal. Chem.*, **385** (1995) 85.
90. U. F. Franck, *Z. Electrochem.*, **62** (1958) 649.
91. O. Lev, M. Sheintuch, L. M. Pismen and C. Yarnitzky, *Nature*, **336** (1988) 488.
92. O. Lev, M. Sheintuch, H. Yarnitsky and L. M. Pismen, *Chem. Eng. Sci.*, **45** (1990) 839.
93. G. Flätgen and K. Krischer, *Phys. Rev. E*, **51** (1995) 3997.
94. E. Nehrer and B. Sakmann, *Nature*, **260** (1976) 799.
95. E. Nehrer and B. Sakmann, *Spektrum der Wissenschaft*, **5** (1992) 48.
96. G. Flätgen, Doctoral Thesis (Freie Universität Berlin, 1995).
97. R. Otterstedt, P. J. Plath, N. I. Jaeger, J. C. Sayer and J. L. Hudson, *Chem. Eng. Sci.*, **51** (1996) 1747.
98. G. Flätgen and K. Krischer, *J. Chem. Phys.*, **103** (1995) 5428.
99. R. M. Hexter and G. M. Albrecht, *Spectrochim. Acta*, **35A** (1979) 233.
100. A. Otto, *Phys. Status Solidi*, **26** (1968) 99.
101. A. Otto, *Z. Physik*, **216** (1968) 398.
102. A. Otto, *Phys. Status Solidi*, **42** (1970) K37.
103. E. Kretschmann and H. Raether, *Z. Naturforsch. A*, **23** (1968) 615.
104. E. Kretschmann, *Z. Phys.*, **241** (1971) 313.
105. M. Fleischmann, P. J. Hendra and A. J. McQuillan, *Chem. Phys. Lett.*, **26** (1974) 163.
106. R. P. Van Duyne, *J. Physique*, **38** (1976) 239.
107. *Surface Enhanced Raman Scattering*, eds. R. K. Chang and T. E. Furtak (Plenum Press, New York, 1982).
108. B. Pettinger, A. Tadjeddine and D. M. Kolb, *Chem. Phys. Lett.*, **66** (1979) 544.
109. M. Futamata, P. Borthen, J. Thomassen, D. Schumacher and A. Otto, *Appl. Spectrosc.*, **48** (1994) 252.
110. M. Futamata, *Langmuir*, **11** (1995) 3894.
111. W. Wittke, A. Hatta and A. Otto, *Appl. Phys. A*, **48** (1989) 289.
112. W. Knoll, M. R. Philpott, J. D. Swalen and A. Girlando, *J. Chem. Phys.*, **77** (1982) 2254.
113. A. Nemetz, U. Fernandez and W. Knoll, *J. Appl. Phys.*, **75** (1994) 1582.
114. A. C. R. Pipino, G. C. Schatz and R. P. Van Duyne, *Phys. Rev. B: Condens. Matter*, **49** (1994) 8320.
115. C. Douketis, T. L. Haslett, V. M. Shalaev, Z. Wang and M. Moskovits, *Physica A*, **207** (1994) 352.
116. J. K. Gimzewski, J. K. Sass, R. R. Schlitter and J. Schott, *Europhys. Lett.*, **8** (1989) 435.
117. S. Ushioda, Y. Uehara and M. Kuwahara, *Appl. Surf. Sci.*, **60–61** (1992) 448.
118. H. J. Simon, D. E. Mitchell and J. G. Watson, *Phys. Rev. Lett.*, **33** (1974) 1531.
119. C. K. Chen, A. R. B. De Castro and Y. R. Shen, *Opt. Lett.*, **4** (1979) 393.
120. G. S. Agarwal and S. S. Jha, *Phys. Rev. B*, **26** (1982) 482.

121. R. M. Corn, M. Romagnoli, M. D. Levenson and M. R. Philpott, *Chem. Phys. Lett.*, **106** (1984) 30.
122. J. C. Quail and H. J. Simon, *Phys. Rev. B*, **31** (1985) 4036.
123. H. J. Simon and Z. Chen, *Phys. Rev. B*, **39** (1989) 3077.
124. O. A. Aktsipetrov, A. A. Nikulin, V. I. Panov, S. I. Vasil'ev and A. V. Petukhov, *Solid State Commun.*, **76** (1990) 55.
125. X. Wang and H. J. Simon, *Opt. Lett.*, **16** (1991) 1475.
126. H. G. Bingler, H. Brunner, M. Klenke, A. Leitner, F. R. Aussenegg and A. Wokaun, *J. Chem. Phys.*, **99** (1993) 7499.
127. H. Akhouayri, M. Neviere, P. Vincent and R. Reinisch, Conference: FRISNO 2. 2nd French-Israeli Symposium on Nonlinear Optics. 25–29 May 1992, **5** (1993) 127.
128. G. Blau, J. L. Coutaz and R. Reinisch, *Opt. Lett.*, **18** (1993) 1352.
129. L. Kuang and H. J. Simon, *Phys. Lett. A*, **197** (1995) 257.
130. Z. Chen and Z. Zhang, *J. Appl. Phys.*, **69** (1991) 7406.
131. U. C. Fischer, U. Duerig and D. W. Pohl, in *Proceedings on "Physical Aspects of Microscopic Characterization of Materials,"* eds. J. Kirschner, K. Murata and J. A. Venables (Scanning Microscopy International, Inc., O'Hare, IL, 1987) p. 2.
132. O. Marti, H. Bielefeldt, B. Hecht, S. Herminghaus, P. Leiderer and J. Mlynek, *Opt. Commun.*, **96** (1993) 225.
133. F. D. Fornel, E. Lesniewska, L. Salomon and J. P. Goudonnet, *Opt. Commun.*, **102** (1993) 1.
134. J. P. Goudonnet, F. d. Fornel, L. Salomon, P. M. Adam and E. Bourillot, *Inst. Phys. Conf. Ser.*, **135** (1994) 229.
135. R. Berndt, A. Baratoff and J. K. Gimzewski, in *Scanning Tunneling Microscopy and Related Methods*, eds. R. J. Behm, N. Garcia and H. Rohrer (Kluwer, Dordrecht, 1990) p. 17.
136. E. Aust, W. Hickel, H. Knobloch, H. Orendi and W. Knoll, *Mol. Cryst. Liq. Cryst. Sci. Technol.*, Sect. A **227** (1993) 49.
137. H. Knobloch and W. Knoll, *J. Chem. Phys.*, **94** (1991) 835.
138. H. Knobloch and W. Knoll, *Makromol. Chem., Macromol. Symp.*, **46** (1991) 389.
139. B. Rothenhaeusler, C. Duschl and W. Knoll, *Thin Solid Films*, **159** (1988) 323.
140. B. Rothenhaeusler and W. Knoll, *Appl. Phys. Lett.*, **52** (1988) 1554.
141. E. M. Yeatman and E. A. Ash, *Proc. Int. Soc. Opt. Eng.*, **897** (1988) 100.
142. E. M. Yeatman, *Biosens. Bioelectron.*, **6** (1996) 635.
143. H. Räther, *Surface Plasmons* (Springer, Berlin, Heidelberg, New York, 1988).
144. Note that a light-scattering process simultaneously requires the annihilation of the incoming photon and the creation of the scattered photon; it therefore involves two generally distinct photons, and its enhancement might be about 10^4 times enlarged, but it is still a linear optical process.
145. C. E. Berger, R. P. Kooyman and J. Greve, *Rev. Sci. Instrum.*, **65** (1994) 2829.
146. J. P. Thost, W. Krieger, N. Kroo, Z. Szentirmay and H. Walther, *Opt. Commun.*, **103** (1993) 194.
147. D. M. Kolb, *J. Physique* (Paris), **38** (1977) C5.

148. J. G. Gordon II and S. Ernst, *Surf. Sci.*, **101** (1980) 499.
149. A. Tadjeddine, *J. Electroanal. Chem.*, **169** (1984) 129.
150. P. G. Dzhavakhidze, A. A. Kornyshev, A. Tadjeddine and M. I. Urbakh, *Electrochim. Acta*, **34** (1989) 1677.
151. A. Tadjeddine, *Electrochim. Acta*, **34** (1989) 29.
152. A. Tadjeddine, A. Rahmani, G. Piazza and M. Costa, *Electrochim. Acta*, **34** (1989) 1681.
153. B. Pettinger and C. Bilger, to be published.
154. A. von Oertzen, Doctoral Thesis (Freie Universität Berlin, 1992).
155. S. Jakubith, H. H. Rotermund, W. Engel, A. von Oertzen and G. Ertl, *Phys. Rev. Lett.*, **65** (1990) 3013.
156. S. Nettesheim, Doctoral Thesis (Freie Universität Berlin, 1993).
157. M. D. Graham, I. G. Kevrekidis, K. Asakura, J. Lauterbach, K. Krischer, H. H. Rotermund and G. Ertl, *Science*, **264** (1994) 80.
158. H. H. Rotermund, *Inst. Phys. Conf. Ser.*, **147** (1995) 215.
159. G. Haas, R. U. Franz, H. H. Rotermund, R. M. Tromp and G. Ertl, *Surf. Sci.*, **352–354** (1995) 1003.
160. G. Flätgen, K. Krischer and G. Ertl, *J. Electroanal. Chem.*, **409** (1996) 183.

5

POTENTIAL CONTROLLED ORDERING IN ORGANIC MONOLAYERS AT ELECTRODE–ELECTROLYTE INTERFACES

N. J. TAO

Department of Physics, Florida International University, Miami, FL 33199

5.1 INTRODUCTION

Organic monolayers are interesting not only because they are attractive for a wide range of applications, but also because they are nice model systems for studying two-dimensional physical phenomena, such as phase transition [1]. Organic monolayers are usually prepared at three interfaces: solid–vacuum (gas), liquid–air, and solid–liquid interfaces. At solid–vacuum interfaces, methods including thermal evaporation and molecular beam epitaxy have been developed [2]. At liquid–air interfaces, the Langmuir technique can prepare organic films with precisely controlled molecular orientations [3]. The films are then transferred onto solid substrates forming the so-called Langmuir–Blodgett films. At solid–liquid interfaces, one simple strategy is to let molecules adsorb onto a substrate from a solution and assemble into a monolayer film by themselves [1,4] The success of the self-assembly techniques has largely relied on covalent or coordination chemistry of the organic molecules and the substrate. Two widely studied systems are silane compounds on SiO_2 and alkanethiols on Au [1]. A related but more sophisticated strategy is to form a film on a conductive

---

Imaging of Surfaces and Interfaces (Frontiers of Electrochemistry, Volume 5).
Edited by Jacek Lipkowski and Philip N. Ross.
ISBN 0-471-24672-7. © 1999 Wiley-VCH, Inc.

or semiconductive substrate under electrochemical conditions [5–9] This strategy is attractive because the substrate potential (or surface charge density) can be varied continuously over a wide range, which can be used to flexibly control the growth and structure of the film. For example, if certain molecules do not self-assemble into a film, an appropriate substrate potential may persuade them to assemble into a film. Once a film has been assembled, the potential may induce various molecular rearrangements in the film via structural phase transitions that may lead to a desirable molecular packing structure for an application. The potential may also be used to induce electrochemical reactions that lead to polymerization of the film [10].

The most fundamental tasks in studying organic monolayers in the electrochemical environment are to understand the formation, molecular packing structure, and reaction of the monolayers as well as their dependence on the substrate potential. These tasks have been primarily carried out by measuring thermodynamic quantities, such as capacitance, charge density, surface tension, and surface excess, and by optical spectroscopies, such as infrared, UV–vis reflectance, surface enhanced Raman, and second harmonic generation [5, 6, 8]. These methods do not directly provide molecular packing structures of the films. They are, however, responsible for most of our current knowledge about the organic monolayers. Direct structural information has been obtained in vacuum by electron diffraction techniques after the monolayers are prepared in the electrochemical environment [7,9,11]. With this approach molecular packing structures of many organic monolayers have been determined in an unambiguous way for the first time. However, the ex situ approach often raises the question whether the transfer from the electrochemical environment to vacuum alters the structures of the monolayers, and it cannot *continuously* monitor various dynamics processes such as nucleation and growth. Two important in situ techniques have become available recently. One is the X-ray techniques using synchrotron radiation sources, which can precisely determine the atomic-scale structure of solid–liquid interfaces in reciprocal space [12]. The other uses the Scanning Tunneling (STM) and Atomic Force Microscopes (AFM), which can image individual molecules in the electrochemical environment in real space [13–14]. Recently in situ STM has been applied to image a number of individual molecules and organic monolayers. Systems that have been studied include DNA [15–17], DNA bases [18–27], and related molecules [28], [Ru(bpy)$_2$(bpy – CH2)$_x$ – bpy]$^{2+}$ [29], bipyridines [30–31], 1,10′ phenanthroline [32] tetramethylthiourea [33], organomercaptans [34], phenol [35], porphyrins [36–39] and benzene [40] on graphite, platinum, gold, or iodine-modified gold substrates. In situ AFM has been less successful but several molecules have been convincingly imaged with AFM, and a comparative STM and AFM study has been shown to provide additional structural information [24–25]. While we focus on organic monolayers in the electrochemical environment, we note that STM and AFM have been applied to image organic molecules in air [41–48], in vacuum [49–51], and in drops of organic solvent [52–62]. Some of these works are reviewed in references [63–65].

In this chapter, we demonstrate that important phenomena: adsorption, nucle-

5.2 ON GRAPHITE BASAL PLANE

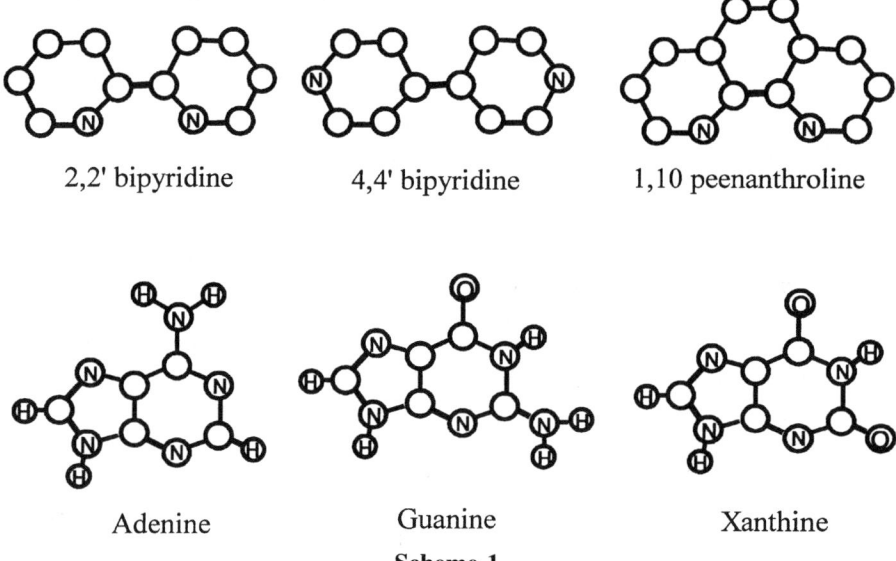

Scheme 1.

ation and growth, structural phase transitions, as well as electron transfer reactions, can be studied with STM and AFM in real time with molecular resolution using selected examples on heterocyclic molecules (scheme 1). We further restrict ourselves to images obtained primarily in our lab because of the difficulty of reproducing high quality images from published works. We compare the results on graphite basal plane and on gold substrates to illustrate the importance of adsorbate–substrate interactions in the molecular packing structures of the adsorbed monolayers. We organize the chapter as follows. In Sections 5.2 and 5.3, we describe the results on graphite basal plane and on Au(111), respectively. In Section 5.4, we discuss some of the important results on other surfaces. In Section 5.5 we turn to some technical aspects of STM and AFM by presenting experimental details used in our experiments, comparing the STM and AFM images of the same molecules and discussing the STM imaging mechanisms. In the end we summarize some of the important findings.

5.2 ON GRAPHITE BASAL PLANE

DNA bases are among the first organic molecules that have been imaged successfully by STM [66–67]. In the pioneering studies, graphite and MoS_2 were used as substrates because clean and atomically flat surfaces could be easily obtained by peeling off the top layers of the layered materials. The samples were prepared by placing drops of aqueous solutions of DNA bases onto a graphite, which was then baked before imaging in air. Using this approach,

Heckl et al. [67] and Allen et al. [66] obtained submolecular resolution images of the four DNA bases on the basal plane of graphite. However, interpreting the high-resolution images was not trivial because the STM images depended sensitively on the registry of the adsorbates to the graphite substrate (Section 5.5.2). Using MoS_2 as substrate, Heckl et al. [67] were able to easily identify each individual guanine molecules and concluded that the hydrogen bonds between the molecules were responsible for the ordered molecular packing. More recently, Poler et al. [68] imaged guanine monolayers in air with Scanning Thermopower Microscope [69]. Both hydrogen bonds and water molecules incorporated into the monolayer were proposed to fit their data.

Srinivasan et al. have studied guanine [18–19] and adenine [20] on graphite electrode under electrochemical conditions with STM. They found that the molecules condensed into ordered structures when the electrode potential was greater than ~0.1 V (Ag/AgCl). In the case of adenine, the ordered structure was found to depend on the electrode potential and on the anions in the electrolytes. These works focused primarily on the molecular packing structure of the monolayers. In addition to molecular packing structure, we have studied the nucleation and growth processes [25] and electrochemical reactions [26, 28] of several purine bases and derivatives with both STM and AFM. Here we review some of these results.

5.2.1 Formation of Guanine Monolayer

In order to study the entire formation process of a monolayer film, we started the experiment by imaging a graphite substrate in 0.1 M NaCl. After obtaining a clear substrate image, we introduced carefully a drop of solution containing sample molecules (e.g., guanine) into the STM or AFM sample cell without interrupting the scanning. We then monitored the nucleation and growth processes in real time by recording the images continuously. A small portion of the images is shown in Figure 5.1. Before introducing guanine into the cell, the substrate is clean and essentially flat (except a few surface steps) (Fig. 5.1a). After introducing guanine into the cell, the molecules begin to adsorb onto the surface and condense into small islands (Fig. 5.1b). These islands grow gradually and cover the entire substrate within ~20 minutes (Fig. 1c–d). The consecutive images can be viewed like a movie that shows the detailed nucleation and growth processes. For example, the islands are compact, rather than fractal as predicted by the diffusion limited aggregation model [70]. This observation indicates that the growth by direct incorporation is dominant in the formation of the monolayer [71]. The smallest observed nuclei are ~30 nm in dimensions, corresponding to several hundred guanine molecules, which is consistent with the estimated critical nuclei for two-dimensional condensation on an electrode based on classical thermodynamic theories [5, 8].

By adjusting the interfacial electrochemical potential, the nucleation and growth processes can be controlled flexibly. Figure 5.2a shows a guanine monolayer on graphite in 0.1 M NaCl. Lowering the potential to –0.33 V (vs.

5.2 ON GRAPHITE BASAL PLANE

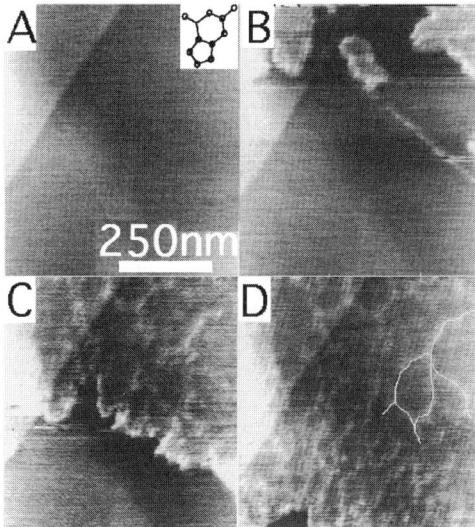

Figure 5.1. AFM images of nucleation and growth of a guanine monolayer on graphite from 0.1 M NaCl solution obtained (a) 0, (b) 7.5, (c) 13.5, and (d) 17.5 min. after introducing a drop of 0.1 mM guanine into the solution [30].

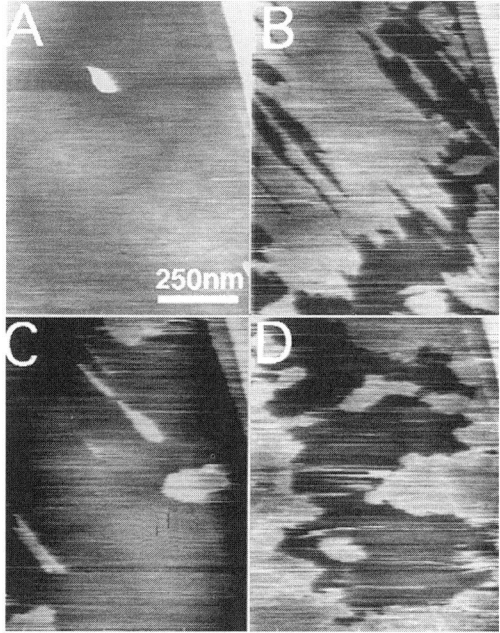

Figure 5.2. (a, c) Substrate potential controlled dissolution and (b) re-condensation of a guanine monolayer on graphite in 0.1 M NaCl. (a) The monolayer at rest potential; (b and c) captured after stepping the substrate potential to −0.33 V (Ag/AgCl in 3 M KCl); (d) obtained after stepping the potential to 0.42 V [30].

Ag/AgCl reference electrode), the monolayer starts to dissolve by forming a few small pits within a few seconds. The pits quickly increase both in size and number density (Fig. 5.2b), and the entire film is completely dissolved within a few minutes (Fig. 5.2c). The process is entirely reversible. About five minutes after raising the potential to +0.42 V, the molecules recondense into a monolayer film (Fig. 5.2d). The potential dependent nucleation and growth observed by AFM agree qualitatively with the interfacial capacitance measurement by Srinivasan et al. (Fig. 5.3) [18]. The measurement shows that the capacitance steps up when lowering to ~ −0.1 V, corresponding to dissolution, and steps up when raising the potential to ~0.1 V, corresponding to condensation. The lack of an exact agreement in the potentials is partly due to the AFM tip that can affect the condensation and dissolution processes and partly due to the dependence of the potentials on the potential sweep rate in the capacitance measurement.

We have studied other molecules, adenine [25] and xanthine [28], and found that they also adsorb onto graphite from aqueous solutions and condense into ordered monolayers in a similar fashion to that of guanine.

We note that the condensation of organic molecules into monolayers was first discovered nearly 40 years ago at the mercury–water interface [72, 73]. With the invention of STM/AFM, the nucleation and growth processes of the condensation can be finally visualized on the molecular scale at solid–water interfaces.

5.2.2 Molecular Packing Structure—The Role of Hydrogen Bonds

The molecular packing structure of the monolayers can be obtained from molecular resolution AFM images. Figure 5.4a shows an AFM image of guanine monolayer grown on graphite in 0.1 M NaCl containing 0.1 mM guanine. The image clearly reveals an ordered two-dimensional array with lattice constants: a = 10.8 ± 0.3 Å, b = 9.3 ± 0.3 Å and $\gamma = 88°\pm2°$, which are in sharp contrast to the hexagonal lattice of the underlying graphite honeycomb lattice revealed by

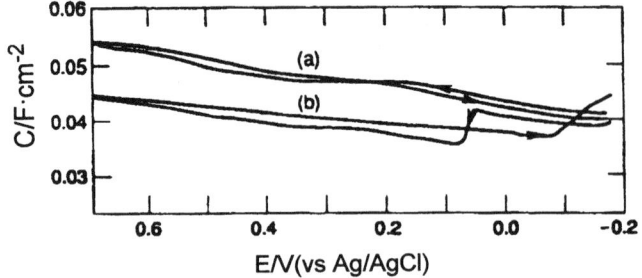

Figure 5.3. Differential capacitance of graphite basal plane in (a) 0.1 M NaCl and (b) 0.1 M NaCl saturated with guanine. The steps in the capacitance represent condensation or dissolution at the interface. The arrows point the direction of potential scan [18].

5.2 ON GRAPHITE BASAL PLANE

Figure 5.4. AFM images of a guanine monolayer on graphite in 0.1 M NaCl. A unit cell is outlined in B. Reproduced from Tao et al. [30].

AFM (inset of Fig. 5.4b). The lattice constants agree reasonably well with some of the previous STM studies [67]. A higher-resolution AFM image (Fig. 5.4b) shows that lattice consists of horseshoe-like features. Interpreting each horseshoe as a guanine molecule, the image can be nicely explained by a model shown in Figure 5.5. In the model, the molecules form zigzag chains in which each molecule is linked to two neighbors via four hydrogen bonds: two N1—H...O6 and two N2—H...N3. The zigzag chains are closely packed together to form

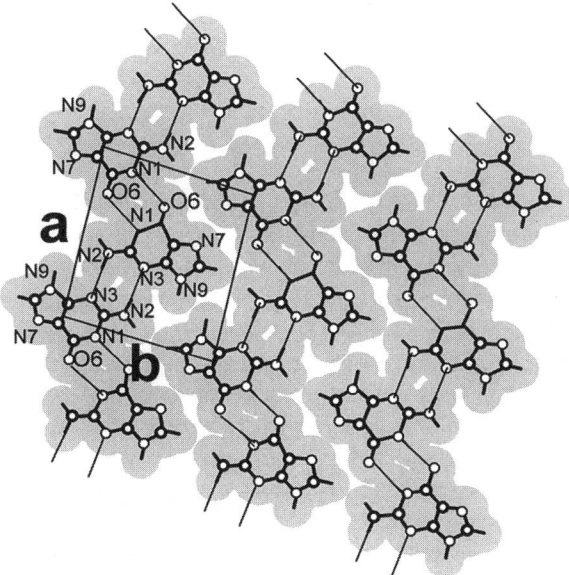

Figure 5.5. Hydrogen bonded network model for the molecular packing structure of guanine monolayer on graphite in 0.1 M NaCl [30].

a two-dimensional array. Using the coordinates of each atom in the molecule determined by X-ray crystallography [74], van der Waals radii of each atom (r_H = 1.2 Å, r_O = 1.5 Å, r_C = 1.7 Å and r_N = 1.5 Å) and hydrogen bond lengths, 2.9 Å for N1—H...O6 and N2—H...N3, the model produces lattice constants of a = 10.5 Å, b = 9.8 Å, and γ = 86°. Given the fact that both van der Waals and hydrogen bonds are susceptible to stretching and bending compared to covalent bonds, the model agrees with the experimental lattice constants reasonably well. Hydrogen bonds between guanine molecules are also responsible for the formation of layered structure in three-dimensional crystals of guanine as revealed by X-ray crystallography [75]. The model shown in Figure 5.5 is essentially the same as that proposed by Heckl et al. [67], except that all the zigzag hydrogen bonded chains are running in the same direction. Assuming that adjacent zigzag chains align in opposite directions, the model produces similar lattice constants. So based on the STM or AFM alone, it is difficult to determine whether zigzag chains alternate their orientations.

In order to confirm the role of the hydrogen bonds in the molecular packing structure of the guanine monolayer, we have carried out the study in alkaline and acidic solutions. At alkaline pH, the hydrogen at N1 of guanine is removed, so we do not expect the hydrogen bonded chains in the model (Fig. 5.5) to be stable since it involves hydrogen bond, N1—H...O6. We started the experiment by imaging a guanine monolayer in 0.1 mM guanine + 0.1 M NaCl. After the molecular packing structure shown in Fig. 5.4a and b was observed, we added 10 μL 0.1 M NaOH at a time into the cell that contains ~100 μL 0.1 M NaCl while scanning. The guanine monolayer dissolves 1–2 min. after injecting 1–3 drops of 10 μL 0.1 M NaOH into the cell (Fig. 5.6a–c). This observation

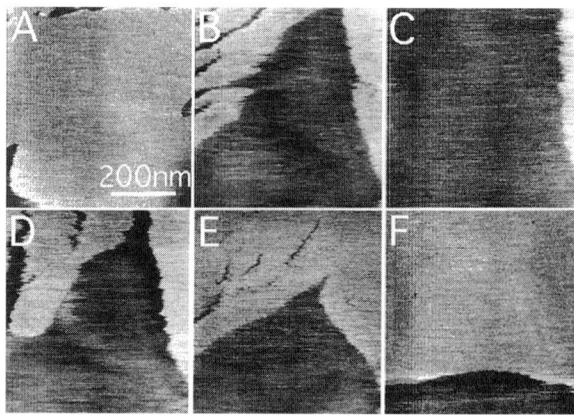

Figure 5.6. (a–c) AFM measurement disassembly and (d–f) reassembly of guanine monolayer on graphite as a drop of 0.1 M NaOH is introduced to the solution cell. The substrate potential was held at rest. The time lapse between two successive images is about 1 min. [30].

5.2 ON GRAPHITE BASAL PLANE

shows that the hydrogen bonded network self-assembled in the neutral solution is not stable in the alkaline solution. However, a new monolayer begins to assemble and gradually cover the entire surface a few minutes after the dissolution (Fig. 5.6d–f). Zooming in on the new monolayer, the AFM images reveal a completely different molecular packing structure in which the molecules organize into parallel stripes with each stripe consisting of three columns of blob-like features (Fig. 5.7a–b). The lattice constants of the structure determined from the images are: a = 6.9 ± 0.3 Å, b = 21.5 ± 1 Å, and γ = 86 ± 2°. This structure has also been observed in the monolayers prepared by directly exposing a graphite substrate to 0.1 M NaOH containing 0.1 mM guanine. If interpreting each blob-like feature as a guanine molecule, a model is proposed as shown in Fig. 5.8. In this model, the molecules form parallel chains with three rows of molecules in each chain. The three rows of molecules are linked together via linear hydrogen bonds, N2—H...N3 and N2—H...N1, and nonlinear hydrogen bonds, N2—H...O6 and N2—H...N7. We note that although the deprotonated N1 cannot form a hydrogen bond with O6, they can form hydrogen bonds with N2 in the amine groups whose hydrogen cannot be removed in the alkaline solution. The hydrogen bonded chains are held together via van der Waals attractions to form a two-dimensional lattice. This model produces lattice constants: a = 7.2 Å, b = 20.6 Å, γ = 88°, which are in good agreement with the experimental values.

We have conducted a similar experiment with 0.1 M $HClO_4$ and found that the monolayer formed in the neutral solution also dissolves after adding a few drops of 0.1 M $HClO_4$ into the cell. However, no new film with an ordered molecular packing grows back. We have also failed to observe any ordered guanine monolayers in 0.1 M $HClO_4$ containing saturated guanine. These results are understandable by considering the formation of hydrogen bonds. In acidic solutions, N1 and N3 are protonated, which prevents them from forming hydrogen bonds with the amine groups of the neighboring molecules, so neither the hydrogen bonded network in the neutral solution nor that in the alkaline solution is expected to be possible. This pH dependency, together with the fact that

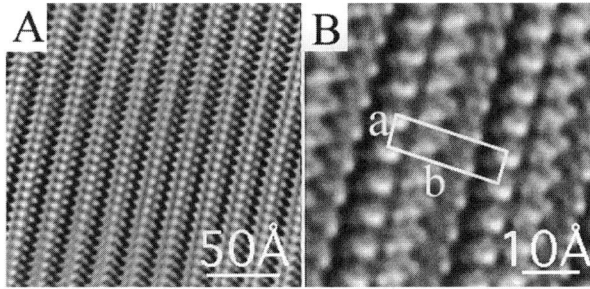

Figure 5.7. AFM images of the monolayer guanine monolayer on graphite in 0.1 M NaOH + 0.1 mM guanine at rest potential [30].

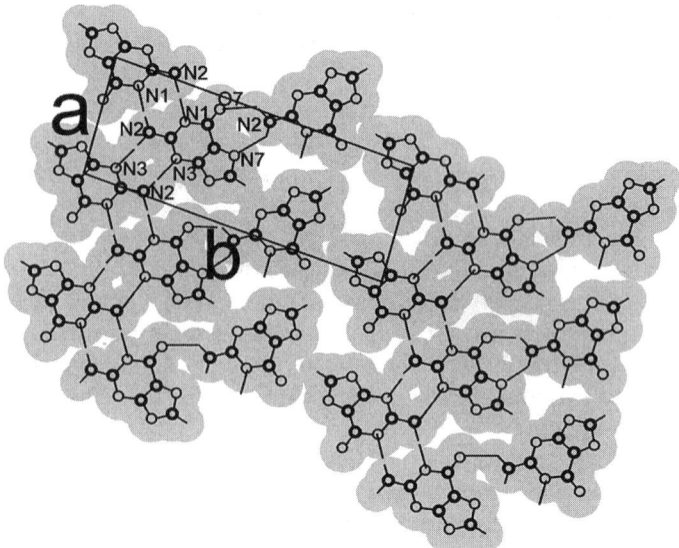

Figure 5.8. Hydrogen bonded network model for the molecular packing structure of guanine monolayer on graphite in 0.1 M NaOH [30].

we have never observed isolated guanine molecules on graphite, indicates that the guanine-graphite interaction is weak and it is the hydrogen bonding between guanine molecules that is responsible for the ordered monolayers.

Similar to guanine, AFM and STM images of ordered monolayers of other purines, such as adenine and xanthine, can also be explained based on hydrogen bonded networks [23–28]. The role of hydrogen bonding in the compact monolayers of purine and pyrimidine derivatives on mercury has been previously proposed based on molecular areas extracted from interfacial capacitance and tension data and analogy to the three-dimensional crystal structure of the molecules [76].

5.2.3 Superperiodic Structures

For guanine monolayer in neutral solutions, we have found two different structures that are separated by a domain boundary (Fig. 5.9a). The lattice constants determined from higher resolution STM images are: $a = 9.8 \pm 0.3$ Å, $b = 18.5 \pm 0.5$ Å, and $\gamma = 88° \pm 2°$ for structure II (Fig. 5.9b), and $a = 8.5 \pm 0.2$ Å, $b = 33 \pm 1$ Å, and $\gamma = 90 \pm 2°$ for structure I (Fig. 5.9c), respectively. Since the periodicities in the b-directions in the STM images are much larger than those determined from the AFM images, we refer these structures as superperiodic structures. The superperiodic structures can be understood by considering the registry of the molecular adsorbate lattice to the substrate lattice as modeled in Figures 5.10 and 5.11. The orientations of the adsorbate lattices in relation to

5.2 ON GRAPHITE BASAL PLANE

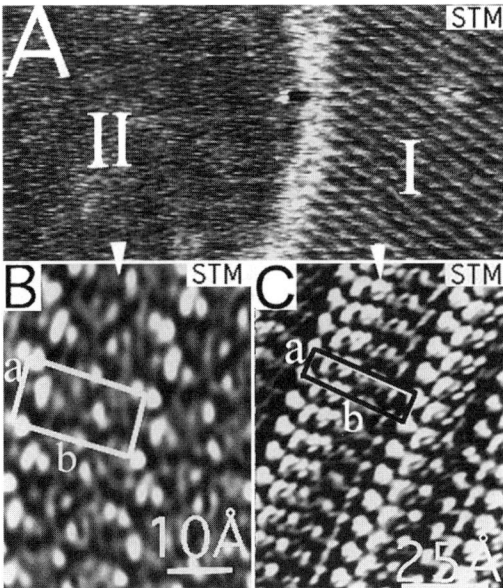

Figure 5.9. (a) STM image of two coexisting superperiodic structures in guanine monolayer grown from 0.1 M NaCl. (b) and (c) Higher-resolution STM images reveal more details of the two structures.

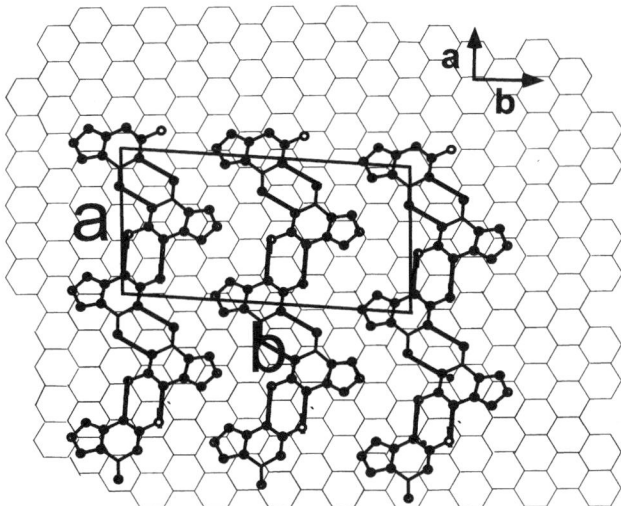

Figure 5.10. Registry of the guanine adsorbate lattice with respect to the graphite substrate lattice in superperiodic structure II.

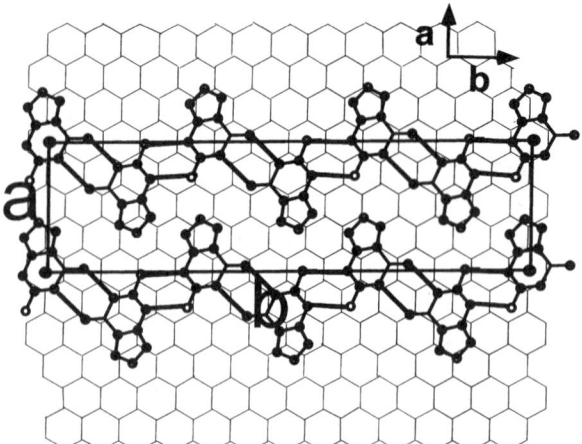

Figure 5.11. Registry of the guanine adsorbate lattice with respect to the graphite substrate lattice in superperiodic structure I.

the substrate lattice used in the models were determined by comparing molecular resolution images of the monolayer with atomic resolution images of the substrate obtained after dissolving the monolayer at a low potential. For structure II, the molecules adsorb at identical substrate lattice sites for every unit cell in the a-direction, but they repeat the same adsorption sites for every two unit cells in the b-direction (Fig. 5.10). This model produces lattice constants: $a = 9.8$ Å, $b = 19.1$ Å, and $\gamma = 87°$, in good agreement with the measured data. Structure I can be explained in a similar way as shown in Fig. 5.11. In the a-direction, the molecules occupy identical substrate lattice sites for every unit cell, but in the b-direction they repeat identical sites once for every three unit cells. The corresponding lattice constants: $a = 8.5$ Å, $b = 32.0$ Å, and $\gamma = 90°$, which also match the experimental data. The molecular packing in structures I and II are rather similar and can be easily missed by the AFM for the reason that will be discussed later (Fig. 5.12). The spacing in structure II is about 0.8 Å greater than that in structure I along each zigzag chain, but it is ~1 Å smaller between two adjacent chains. The difference may be interpreted as a rotation in the hydrogen bonds by 9°, which is possible since hydrogen bonds are only weakly directional [74].

We note that superperiodic structures have been observed in other systems, such as underpotential-deposited Pb on Au(111) [78] and monolayer iodine on Au(111) [79]. In those systems, the superperiodic structures have been attributed to the height variations in the adsorbed layers due to different adsorption sites. However, because of the planar geometry and rigid structure of the heterocyclic molecules, the height variation due to different adsorption sites in the present case is much smaller than the observed ~0.3 Å in STM, which explains why the superperiodic structures are much harder to observe with AFM. Since STM

5.2 ON GRAPHITE BASAL PLANE

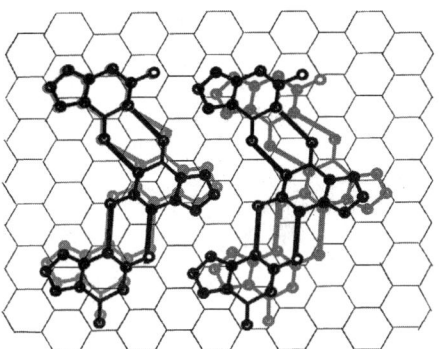

Figure 5.12. Comparison of molecular packing in superperiodic structures I (black lines) and II (gray lines).

probes local electron density of states (LDOS) [80–81] this result indicates that the modulation in the adsorbate LDOS by the substrate is responsible for the superperiodic structures [82].

We have observed different superperiodic structures in xanthine and adenine monolayers [25,28]. These superperiodic structures can also be explained based on variations in the LDOS due to different adsorption sites. Transitions between the different structures can be induced by the substrate potential. An example of such potential-induced phase transition is described next.

5.2.4 Potential-Induced Structural Phase Transitions

We have observed two superperiodic structures that can coexist at the rest potential in xanthine monolayers, which are rather similar to guanine monolayers. Figure 5.13a is an STM image of an xanthine monolayer in 0.1 M NaCl at rest potential. The image shows that the film consists of domains of various sizes. Higher magnification STM images reveal two distinctive superperiodic structures (insets of Fig. 5.13a and b). Domains 1, 2, 4, and 5 have parallel stripes that are separated by ~100 Å (inset of Figs. 5.13a), but domain 3 has a rectangular superperiodic structure (inset of Fig. 5.13b). The molecules in all the domains pack into a similar structure but with different orientations relative to the substrate lattice. This kind of adsorbate lattice rotation to misalign itself with the substrate has been observed in many monolayer films in vacuum [83] and described by several theories [84].

By varying the substrate potential, the orientation of the adsorbate lattice can be changed via a phase transition. By increasing the potential, domains 1 and 2 begin to shrink (Fig. 5.13d) and eventually disappear (Fig. 5.13e and f). Then domain 3 expands outwards (Fig. 5.13f) and covers the entire surface, which shows that the rotated phase in domain 3 is more stable at high potentials. The mechanism of this phase transition is not yet understood. We note that the phase

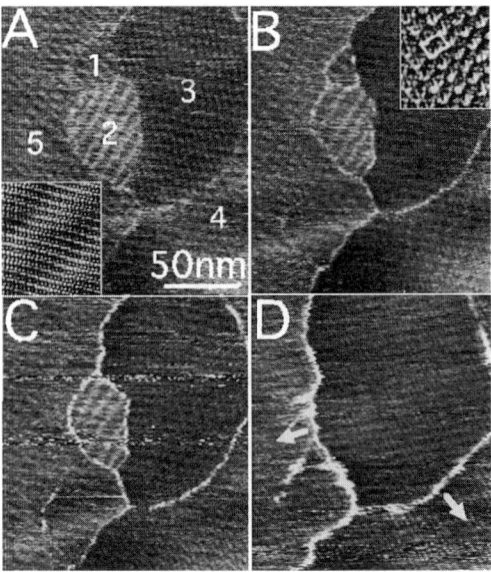

Figure 5.13. A potential-induced phase transition between superperiodic structures (insets in a and b) in a xanthine monolayer. (a) is an STM image at V = 0.05 V (vs. Ag/AgCl). (b), (c), and (d) were recorded ~10, 30, and 50 s after the potential is increased to 0.8 V [30].

transition does not show up in the voltammograms and the capacitance plots, probably because it involves only a change in the adsorbate lattice orientation.

5.2.4 Electrochemical Reactions

Raising the potential to 0.6 – 1.2 V, a number of heterocyclic molecules are oxidized on graphite or glassy carbon electrodes, which have been studied because of their relevance to the reactions in biological systems [85–86]. We have studied the oxidations of several molecules, and describe some of the results for xanthine here. Xanthine is oxidized on graphite electrode at ~0.9 V vs. Ag/AgCl (Fig. 5.14). Based on an electrolysis study, it has been proposed that the molecule is oxidized via a four-electron transfer and becomes a uric acid diimine that can be further oxidized at high potentials [85–86]. From the area of the reaction peak in the cyclic voltammogram, the total amount of charge transfer involved in the oxidation is estimated to be ~160 $\mu C/cm^2$. Varying the sweep rate, the peak height changes proportionally while the charge remains constant, which indicates that the reaction is not diffusion controlled and all the adsorbed molecules are oxidized during each potential cycle. Using the molecular packing density determined from the AFM images, the oxidation has been found to involve ~4 electrons per molecule, which supports the electrolysis result [85–86].

5.2 ON GRAPHITE BASAL PLANE

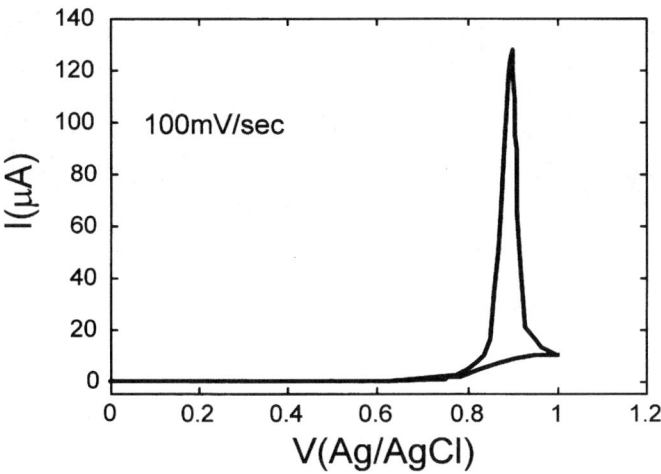

Figure 5.14. A cyclic voltammogram of graphite in 0.4 mM xanthine + 0.1 M NaCl. The sweep rate was 0.1 V/s. The electrode area was ~0.38 cm² [28].

By slowly raising the potential to the oxidation, we have studied the oxidation kinetics by recording the STM and AFM images continuously. Before the electrode potential is brought close to the oxidation potential, the image (Fig. 5.15A) shows a xanthine monolayer consisting of domains. Raising the electrode potential to the oxidation potential, the domain boundaries brighten up (Fig. 5.15b), showing that the reaction takes place preferentially at the defect sites. The brightened area expands and eventually covers the entire surface as the entire monolayer becomes oxidized (Fig. 5.15c). This observation demonstrates that the reaction rate depends strongly on the density and distribution of the defects in the monolayer.

The temporal resolution for studying the reaction dynamics using STM and AFM is typically in the order of seconds due to the limited scan rate of the instruments. To overcome this difficulty, we have recently used a potential-pulse method to study the oxidation kinetics. In this method, the reaction is studied by applying a sequence of potential pulses to the substrate. Figure 5.16 shows

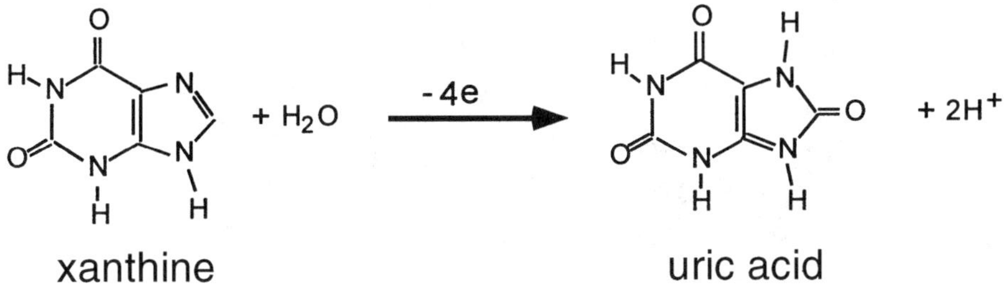

xanthine + H$_2$O $\xrightarrow{-4e}$ uric acid + 2H$^+$

Scheme 2

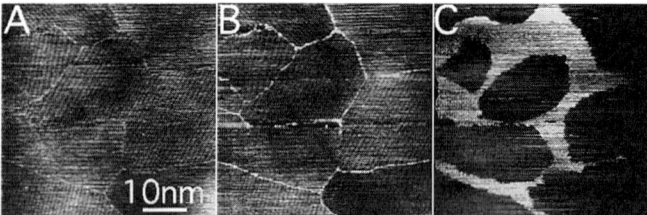

Figure 5.15. STM measurement of oxidation of xanthine on graphite in 0.1 M NaCl. (a) Image obtained before the reaction and (b) and (c) images taken during the reaction. The bright and dark areas are reacted and unreacted areas, respectively, in (c) [30].

STM images of xanthine (a) before and (b) after a 1 V potential pulse with a duration of 10 ms was applied to the substrate. The oxidized molecules appear to be brighter (pointed by an arrow) and less well ordered than the reactant molecules. After the monolayer is completely oxidized, the oxidized molecules are also closely packed but adapt a different ordered structure [28]. By changing the pulse height and duration, detailed dynamic information of the reaction can be studied with a temporal resolution of 1 ms. Higher temporal resolution is possible using potentiostat and electrochemical cell with shorter response times.

5.3 ON Au(111)

Au(111) is another surface well studied through a variety of techniques including STM/AFM and electrochemical techniques. The surface has a $23 \times \sqrt{3}$ reconstruction in which an extra row of Au atoms is squeezed into the surface layer for every 22 rows of atoms resulting in a periodic alternation between face-centered-cubic (fcc) stacking and hexagonal-close-packed (hcp) stacking [87–90]. In the STM [91–95] or AFM images [96] (Fig. 5.17), the transitions

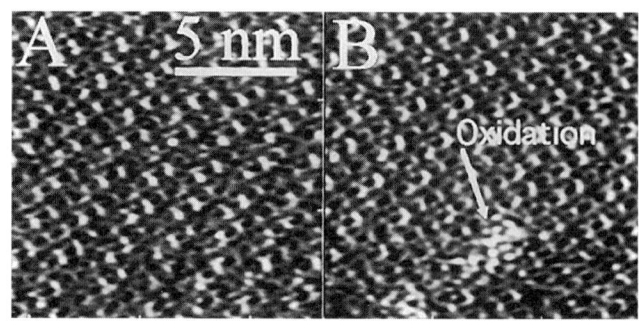

Figure 5.16. A xanthine monolayer on graphite in 0.1 M NaCl (a) before and (b) after a 1 V potential pulse (10 ms) was applied.

5.3 ON Au(111)

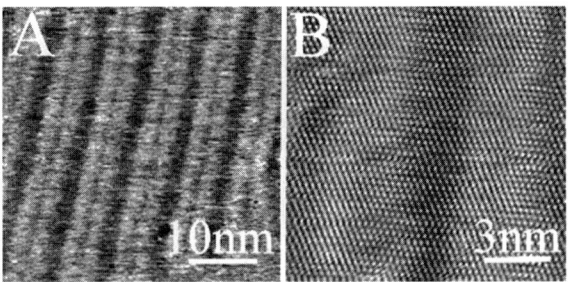

Figure 5.17. Reconstructed Au(111) in 0.1 NaClO$_4$ at 0.04 V (Ag/AgCl) revealed by STM.

between the fcc and hcp stackings are revealed as two parallel stripes in each unit cell. In the absence of strong adsorption, the reconstructed surface is stable when the potential is below the potential of zero charge (pzc). Increasing the potential above pzc, the surface transforms into the ideal 1 × 1 structure and the extra atoms released from the reconstructed surface form islands and diffuse to step edges of the surface [93–97]. More recent X-ray [98] and second harmonic generation experiment [99] have shown that the adsorption of organic molecules can either lift or stabilize the reconstruction.

Au(111) has been used as substrate to study molecular adsorption of many organic molecules. Lindsay et al. imaged DNA and DNA bases on Au(111) in aqueous solutions [15–16]. Wandlowski et al. studied the adsorption of cytosine and uracil on Au(111), Au(110), and Au(100) [21–22]. By combining the electrochemical techniques, current-potential, capacitance-potential, and current-time, with the STM, they studied the kinetics of the condensation and dissolution of the monolayers. For uracil, they found that the molecules form ordered physisorbed and chemisorbed adalayers, depending on the electrode potential. At low potentials, uracil molecules form a hydrogen-bonded network in which the planar molecules lie flat on the surfaces, which is in analogy to DNA bases on graphite. At high potentials, the molecules deprotonate and form another ordered structure in which the individual molecules stand vertically on the surfaces. Richard and Gewirth monitored the electrooxidation of phenoxide to oligophenol on Au(111) in alkaline solutions [35]. Prior to oxidation, phenoxide forms a ($\sqrt{3} \times \sqrt{3}$)R30° with the molecule end-on through the O atom. The oxidation products primarily consist of monomers, dimers, and trimers and are disordered on the surface. Bunge et al. examined tetramethylthiourea adsorption on Au(111) with cyclic voltammetry, capacitance measurements, and in situ STM, and they concluded that the molecules pack into a 3 × 3 structure [33]. Dakkouri et al. applied STM to cysteine on Au(111) and observed a ($\sqrt{3} \times \sqrt{3}$)R30° structure [34].

We studied a number heterocyclic molecules on Au(111) [30–32]. In contrast to the results on graphite, where the hydrogen bonding between the adsorbate molecules plays a dominant role in the formation of the ordered monolayers, the adsorbate–substrate interaction via the nitrogen σ lone pair electrons of the heterocyclic molecules is strong at high substrate potentials, which enables the planar molecules to stand vertically on the substrate [99]. The vertically standing molecules self-assemble into polymer-like chains in which the individual molecules stack via the π–π interactions [23]. The individual chains are randomly oriented at low substrate potentials, but they begin to line up in parallel as the potential reaches a critical value via a first order phase transition [30–31]. The phase transition resembles the well-known nematic-isotropic phase transitions in liquid crystal materials in three dimensions [100]. At very negative potentials, however, the molecules are believed to lie flat on the surface. We use 2,2' bipyridine (22BPY) as an example to illustrate these processes, and summarize results on phenanthroline, adenine, and guanine as a comparison.

5.3.1 Thermodynamic Data

Lipkowski et al. have developed a procedure for extracting important thermodynamics quantities, such as charge density, surface excess, and Gibbs adsorption energy using cyclic voltammetry, ac voltammetry, chronocoulommetry [8]. They applied the procedure to 22BPY adsorption on Au(111) [99]. Based on the thermodynamic quantities (Figs. 5.18 and 5.19), they concluded that the planar molecule assumes a flat orientation at the negatively charged surface. At the positively charged surface, the molecule assumes a vertical orientation in which one of the nitrogen atoms face the Au surface. Because of a gradual change in the surface excess and multiple peaks on the capacitance curves, they believed

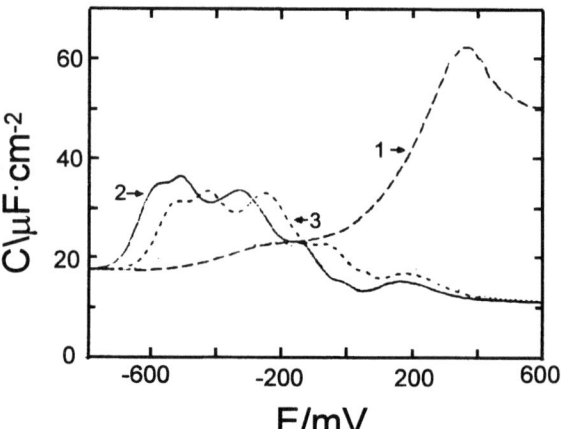

Figure 5.18. Differential capacitance curves for the pure 0.1 M $KClO_4$ solution (curve 1 and dashed line) and for the supporting electrolyte with 10^{-3} M and 5.5×10^{-4} M concentrations of 22BPY. The curves were recorded using a voltage sweep in the positive direction at the sweep rate 5 mV/s and the ac modulation frequency 25 Hz [93].

5.3 ON Au(111)

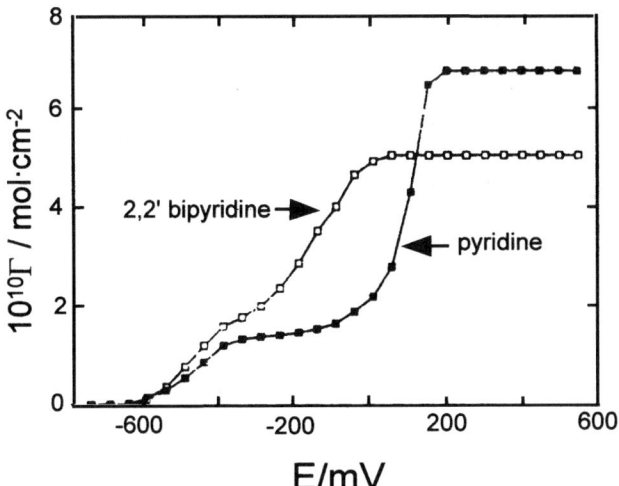

Figure 5.19. Surface excess versus electrode potential plots for 1×10^{-3} M solutions of 22BPY (open points) and pyridine (black points) in 0.1 M KClO$_4$ [93].

that the transition from the flat to the vertical orientation is gradual and goes through a series of intermediate states. In addition, they found the adsorption of 22BPY has little effect on the substrate $23 \times \sqrt{3}$ reconstruction based on the second harmonic generation spectroscopy data.

5.3.2 Formation of 22BPY Monolayer

We have studied the formation of 22BPY monolayer on Au(111) with STM by imaging a reconstructed Au(111) in a bare 0.1M NaClO$_4$ at rest potential. While scanning, we then introduced a drop of 1 mM 22BPY + 0.1M NaClO$_4$ into the solution cell and monitored the adsorption of 22BPY onto the surface. Figure 5.20a is an STM image that shows a reconstructed Au(111) in 0.1 M NaClO$_4$ before the introduction of 22BPY. A few minutes after introducing 22BPY into the solution cell, 22BPY molecules begin to adsorb on the surface as pointed by an arrow in Figure 5.20b. The adsorbed molecules appear as polymer-like chains, which are described in detail in Section 5.3.3. As more 22BPY molecules adsorb onto the surface, it becomes clear that the molecules adsorb preferentially along the stripes of the reconstruction (Fig. 5.20c).

In contrast to those on graphite, 22BPY, guanine and adenine monolayers on Au(111) do not start from a few nucleation islands and then grow and coalesce into a continuous film. The difference reflects the difference in the adsorbate–substrate and adsorbate–adsorbate interactions between the molecules on the two surfaces. On graphite, each molecule is weakly bound to the substrate and the hydrogen bonding between the molecules favors the molecules to condense into two dimensional arrays by forming hydrogen bonded networks. While on gold

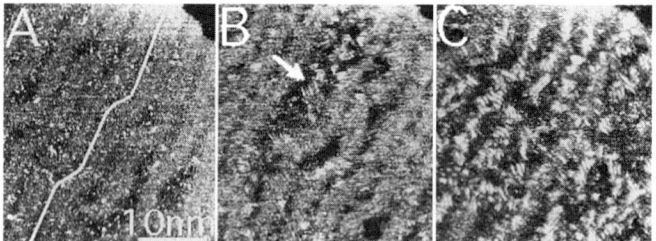

Figure 5.20. STM images of a reconstructed Au(111) in 0.1 M NaClO$_4$ (a) 5 minutes, (b) 9 minutes, and (c) 23 minutes after introducing a drop of 1 mM 22BPY + 0.1 M NaClO$_4$ into the cell. The line in (a) marks the direction of the reconstruction stripe and the arrow in (b) points the formation of 22BPY polymer-like chains [30].

surfaces, the adsorbate–substrate interaction is much stronger at positive potentials than the adsorbate–adsorbate interactions, therefore the tendency of forming compact islands is weaker.

5.3.3 Molecular Packing Structure

The 22BPY chains can exist in either a disordered or an ordered phases, depending on the potential (see Section 5.3.4). In the ordered phase (Fig. 5.21), the chains organize into domains within which they pack closely in parallel along one of the three equivalent Au(111) directions. The separation between two adjacent parallel chains in the ordered phase is measured to be 0.90 ± 0.02 nm. In the disordered phase, the chains pack loosely and their orientations do not strictly follow the Au(111) lattice directions. Along each chain, the individual molecules arrange periodically with a repeat distance of 0.38 ± 0.01 nm. The STM images confirm that 22BPY stands vertically on the Au(111) substrate with the two nitrogen atoms facing the substrate surface [99]. In addition, they

Figure 5.21. STM images of an ordered 22BPY monolayer on Au(111) in 0.1 M NaClO$_4$ [30].

5.3 ON Au(111)

reveal that the vertically standing molecules self-assemble into chains in which the individual molecules stack via the π–π interaction. The tendency of planar heterocyclic molecules to stack into polymeric aggregates in aqueous solutions has been studied by NMR [101], but only aggregates consisting of a few 22BPY molecules are stable in bulk solution at room temperature. However, with the help of a surface, long chains of stacked molecules are stable at room temperature. The images also show that each molecule is tilted from the chain direction by 60 ± 5°. By taking into account this ~ 60° tilt, the perpendicular distance between two adjacent molecules is calculated to be 0.33 nm, which corresponds well with the typical distance between two stacked planar molecules found in many other systems [74]. The 60° tilt may be attributed to 22BPY molecules that avoid perfect alignment with their neighbors by shifting aside ~0.19 nm (see Fig. 5.22). This behavior of avoiding perfect alignment between

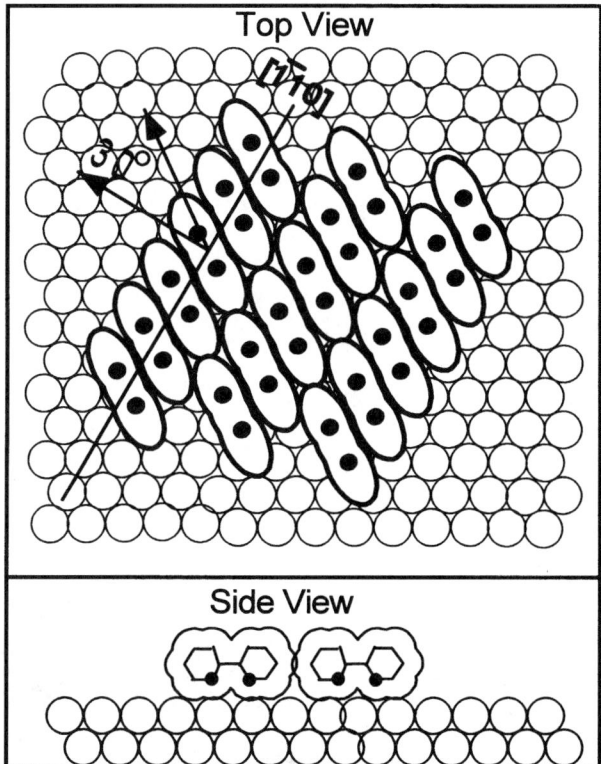

Figure 5.22. Proposed molecular packing of the 22BPY monolayer. Each molecule stands vertically with the two nitrogen atoms (filled circles) facing the surface. The individual molecules stack into polymer-like chains that align in parallel to form a two-dimensional lattice. The 22BPY dimensions from X-crystallography are used in this model [30].

two adjacent stacking molecules has been observed in nucleic acid bases and many aromatic molecules [74]. A recent model of π–π interaction between planar molecules shows that the misalignment is a result of competition between the π–σ attraction and π–π repulsion [101].

High-resolution STM reveals each 22BPY molecule as two blobs separated by ~0.3 nm (Fig. 5.21). This distance corresponds to the separation between the two nitrogen atoms in the molecule, so electron tunneling through the nitrogen atoms dominates the STM image. This observation is reasonable because the nitrogen atoms interact directly with the substrate, which provides an electronic coupling between the molecular and the substrate states near the Fermi level. In a recent study of Cu-TBP-porphyrin, only the part of the molecule that is in contact with the substrate is observed in the STM image [102].

Similar to 22BPY, several other heterocyclic molecules, such as phenanthroline, adenine, and guanine molecules, stack also into polymer-like chains with the individual molecules standing vertically on the surface. Figures 5.23 shows STM images (a) adenine and (b) guanine obtained in 1 mM adenine + 0.1 M NaClO$_4$ and 0.1 mM guanine + 0.1 M NaClO$_4$, respectively.

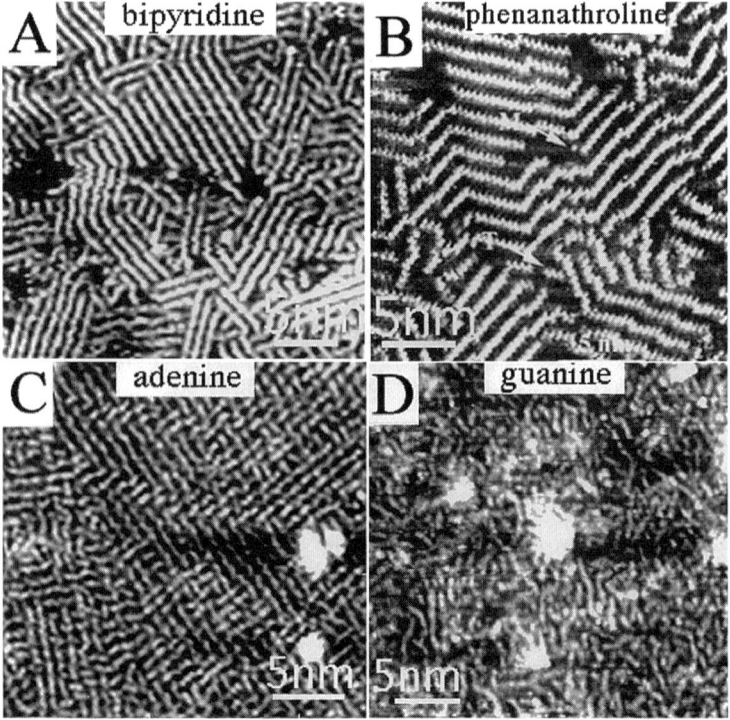

Figure 5.23. STM images of (a) 22BPY, (b) phenanathroline, (c) adenine, and (d) guanine adsorbed on Au(111) from 0.1 M NaClO$_4$.

5.3.4 Structural Phase Transition

The orientational order of the 22BPY chains depends on the substrate potential. At low substrate potentials, the chains are randomly oriented, but they line up in parallel as the potential is raised to a critical value via a first-order disorder–order phase transition. The detailed dynamic phase transition process has been studied with STM [30–31]. A small portion of the images that show the time sequence is described as follows.

Figure 5.24a is an STM image of a 22BPY monolayer in 0.1 M NaClO$_4$ at 0.1 V, which shows essentially randomly oriented polymeric chains. Increasing the potential to 0.25 V, the chains begin to align in parallel, and pits are also forming during the process (Fig. 5.24b). As more chains align in parallel at higher potentials, the size and number of pits increase (Fig. 5.24c). The chains in each domain align along one of the three equivalent Au(111) lattice directions. The formation of pits is because as the chains line up in parallel they also diffuse towards each other and pack closely. This indicates an attractive force between the parallel chains, which increases as the potential increases. We believe that this potential dependent force is responsible for the potential-induced phase transition. A recent self-consistent density functional calculation suggests that the potential dependent attraction between parallel chains is due to

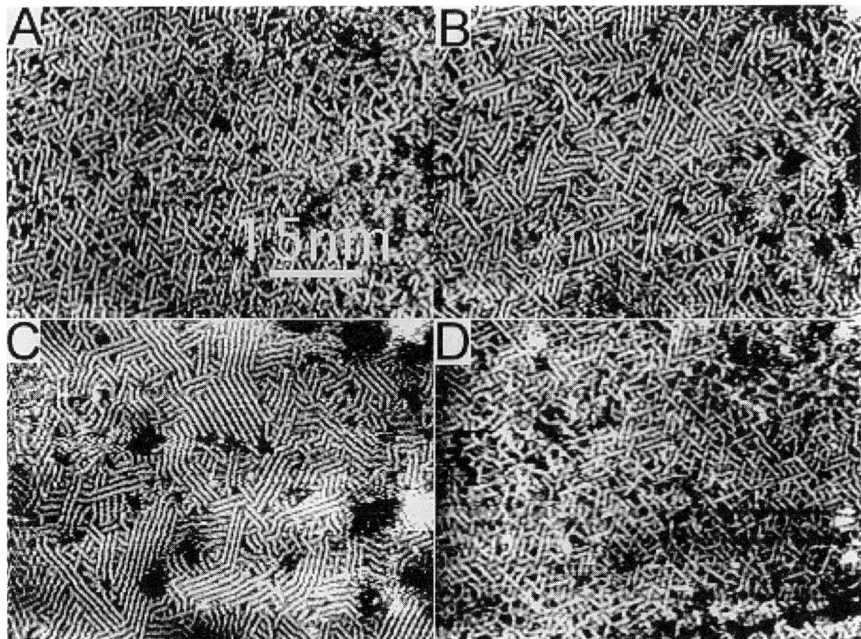

Figure 5.24. Potential-induced reversible disorder–order phase transition in a 22BPY monolayer on Au(111) in 0.1 M NaClO$_4$. The images were obtained at potentials of (a) 0.10 V, (b) 0.25 V, (c) 0.37 V, and (d) 0.05 V.

a substrate-mediated effective force that arises as an adsorbed 22BPY perturbs its surrounding local surface potential, which affects the adsorption energy of the nearby molecules [31]. Lowering the potential, the disordered phase recovers (Fig. 5.24d). It starts as the chains near the edges of the pits diffuse into the pits and become randomly reoriented because of the extra space in the pits. The space freed by the diffusion of the chains into the pits allows nearby chains to become randomly oriented. In this fashion, the disordered phase eventually spreads out and covers the entire surface (Fig. 5.24d).

We have observed a similar potential-induced order–disorder phase transition in phenanthroline monolayers [32], but considerably different behaviors for the adenine and guanine. In the adenine monolayers, the polymeric chains pack in parallel along one of the three equivalent Au(111) lattice directions, which is remarkably similar to that of 22BPY chains in the ordered phase, but this ordered phase cannot be driven into a disordered phase when varying the potential between -0.2 V and 0.9 V. At lower concentrations (1–10 μM), the adenine layer is less compact but the chains still tend to align in parallel. This observation indicates an attractive force between the parallel adenine chains that holds them together. When decreasing the potential, the chains may suddenly rotate *collectively* but they do not become randomly oriented as in the case of 22BPY and phenanthroline. However, the order–disorder transition can be triggered by varying the adenine concentration. At very low concentrations (below ~1 μM), the adenine chains are very short, consisting of only a few stacked molecules, and they are essentially randomly oriented on the surface. The existence of this disordered phase for short chains can be understood in terms of the Onsager theory [100]. According to the theory, the existence of ordered and disordered phases in the systems of polymer rods are because of the competition between entropy, which favors random orientation, and steric repulsion, which favors parallel orientation. Shorter chains are easier to become randomly oriented without causing steric repulsion than the longer ones and therefore favor the disordered phase. Guanine stacks also into polymer-like chains, but the chains are randomly oriented, similar to the disordered phases of 22BPY, phenanthroline and adenine. Increasing the potential form -0.2 V to 0.9 V, the disordered phase did not transform into an ordered phase. We cannot drive the disordered phase into an ordered phase by increasing the guanine concentration either because of low solubility of guanine in aqueous solutions.

5.4 ON OTHER SURFACES

In addition to graphite and Au substrates, several STM studies of organic molecules on other metallic surfaces or modified metallic surfaces have been reported under electrochemical conditions. Yau et al. studied benzene chemisorption on Rh(111) and Pt(111) in HF solution [40]. They found a c(2$\sqrt{3}\times 3$)rect structure for the adsorbed benzene on both substrates in the double-layer charging region, but the degree of ordering of the structure on Rh(111) was higher than that on Pt(111). At negative potentials, the c($2\sqrt{3}\times3$)rect structure on Rh(111) transformed into the 3×3 structure, which was previously observed in UHV for benzene coadsorbed with CO on Pt. They proposed that the role of CO coadsorption might be replaced by water molecules or hydronium cations in the electrolyte. The c($2\sqrt{3}\times3$)rect structure on Pt(111), however, transformed into a ($\sqrt{21}\times\sqrt{21}$)R10.9° at negative potentials. Yau et al. also studied several other aromatic molecules, such as naphthalene, naphthoquinones, and anthracene on Rh(111) and Pt(111) in aqueous HF solutions [103]. On Rh(111), naphthalene and naphthoquinone form a long-range ordered ($3\sqrt{3}\times3\sqrt{3}$)R30° structure, while anthracene forms only a disordered structure. On Pt(111) long-range ordering is absent for the three molecules.

Itaya's group has studied adsorption of 5, 10, 15, 20-tetrakis(N-methylpyridinum-4-y)-21H, 23H-prophine, crystal violet, and methyl-pyridinium phenylenedivinylene on iodine-modified Au and Ag electrodes [37, 39, 104–105]. They found that the presence of the iodine layers allows the molecules to form highly ordered structures on the surfaces. This approach may be important for attaching other important organic molecules including large biological molecules to the surfaces for STM imaging. Li and Tao have studied heterocyclic molecules, such as 22BPY and phenanthroline on a Cu layer on Au(111) prepared by underpotential deposition [106]. Their results show that the molecules are more strongly adsorbed on the surface and pack into ordered structures with a longer range than those on bare Au(111).

5.5 TECHNICAL ASPECTS OF ELECTROCHEMICAL STM AND AFM

In the previous sections, we focused on the results of STM/AFM, now we turn to some technical aspects of electrochemical STM/AFM. First, we present some experimental details involved in obtaining the STM and AFM images shown in the previous sections. Then we compare STM and AFM images of some molecules. Finally we discuss the STM imaging mechanism for organic molecules.

5.5.1 Experimental Details of STM and AFM

5.5.1.1 Substrates Highly oriented pyrolytic graphite (HOPG) substrates of various grades are commerically available from Advanced Ceramics. The ZYH grade HOPG has a lower overall surface flatness than the more expensive ZYA and ZYB grade substrates, but it is just as good for the STM and AFM studies discussed in Section 5.2 because only a small area is scanned by STM and AFM. In our experiment, a fresh atomically flat graphite surface was created by peeling off the top layers with an adhesive tape, and covered immediately with a drop of electrolyte. A drawback for using HOPG as substrate to image chain-like molecules is the various surface defects that mimic isolated polymer chains, which has resulted in a large number of published artifacts in the early years of STM applications [107]. This kind of artifacts is not a problem in the study of organic monolayers in which the molecules are densely packed and often ordered. For imaging ordered adsorbates, the superperiodic structures (Moiré patterns) due to orientational misalignment of the very top layer of graphite could also lead to misinterpretation [108]. This misinterpretation can be easily avoided if one is cautious about rare images because the Moiré patterns are quite rare. In addition, the Moiré patterns have a hexagonal symmetry that can be distinguished from many adsorbate structures.

We prepared the Au(111) substrates by epitaxially growing gold on mica in vacuum (pressure $< 1 \times 10^{-8}$ torr) using a procedure described in DeRose et al [109]. First, we placed freshly cleaved mica sheets of $\sim 9 \times 12$ mm^2 in a vacuum chamber and baked at $\sim$380 C$^\circ$ for 12 hours. We then evaporate gold onto the mica sheets to form films of $\sim$1500 Å at a rate of 1–2 Å/sec. We allowed the films to anneal at $\sim$400 C$^\circ$ in vacuum for four or five hours before cooling them to room temperature. Before each STM or AFM experiment, we briefly annealed a gold substrate in a H$_2$ flame and covered immediately with a drop of electrolyte. Au(111) has also been prepared by melting a gold wire into a bead in a H$_2$ + O$_2$ flame [110]. For other orientations, Au single crystals are normally used [22, 94, 97]. The surfaces can be oriented by Laue backscattering to better than 1° and are polished with fine diamond paste and then also annealed in a flame before each STM experiment.

5.5.1.2 STM Tip Fabrication W and Pt$_{0.8}$Ir$_{0.2}$ wires are common materials used to fabricate STM tips for ECSTM. These tips can be electrochemically etched from concentrated (e.g., between 1 M to 10 M) NaOH solutions or simply by cutting the wires with a wire cutter [111]. The tips are coated with Apiezon wax or nail polish to reduce ionic conduction (leakage current) [112]. Well-coated tips are especially important for imaging organic molecules, because a small tunneling current is often needed to minimize the tip perturbation on weakly adsorbed layers. Pt$_{0.8}$Ir$_{0.2}$ tips are more stable in air than W tips, but the leakage current tends to increase as the potential decreases. The W tips have less leakage current at low potentials but more leakage current at high potentials than do Pt$_{0.8}$Ir$_{0.2}$ tips. So for investigating a surface phenomenon over a wide potential range, both tips are sometimes used.

5.5.1.3 Electrochemical Environment In addition to clean substrates and well-coated STM tips, a clean electrochemical environment is essential for successful and meaningful electrochemical STM or AFM experiments. We used Teflon electrochemical cells for our STM and AFM (Pico-SPM, Molecular Imaging Co.). They were thoroughly cleaned in 70% H$_2$SO$_4$ + 30% H$_2$O$_2$ before each experiment. For electrochemical potential control, we used a fresh Pt wire as the counter electrode, and a Ag, Pt, or oxidized Au wire as quasi-reference electrode. The electrodes, however, need to be calibrated immedi-

5.5 TECHNICAL ASPECTS OF ELECTROCHEMICAL STM AND AFM

ately before and after each experiment. More conventional reference electrodes have been reported in STM experiments, but the risk of contamination and complexity of the electrodes introduce extra difficulties in the experimental setup. Our experiments were operated in a homemade N_2 chamber when using Nanoscope III (Digital Instruments Co.) or more recently in the equipped glass N_2 chamber of a Pico-SPM.

5.5.2 STM and AFM Comparison

To a first-order approximation, STM probes the sample LDOS near the substrate Fermi level for a small tip-substrate bias voltage, while AFM depends on the tip-sample interaction [81]. Although both can achieve molecular resolution, a comparative STM and AFM study of the same sample can provide additional information about the studied samples, and it also reduces the chance of misinterpreting the images. We have imaged the molecules discussed previously with both STM and AFM. On Au(111) where the molecules stand vertically on the surface, the STM and AFM images are remarkably similar (Fig. 5.25). However, on graphite where the molecules lie flat on the surface, the STM and AFM images are significantly different. Figure 5.26 compares the STM and AFM images of xanthine and guanine on graphite. In the AFM images (a and c) both xanthine and guanine appear as blobs. Although internal features of the molecules are occasionally resolved by AFM, Figure 5.26a and c represent typical AFM images. In sharp contrast, the STM images (b and d) reveal much

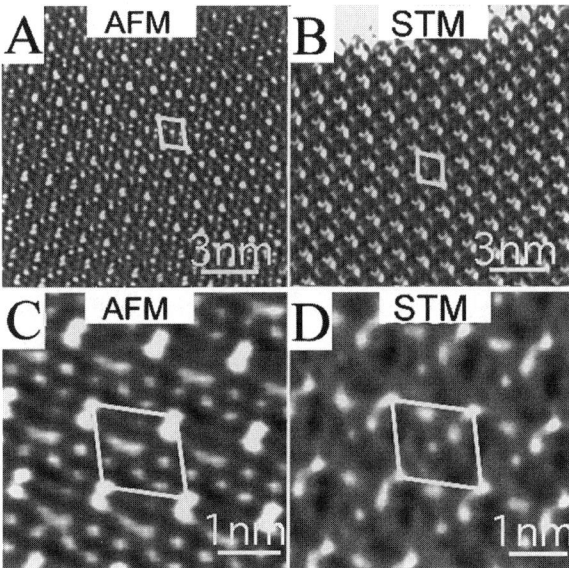

Figure 5.25. (a and c) AFM and (b and d) STM for a 4, 4′ bipyridine (44BPY) in 1 mM 44BPY + 0.1 M NaClO4 at rest potential [31].

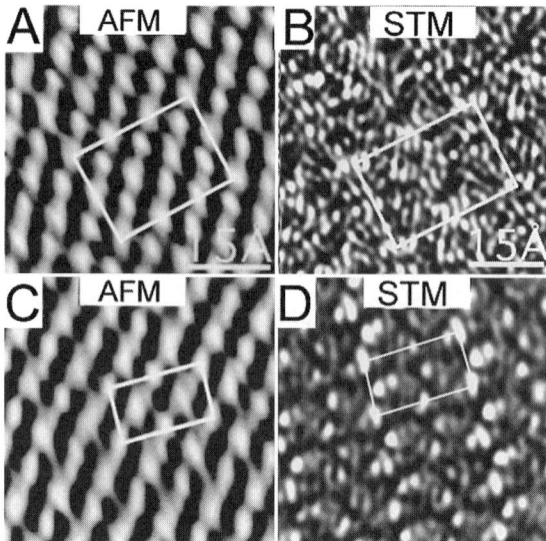

Figure 5.26. Comparison of (a and c) AFM and (b and d) STM for xanthine (a and b) and guanine (c and d) monolayers on graphite grown from 0.1 M NaCl.

more complex blob-like features. Since the dimensions of the blobs are much smaller than the dimensions of the molecules, these blobs reflect the internal electronic density of states of the molecules.

It is also interesting to compare the xanthine (a and b) and guanine (c and d) images. The AFM images of xanthine and guanine are nearly identical, which is expected because the two molecules have almost the same structure (the NH_2 group in guanine is replaced by the O in xanthine, see Scheme 1). However, their STM images are significantly different, which reflect the difference in their electronic states (they are chemically different). This comparison demonstrates that molecules with similar structures may be identified by STM. Although STM seems to have higher resolution than AFM for the small organic molecules studied in our lab, it is more prone to misinterpretation than AFM. This is because STM often resolves more complicated patterns, and it probes the electronic states of the molecules rather than geometrical positions of each atom. Misinterpretation can also occur in the case of AFM because of unknown tip geometry and strong tip-sample interactions that may change both the sample and the tip geometry during an experiment. But the lattice constants of a periodic structure observed in AFM are often not particularly sensitive to detailed tip-sample interactions or tip geometry. It has been shown that periodic surface lattices can be observed with a blunt tip that has many atoms in contact with the sample surface [113]. As the tip scans across a surface, the periodic modulation in the force on the tip reflects the periodicity of the sample surface lattice. So, the risk of misinterpreting an AFM image can be reduced if we only rely on periodic structures. The lattice constants of the periodic structure are

5.5.3 STM Imaging Mechanisms

As already discussed, STM is capable of revealing the internal features of molecules. These features clearly contain important electronic and structural information about the molecules, but they have not been widely used to extract such information because of the lack of a complete understanding of STM imaging mechanism. Shortly after its invention, STM was successfully applied to imaging organic molecules, including large biological molecules. Two important questions have been repeatedly asked since then. How can electrons tunnel through large "insulating" organic molecules? Also, if electrons can tunnel through the molecules, how does the tunneling current that probes sample LDOS near the substrate Fermi level reflect the image of the sample molecules whose molecular energy levels are usually far away from the Fermi level?

In order to answer these questions many models have been proposed [52, 114–128]. For inert gas atoms adsorbed on metal surfaces, it has been shown that the images are due to a small mixture of the unoccupied atomic states and the substrate states near the Fermi level [115]. For alkanes on graphite, Spong et al. suggested that adsorbate-induced local variation in the work function is responsible for the STM images [52]. For large molecules, Yuan et al. proposed that the measured current is due to an ion-assisted conduction in the water layer that covers the molecules [124]. Tang et al. proposed another non-tunneling mechanism based on field-induced conduction mediated by impurity [125]. Joachim et al. [116] and Fisher et al. [120] showed that resonant tunneling through the tail of the molecular states may be enough to interpret the current measured in STM. Kuznetsov et al. proposed that a strong coupling between the electronic states of a large molecule and the environmental fluctuation can bring a molecular state close to the substrate Fermi level for resonant tunneling to take place [121]. Lindsay et al. showed that a tip force applied on the imaged molecules can significantly shift the molecular states thus inducing resonant tunneling [114]. In the electrochemical environment, Schmickler [122] developed a detailed theory that takes into account both the adsorbate–substrate interaction using Newen's resonant state approach and adsorbate–solvent coupling [122]. Repphun and Halbritter proposed a tunneling mechanism through localized states at solid–liquid interfaces [127].

In order to interpret STM images accurately, it is important to calculate the electronic structure of the molecule and the substrate. Using Hückel molecular orbital theory, Chiang et al. calculated STM images of several organic molecules on metal surfaces [126]. While the Hückel theory is crude by the standard of modern quantum chemistry, the method enjoys the advantage of its ability to be carried out on a personal computer. Joachim and Sautet developed an electron scattering quantum chemical procedure to calculate STM images successfully for a number of molecules [116–117]. Ou-Yang et al. developed a formulism for the STM current using the time-dependent first-order perturbation theory for a two-Hamiltonian system and applied it to adenine on graphite [119]. Whangbo et al. calculated STM images of several organic molecules using extended Hückel tight-binding method [82] Fisher et al. carried out a first-principle calculation of benzene on graphite [120]. These computational approaches demonstrate the success as well as difficulties of each approach.

To understand imaging mechanism heterocyclic molecules presented in the previous sections, we have undertaken a combined experimental and theoretical study of guanine on graphite [128]. Some of the results are reviewed here.

5.5.3.1 Role of Tip-Molecule Interaction in STM Imaging Because the tip force applied to the imaged molecules can significantly change the electronic states of the molecules and therefore the STM images, it is of great importance to determine the tip–molecule interaction. In air or in UHV, the interactions have been directly measured by Dürig et al. [129] and studied by simultaneous STM and AFM in large organic molecules [130–131]. However these techniques cannot be easily extended to measuring the interaction in the electrochemical STM. We extracted the tip–molecule interaction of guanine on graphite in air and in the electrochemical environment by determining the deformation caused by the tip on the organic monolayers on graphite. The details of the procedure were published elsewhere [128]. The average force on each molecule as a function of the STM tunneling current with a fixed tip-substrate bias voltage is plotted in Figure 5.27. The filled and open circles are data obtained in air and in 0.1 M NaCl, respectively. The force in air is much greater than that in the aqueous solution, which is probably due to contamination or capillary force in air. In both air and solution, the force decreases as the tunneling current decreases as expected. But even at the smallest tunneling current, 3 pA, the force is still in the order of a few nN per molecule. We also determined the force as a function of tip-substrate bias voltage with a fixed tunneling current, which shows a few nN force at even a rather large tip-substrate bias (Fig. 5.28).

5.5.3.2 Theoretical Calculations We performed a first-principle calculation of guanine on graphite using the local density approximation (LDA) of the density functional theory [132–135]. The electrons in the chemically active outer shell of each atom in guanine are considered while the inert core electrons (1 S shell for C, N, and O) are replaced by pseudopotentials [136] that have been extensively tested under several atomic configurations [137]. The graphite substrate was represented by a jellium slab with appropriate charge density. The electronic interaction of the entire guanine + graphite system was treated self-consistently. The STM image was calculated by determining the LDOS at the Fermi level as a function of force on the molecule. Without the tip force, we found that the substrate Fermi level is at 2.3 eV above the HOMO and 1.7 eV below the LUMO, which means a greater contribution to the STM image from LUMO than from HOMO. However, the LDOS at the Fermi level is very small without force. Increasing the tip force to 2.5 nN, the LDOS increases

5.5 TECHNICAL ASPECTS OF ELECTROCHEMICAL STM AND AFM

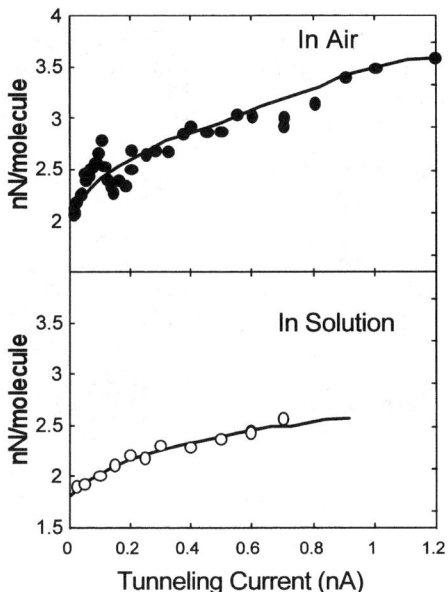

Figure 5.27. Average force per guanine molecule as a function of tunneling current with the tip-substrate bias fixed at -0.1 V in air (top) and in 0.1 M NaCl (bottom). The solid curves are guide for eyes [122].

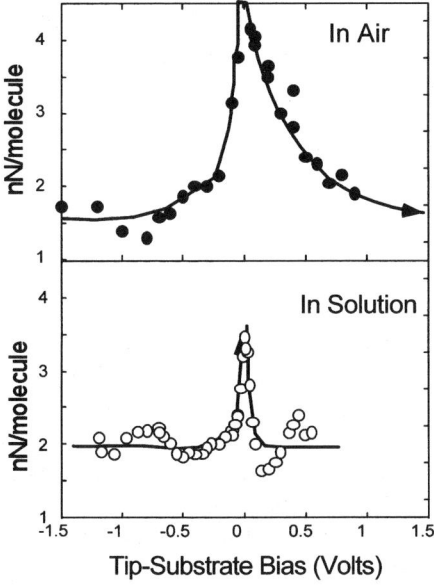

Figure 5.28. Average force per guanine molecule as a function of tip-substrate bias with the tunneling current fixed at 0.1 nA in air (top) and in 0.1 M NaCl (bottom). The solid curves are guide for eyes [122].

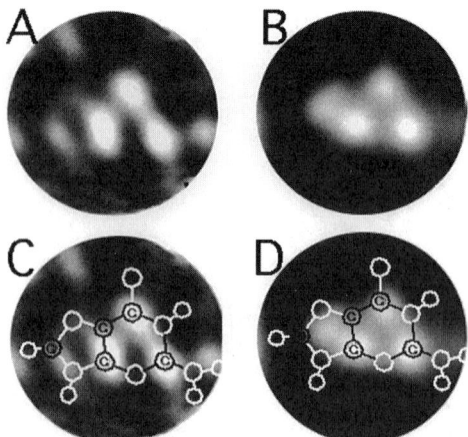

Figure 5.29. (a and c) Measured and (b and d) calculated STM images of guanine on graphite. Ball-stick model of guanine is superimposed on the (c) measure and (d) calculated images [122].

significantly because the LUMO is shifted downward by 0.4 eV. This can be understood by the following consideration. When the molecule is being pushed towards the substrate, the electrons in the molecule are affected more by the negative potential of the substrate, resulting in a downward shift of the energy levels of molecule. This brings the LUMO closer to the substrate Fermi level, therefore leading to more tunneling current. The calculated STM image agrees reasonably well with the experimental STM image (Fig. 5.29).

This study demonstrates the importance of the tip force in the STM image of guanine on graphite [114]. For large and more flexible molecules, we expect the effect of tip force to be stronger and self-consistent calculations to become more important for understanding STM images.

5.6 SUMMARY

We have shown that important phenomena, such as adsorption, nucleation, growth, structural phase transitions, and electrochemical reactions, in organic monolayers at electrode-electrolyte interfaces can be studied by in situ STM and AFM using selected examples on heterocyclic molecules at graphite-water and Au(111)-water interfaces. The number of organic molecules that have been studied by STM and AFM are expanding rapidly in recent years, and it is clear that the technique will play a greater role in understanding molecular adsorption at solid–liquid interfaces.

On graphite substrate, the heterocyclic molecules adsorb onto the interfaces and condense into monolayers via nucleation and growth processes that can be flexibly controlled by the substrate potential. The molecules pack into two-dimensional hydrogen bonded networks. Changing the solution pH changes the degree of protonation or deprotonation of the nitrogen atoms, which changes the hydrogen bonded network. By increasing the potential, a phase transition corresponding to the change in the orientation of the adsorbate lattices with respect to the underlying graphite lattice was observed. At even higher potentials, electron transfer reactions take place starting preferentially from defects in the monolayers.

On Au(111), the molecules lie flat on the surface at negative potentials and stand vertically with nitrogen atoms facing the surface at positive potentials. The vertically standing molecules form polymer-like chains in which the individual molecules stack with a repeat distance of ~3.5 Å. The chains can exist either in a disordered phase in which they orient randomly or an ordered phase in which they pack closely in parallel and align along one of the three equivalent Au(111) lattice directions. For both 22BPY and phenanthroline, the chains are in the disordered phase at low potentials and transform into the ordered phase if the substrate reaches a critical potential. This disorder–order phase transition is driven by a potential dependent attractive force between the chains. The completely different behaviors for the molecules on graphite and Au surfaces are because of the different adsorbate–adsorbate and adsorbate–substrate interactions on the two surfaces. On graphite, the molecules lie flat on the surface and interact with the substrate via weak π-electrons, which favors the formation of hydrogen bonded networks between the adsorbed molecules. On Au or electrodeposited Cu surfaces, the σ-electron lone pair electrons are strongly attracted to the metal surfaces, especially at positive charges, which enables the molecules to overcome the tendency of forming hydrogen bonded networks.

For the heterocyclic molecules on graphite, we found that STM can resolve the internal features of each molecule while AFM usually reveals each molecule only as a blob. To understand the STM images, self-consistent calculations of molecular and substrate electronic states with the consideration of tip-sample interaction are crucial to achieve understanding of STM images.

ACKNOWLEDGMENTS

Financial support is acknowledged through grants from the Cottrell College Science Award (CC3608), Petroleum Research Fund (28163-GB7), AFSOR (F49620-96-1-0346), and NIH (GM-08205). We would like to thank X. Wang and Th. Wandlowski for critical comments and discussions, C. Z. Li, F. Cunha, and Q. Jing, and B. Doung and John D'Agnese for help in the lab.

REFERENCES

1. A. Ulman, *Introduction to Ultrathin Organic Films* (Academic Press, Boston, 1991).

2. G. E. Collins, V. S. Williams, L-K. Chau, K. W. Nebesny, C. England, P. A. Lee, T. Lowe, Q. Fernando, N. R. Armstrong, *Synthetic Metals* **54,** 351 (1993).
3. G. Roberts, *Langmuir–Blodgett Films.* (Plenum Press, New York, 1990).
4. L. H. Dubois, R. G. Nuzzo, *Annu. Rev. Phys. Chem.* **43,** 437 (1992).
5. Cl. Buess-Herman, *J. Electroanal. Chem.* **186,** 27 (1985).
6. R. de Levie, *Chem. Rev.* **88,** 599 (1988).
7. A. T. Hubbard, *Chem. Rev.* **88,** 633 (1988).
8. J. Lipkowski, P. N. Ross, *Adsorption of Molecules at Metal Electrodes* (VCH, New York, 1992).
9. C. Shannon, D. G. Frank, A. T. Hubbard, *Annu. Rev. Phys. Chem.* **42,** 393 (1991).
10. H. Meier, *Organic Semiconductors* (Verlag Chemic, Weinheim H, 1974).
11. M. P. Soriaga, *Prog. Surf. Sci.* **39,** 525 (1992).
12. M. F. Toney, O. R. Melroy, in *In-Situ Studies of Electrochemical Interfaces*, H. D. Abruna, ed. (VCH Verlag Chemical, Berlin, 1991); M. G. Samant, M. F. Toney, G. L. Borges, L. Blum, O. R. Melroy, *Surf. Sci.* **92,** 220 (1988); B. Ocko, J. Wang, A. Davenport, H. Isaacs, *Phys. Rev. Lett.* **65,** 1466 (1990).
13. H. Y. Liu, F. R. F. Fan, C. W. Lin, A. Bard, *J. Am. Chem. Soc.* **108,** 3838 (1986).
14. R. Sonnenfeld, P. K. Hansma, *Science,* **232,** 211 (1986).
15. S. M. Lindsay, B. Barris, *J. Vac. Sci. Technol.* **A6,** 544 (1988).
16. S. M. Lindsay, T. Thundat, L. Nagahara, U. Knipping, R. L. Rill, *Science* **244,** 1063 (1989).
17. L. A. Nagahara, T. Thundat, P. I. Oden, S. M. Lindsay, R. L. Rill, *Ultramicroscopy* **33,** 107 (1990).
18. R. Srinivasan, J. C. Murphy, R. FainChtein. *J. Electroanal. Chem.* **312,** 293 (1991).
19. R. Srinivasan, J. C. Murphy, *Ultramicroscopy* **42–44,** 453 (1992).
20. R. Srinivasan, P. Gopalan, *J. Phys. Chem.* **97,** 8770 (1993).
21. M. Holzle, T. Wandlowski, D. M. Kolb, *Surf. Sci.* **335,** 281 (1995).
22. Th. Wandlowski, Th. Dretschkow, A. S. Dakkouri, *Langmuir* **13,** 2843 (1997).
23. N. J. Tao, J. A. DeRose, S. M. Lindsay, *J. Phys. Chem.* **97,** 910 (1993).
24. N. J. Tao, Z. Shi, *Suf. Sci. Lett.* **301,** L217 (1994).
25. N. J. Tao, Z. Shi, *J. Phys. Chem.* **98,** 1464 (1994).
26. N. J. Tao, Z. Shi, *J. Phys. Chem.* **98,** 7422 (1994).
27. Th. Wandlowski, D. Lampner, S. M. Lindsay, *J. Electroanal. Chem.* **404,** 215 (1996).
28. N. J. Tao, Z. Shi, *Suf. Sci. Lett.* **321,** L149 (1994).
29. J. H. Schott, C. R. Arana, H. D. Abruna, H. H. Petoch, G. M. Elliot, H. S. White, *J. Phys. Chem.* **96,** 5222 (1992).
30. N. J. Tao, C. Z. Li, F. Cunha, Q. Jing, Scanning Microscopy, 1997, in press.
31. F. Cunha, N. J. Tao, *Phys. Rev. Lett.* **75,** 2376 (1995); F. Cunha, N. J. Tao, X. W. Wang, O. Jin, B. Duong, D'Agnese, *Langmuir*, **12,** 6410 (1996).
32. F. Cunha, Q. Jin, N. J. Tao, C. Z. Li, *Suf. Sci.* **389,** 19 (1997).
33. E. Bunge, R. J. Nichols, B. Roelfs, H. Meyer, H. Baumgartel, *Langmuir*, **12,** 3060 (1996).

34. A. Dakkaouri, D. M. Kolb, R. Edelstein-Shima, D. Mandler, *Langmuir*, **12,** 2849 (1996); J. Pan, N. J. Tao, S. M. Lindsay, *Langmuir*, **9,** 1556 (1993); F. P. Zamborini, R. M. Crooks, *Langmuir*, **13,** 122 (1997).
35. K. M. Richard, A. A. Gewirth, *J. Phys. Chem.* **99,** 12288 (1995).
36. N. J. Tao, G. Cardenas, F. Cunha, Z. Shi Z, *Langmuir*, **11,** 4445 (1995).
37. M. Kunitake, N. Batina, K. Itaya, *Langmuir*, **11,** 2337 (1995).
38. N. J. Tao, *Phys. Rev. Lett.* **76,** 4066 (1996).
39. K. Ogaki, N. Batina, M. Kunitake, K. Itaya, *J. Phys. Chem.* **100,** 7185 (1996).
40. S. L. Yau, Y. G. Kim, K. Itaya, *J. Am. Chem. Soc.* **118,** 7795 (1996).
41. D. K. Luttrull, J. Graham, J. A. DeRose, D. Gust, T. A. Moore, S. M. Lindsay, *Langmuir*, **8,** 765 (1992); M. Hara, H. Sasabe, W. Knoll, *Thin Solid Films* **273,** 66 (1996).
42. W. M. Heckl, K. M. R. Kallury, M. Thompson, C. Gerber, H. J. K. Horber, G. Binnig, *Langmuir*, **5,** 1433 (1989).
43. C. A. Widrig, C. A. Alves, M. D. Porter, *J. Am. Chem. Soc.* **113,** 2805 (1991); Y. T. Kim, A. J. Bard, *Langmuir* **8,** 1096 (1992); C. Schonenberger, J. Jorritsma, J. A. M. Sondag-Huethorst, L. C. J. Fokkink, *J. Phys. Chem.* **99,** 3259 (1995).
44. Y.-T. Kim, R. L. McCarley, A. J. Bard, *J. Phys. Chem.* **96,** 7416 (1992).
45. S. R. Snyder, H. S. White, S. López, H. D. Abruña, *J. Am. Chem. Soc.* **112,** 1333 (1990).
46. J. J. Breen, G. W. Flynn, *J. Phys. Chem.* **96,** 6825 (1992).
47. G. Liu, P. Penter, C. E. D. Chidsey, D. F. Ogletree, P. Eisenberger, M. Salmeron, *J. Chem. Phys.* **101,** 4301 (1994).
48. R. Viswanathan, J. A. Zasadzinski, D. K. Schwartz, *Science*, **261,** 449 (1993); J. Y. Josefowicz, N. C. Maliszewskyj, S. H. J. Idzak, P. A. Heiney, J. P. McCauley, Jr., A. B. Smith III, *Science*, **260,** 323 (1993).
49. P. H. Lippel, R. J. Wilson, M. D. Miller, Ch. Wöll, S. Chiang, *Phys. Rev. Lett.* **62,** 171 (1989); V. M. Hallmark, S. Chiang, K.-P. Meinhardt, K. Hafner, *Phys. Rev. Lett.* **70,** 3740 (1993); G. E. Poirier, E. D. Pylant, *Science*, **272,** 1145 (1996).
50. S. J. Stranick, M. M. Kamna, P. S. Weiss, *Surf. Sci.* **338,** 41 (1995). S. R. Forrest, P. E. Burrows, E. I. Haskal, F. F. So, *Phys. Rev. B* **49,** 11309 (1994); C. Ludwig, J. Gompf, J. Petersen, R. Strohmaier, W. Eisenmenger, *Z. Phys. B* **93,** 365 (1994).
51. D. Smith, H. Horber, G. Binnig, H. Nejoh, *Nature*, **344,** 641 (1990).
52. J. Spong, H. Mizes, L. LaComb, M. Dovek, J. Frommer, J. Foster, *Nature*, **338,** 137 (1989).
53. W. Mizutani, M. Shigano, Y. Sakakibara, K. Kajimura, M. Ono, S. Tanishima, K. Ohno, N. Toshima, *J. Vac. Sci. Technol.* **A8,** 675 (1990); W. Mizutani, M. Shigano, M. Ono, K. Kajimura, *Appl. Phys. Lett.* **56,** 1974 (1990).
54. N. A. Clark, D. M. Walba, P. D. Beale, *Phys. Rev. Lett.* **70,** 607 (1993).
55. T. McMaster, H. Carr, M. Miles, P. Cairns, V. Morris, *J. Vac. Sci. Technol.* **A8,** 672 (1990).
56. G. Mcgonigal, R. Bernhardt, D. Thompson, *Appl. Phys. Lett.* **57,** 28 (1990).

57. D. M. Cyr, B. Venkataraman, G. W. Flynn, A. Black, G. M. Whitesides, *J. Phys. Chem.* **100,** 13747 (1996).

58. H. Yamada, S. Akamine, C. F. Quate, *Ultramicroscopy*, **42–44,** 1044 (1992); M. Hara, Y. Iwakabe, K. Tochigi, H. Sasabe, A. F. Garito, A. Yamada, *Nature*, **344,** 228 (1990).

59. J. Bucher, H. Roeder, K. Kern, *Surf. Sci.* **289,** 370 (1993).

60. J. P. Rabe, S. Buchholz, *Science*, **253,** 442 (1991); J. P. Rabe, S. Buchholz, *Phys. Rev. Lett.* **66,** 2096 (1991); A. Stabel, R. Heinz, J. P. Rabe, G. Wegner, F. C. De Schryver, D. Corens, W. Dehaen, C. Suling, *J. Phys. Chem.* **99,** 8690 (1995).

61. D. L. Patrick, T. P. Beebe, *Langmuir*, **10,** 298 (1994).

62. A. Wawkuschewski, H.-J. Cantow, S. N. Magonov, *Langmuir*, **9,** 2778 (1993); A. Wawkuschewski, H.-J. Cantow, S. N. Magonov, M. Moller, W. Liang, M. Whangbo, *Adv. Mater.* **5,** 821 (1993).

63. J. Frommer, *Angew. Chem. Int. Ed. Engl.* **31,** 1298 (1992).

64. S. Chiang, in *Scanning Tunneling Microscopy 1*; H. Guntherodt, R. Weisendanger, eds. (Springer-Verlag: Berlin, 1991), Chapter 7.

65. D. M. Cyr, B. Venhataraman, G. W. Flynn, *Chem. Mater.* **8,** 1600 (1996).

66. M. J. Allen, M. Balooch, S. Subbiah, R. J. Tench, R. Balhorn, W. Siekhaus, *Ultramicroscopy*, **42–44,** 1049 (1992).

67. W. M. Heckl, D. P. E. Smith, G. Binng, H. Klagges, T. W. Hansch, J. Maddocks, *Proc. Natl. Acad. Sci. USA* **88,** 8003 (1991); W. M. Heckl, *Thin Solid Films*, **210–211,** 640 (1992).

68. J. C. Poler, R. M. Zimmermann, E. C. Cox, *Langmuir*, **11,** 2689 (1995).

69. C. C. Williams and H. K. Wichramasinghe, *J. Vac. Sci. Technol. B* **9,** 537 (1991).

70. T. A. Witten and L. M. Sander, *Phys. Rev. Lett.* **47,** 1400 (1982).

71. B. Marner and W. Schmickler, *J. Phys. Chem.* **93,** 3186 (1989).

72. W. Lorenz, *Z. Elekstrochem.* **62,** 192 (1958).

73. A. N. Frumkin, B. B. Damaskin, *Dokl. Akad. Nauk SSSR*, **129,** 862 (1959).

74. W. Saenger, *Principles of Nucleic Acid Structure* (Springer, New York, 1984).

75. U. Thewalt, C. E. Bugg, R. E. Marsh, *Acta Cryst. B* **27,** 2358 (1971).

76. M. H. Saffarian, R. Sridhara, R. de Levie, *J. Electroanal. Chem.* **218,** 273 (1987).

77. R. de Levie, T. Wandlowski, *J. Electroanal. Chem.* **366,** 265 (1994).

78. N. J. Tao, J. Pan, Y. Li, P. I. Oden, J. A. DeRose, S. M. Lindsay, *Surf. Sci.* **271,** L338 (1992).

79. X. Gao, M. J. Weaver, *J. Am. Chem. Soc.* **114,** 8544 (1992); W. Haiss, J. K. Sass, X. Gao, M. J. Weaver, *Surf. Sci.* **274,** L593 (1992).

80. J. Tersoff, D. R. Hamman, *Phys. Rev. B* **31,** 805 (1985).

81. J. C. Chen, *Introduction to Scanning Tunneling Microscopy* (Oxford University Press, New York, 1993).

82. D.-K. Seo, J. Ren, M.-H. Whangbo, *Surf. Sci.* **370,** 252 (1997).

83. C. G. Shaw, S. C. Fain, Jr., M. D. Chinn, *Phys. Rev. Lett.* **41,** 955 (1978).

84. A. D. Novaco, T. P. McTague, *Phys. Rev. Lett.* **38,** 1286 (1977).

85. G. Dryhurst, in *Electrochemistry of Biological Molecules* (Academic Press, New York, 1977).
86. G. Dryhurst, G. F. Pace, *J. Electrochem. Soc.* **117,** 1259 (1970).
87. J. Perderau, J. P. Biberian, G. E. Rhead, *J. Phys.* **F4,** 798 (1974).
88. H. Melle, E. Menzel, *Z. Naturforsch,* **33a,** 282 (1978).
89. U. Harten, A. M. Lahee, J. P. Toennies, Ch. Wöll, *Phys. Rev. Lett.* **54,** 2619 (1985).
90. K. Takayanagi, K. Yagi, *Trans. Jpn. Inst. Met.* **24,** 337 (1983).
91. Ch. Wöll, S. Chiang, R. J. Wilson, P. H. Lippel, *Phys. Rev. B* **39,** 7988 (1989).
92. J. V. Barth, H. Brune, G. Ertl, R. J. Behm, *Phys. Rev. B* **42,** 9307 (1990).
93. N. J. Tao, S. M. Lindsay, *J. Appl. Phys.* **70,** 5141 (1991).
94. X. Gao, M. J. Weaver, *J. Chem. Phys.* **95,** 6993 (1991).
95. N. J. Tao, S. M. Lindsay, *Surf. Sci.* **274,** L454 (1992).
96. P. I. Oden, N. J. Tao, S. M. Lindsay, *J. Vac. Sci. Tec.* **11,** 137 (1993).
97. J. Wang, A. J. Davenport, H. S. Isaacs, B. M. Ocko, *Science,* **255,** 1416 (1992); J. Wang, B. M. Ocko, J. Davenport, H. S. Issacs, *Phys. Rev. B* **46,** 10321 (1992).
98. Th. Wandlowski, B. M. Ocko, O. M. Magnussen, S. Wu, J. Lipkowski, *J. Electroanal. Chem.* **409,** 155 (1996).
99. D. Yang, D. Bizzotto, J. Lipkowski, B. Pettinger, S. Mirwald, *J. Phys. Chem.* **98,** 7083 (1994).
100. P. G. de Gennes, J. Prost, *The Physics of Liquid Crystals* (Oxford University Press, Oxford, 1993).
101. C. A. Hunter, J. K. M. Sanders, *J. Am. Chem. Soc.* **112,** 5525 (1990).
102. T. A. Jung, R. R. Schlitter, J. K. Gimzewski, H. Tang, H., C. Joachim, *Science,* **271,** 181 (1996).
103. S. L. Yau, Y.-G. Kim, K. Itaya, *J. Phys. Chem. B* **101,** 3547 (1997).
104. M. Kunitake, N. Batina, K. Itaya, *Langmuir,* **11,** 2337 (1995).
105. K. Ogaki, N. Batina, M. Kunitake, K. Itaya, *J. Phys. Chem.* **100,** 7185 (1996).
106. C. Z. Li, N. J. Tao, unpublished.
107. C. R. Clemmer, T. P. Beebe, *Science,* **251,** 640 (1990).
108. P. I. Oden, T. Thundat, T., L. A. Nagahara, S. M. Lindsay, *Surf. Sci.* **254,** L454 (1991).
109. J. A. DeRose, D. B. Lampner, S. M. Lindsay, N. J. Tao, *J. Vac. Sci. Tec.* **11,** 776 (1993).
110. K. Itaya, S. Sugawara, K. Sashikata, N. Furuya, *J. Vac. Sci. Tec. A* **8,** 515 (1990).
111. J. P. Ibe, P. P. Bey, S. L. Brandow, R. A. Brizzolara, N. A. Burnham, D. P. DiLella, K. P. Lee, C. R. K. Marrian, R. J. Colton, *J. Vac. Sci. Technol. A* **8,** 3570 (1990).
112. L. A. Nagahara, T. Thundat, S. M. Lindsay, *Rev. Sci. Instrum.* **60,** 3128 (1989).
113. N. A. Burnham, R. J. Colton, in *Scanning Tunneling Microscopy, Theory, Techniques and Applications,* D. Bonnel, eds. (VCH, New York, 1992), Chapter 7; F. Ohensorge, G. Binnig, *Science,* **260,** 1451 (1993).
114. S. M. Lindsay, O. F. Sankey, Y. Li, C. Herbst, A. Rupprecht, *J. Phys. Chem.* **94,** 4655 (1990).

115. N. D. Lang, *IBM J. Red. Dev.* **30,** 374 (1986); D. M. Eigler, P. S. Weiss, E. K. Schweizer, N. D. Lang, *Phys. Rev. Lett.* **56,** 1189 (1991).
116. C. Joachim, P. Sautet, in *Scanning Tunneling Microscope and Related Methods* R. J. Behm et al., eds., (Kluwer Academic, Netherlands, 1990).
117. P. Sautet, M.-L. Bocquet, *Surf. Sci. Lett.* **304,** L445 (1994).
118. V. Mujica, M. Kemp, M. A. Ratner, *J. Chem. Phys.* **101,** 6849 (1994).
119. H. Ou-Yang, B. Kallebring, R. A. Marcus, *J. Chem. Phys.* **98,** 7565 (1993).
120. A. J. Fisher, P. E. Blöchl, *Phys. Rev. Lett.* **70,** 3263 (1993).
121. A. M. Kuznetsov, P. Sommer-Larsen, J. Ulstrup, *Surf. Sci.* **275,** 52 (1992).
122. W. Schimickler, *J. Electroanal. Chem.* **296,** 283 (1990); W. Schmickler, C. Widrig, *J. Electroanal. Chem.* **336,** 213 (1992).
123. A. K. Mishra, S. K. Rangarajan, *J. Molecular Structure*, **361,** 101 (1996).
124. J. Y. Yuan, Z. Shao, C. Cao, *Phys. Rev. Lett.* **67,** 863 (1991).
125. S. L. Tang, A. J. McGhie, A. Suna, *Phys. Rev.* **47,** 3850 (1993).
126. D. N. Futaba, S. Chiang, *J. Vac. Sci. Technol. A* **15,** 1295 (1997); V. M. Hallmark, S. Chiang, *Surf. Sci.* **329,** 255 (1995).
127. G. Repphun, J. Halbritter, *J. Vac. Sci. Technol. A* **13,** 1693 (1995).
128. X. W. Wang, N. J. Tao, F. Cunha, *J. Chem. Phys.* **105,** 3747 (1996).
129. R. Dürig, O. Zuger, B. Michel, L. Haussling, H. Ringsdorf, *J. Chem. Phys.* **48,** 1711 (1993).
130. M. Specht, F. Ohnesorge, W. M. Heckl, *Surf. Sci.* **257,** L653 (1991).
131. W. Mizutani, D. Anselmetti, B. Michel, in *Computations for the Nano-Scale*, P. E. Blochl, C. Joachim, A. J. Fisher, eds. (Kluwer Academic Publishers, The Netherlands, 1993) p. 43–48.
132. W. Kohn, L. J. Sham, *Phys. Rev. A* **140,** 1133 (1965).
133. R. O. Jones, O. Gunnarssorn, *Rev. Mod. Phys.* **61,** 689 (1989).
134. S. G. Louie, K.-M. Ho, M. L. Cohen, *Phys. Rev. B* **19,** 1774 (1979).
135. X. Wang, *Surf. Sci.* **322,** 51 (1994).
136. D. R. Hamann, M. Schlüter, C. Chiang, *Phys. Rev. Lett.* **43,** 1494 (1979).
137. S. Fahy, X. W. Wang, and S. G. Louie, *Phys. Rev. B* **42,** 3503 (1990).

6

SCANNING PROBE MICROSCOPY OF ORGANIC THIN FILMS AT ELECTRODE SURFACES

J.-B. D. Green
Naval Research Laboratory, 4555, Overlook Avenue, Washington, DC 20375-5342

C. A. McDermott, M. T. McDermott
Department of Chemistry, University of Alberta, Edmonton, Alberta T6G 2G2, Canada

M. D. Porter
Ames Laboratory-USDOE, Department of Chemistry and Microanalytical Instrumentation Center, Iowa State University, Ames, IA 50011

6.1 SCOPE

A long-standing ambition of electrochemists is the control of reactivity at electrode surfaces [1–6]. At issue is the ability to tailor the architecture of the electrode surface in ways that will optimize the rates and/or selectivities of heterogeneous electron-transfer processes of relevance to technologies such as catalysis, corrosion, energy production and storage, photoelectrochemistry, chemical analysis, and synthesis. In many instances, approaches to modify reactivity exploit an alteration of the molecular architecture of the electrode surface via the deposition of a thin organic film [7,8]. These approaches, however, often contain a high degree of empiricism. That is, a projection of the performance

---

Imaging of Surfaces and Interfaces (Frontiers of Electrochemistry, Volume 5).
Edited by Jacek Lipkowski and Philip N. Ross.
ISBN 0-471-24672-7 © 1999 Wiley-VCH, Inc.

of a modified electrode often assumes a direct translation of the reactivity of a precursor in the three-dimensional environment of solutions to the lower dimensionality for the corresponding immobilized species.

Though the preceding assumption is useful as a starting point in the creation of novel electrode materials, the lack of a detailed understanding of how immobilization can affect reactivity limits the ability of electrochemists to proceed predictively in subsequent design steps for optimization of performance. It is therefore important to develop insights into the fundamental issues that govern the chemistry and physics of electrochemical interfaces by asking questions such as:

1. What is the chemical identity of the immobilized species?
2. What is the mode of attachment, spatial orientation, coverage, and two-dimensional arrangement of the immobilized species?
3. How do the interactions between adsorbate, solvent, supporting electrolyte, and electrode influence the reactivity and stability of the immobilized species?
4. How do electrochemical variables such as applied voltage affect the structure and reactivity of the immobilized species?

These questions are also relevant to numerous other areas of surface and materials science, such as adhesion, biocompatibility, and colloidal stabilization. For example, factors that facilitate the ionization of acidic functionalities of organic materials are crucial to the stabilization of polymer dispersions and colloidal solutions [9]. Similarly, fundamental studies of electrocatalysis probe how adsorbate-substrate interactions perturb the electronic structure of an adsorbate and the subsequent impact of such perturbations on electron transfer kinetics [1–6]. Seeking answers to those questions is therefore one of the most important and yet difficult challenges in interfacial science today.

To address the questions, it is necessary to develop surface analytical techniques that probe the details of the electrochemical interface at a molecular level. Ideally, such a method would produce information that would identify the types, quantities, spatial orientation, two-dimensional arrangement, and microscopic reactivities of adsorbates. In view of these requirements and the general complexity and diversity of electrochemical interfaces, it is not surprising that the ideal surface probe is nonexistent. Thus, the application of an integrated approach that combines a carefully selected set of surface analysis probes represents a more tractable strategy. Such approaches generally employ techniques that can be divided into two broad groups. The first includes techniques that characterize interfaces at a macroscopic level [10], yielding details about the general (i.e., average) population of the immobilized species. For example, techniques such as X-ray photoelectron spectroscopy (XPS) have been used to probe the elemental composition of electrode surfaces, providing semi-quantitative data at submonolayer coverages; approaches have also been devised for the identification of chemical functionalities. However, details concerning the roles of solvent and supporting electrolyte may be compromised by the transfer of an electrode to the ultra high vacuum environment of the XPS sample chamber. Vibrational spectroscopic techniques (e.g., infrared and Raman spectroscopies) open avenues to in situ characterizations as well as provide data directly attributable to the chemical identity and spatial orientation of an organic adsorbate [10]. We note that some vibrational spectroscopies are unable to detect adsorbates at levels as low as XPS and other types of surface analytical techniques, which may prove a hindrance in some applications.

The second class of techniques includes those that probe interfaces at nanometer length scales. These techniques comprise in large part the many forms of scanning tunneling (STM) [11,12] and scanning force (SFM) microscopy [13,14], both of which are the focus of this chapter. These techniques provide a means to explore the two-dimensional spatial arrangements of adsorbates at molecular length scales, which when coupled with macroscopic descriptions, yield a more complete description of the interfacial architecture.

As part of the overall theme of this monograph, this chapter examines the application of STM and SFM as probes of the two-dimensional architecture of organic films employed to modify electrode surfaces. For this purpose, we have divided the remainder of our review into two sections, one focused on STM and the other on SFM. Both sections begin with a brief discussion of the instrumentation and imaging mechanisms for each technique. The next part of each section highlights recent examples of characterizations of organic monolayers and polymeric films. These examples, though by no means exhausting the many applications of STM and SFM to electrochemistry [12], were selected to illustrate the broad-range utility of the techniques in this area. We note that there are many other vibrant areas in electrochemistry where scanning tunneling and scanning force microscopies have proven of value but are not included in this chapter; reviews of these approaches, which include the technique known as scanning electrochemical microscopy, have recently appeared [15]. We conclude by noting briefly the areas in electrochemistry where these techniques are likely to have a major impact.

6.2 STM CHARACTERIZATIONS OF ORGANIC THIN FILMS AT ELECTRODE SURFACES

This section presents an overview of the applications of STM to the characterization of the structure and reactivity of organic thin films at electrode surfaces. After a brief discussion of the instrumental and theoretical aspects of STM, we describe characterizations of spontaneously adsorbed monolayers and other types of ordered organic coatings of interest to electrochemists, followed by a discussion of imaging studies at polymeric films.

6.2.1 Theory and Instrumentation

There have been several in-depth discussions on both the theory and application of STM [11,16,17] as a surface characterization tool. We refer readers to the cited precedence for more exacting discussions of theory (e.g., theories of three-dimensional tunneling currents, densities of surface electronic states, and tunneling through insulating films) and instrumentation (e.g., tip displacement mechanisms and tip preparation). Our goal in this section is to present sufficient background material to facilitate the discussion of the experimental work that follows in subsequent sections.

As depicted in Figure 6.1, STM employs piezoelectric microactuators to position an atomically sharp metal tip and electrically conductive sample sur-

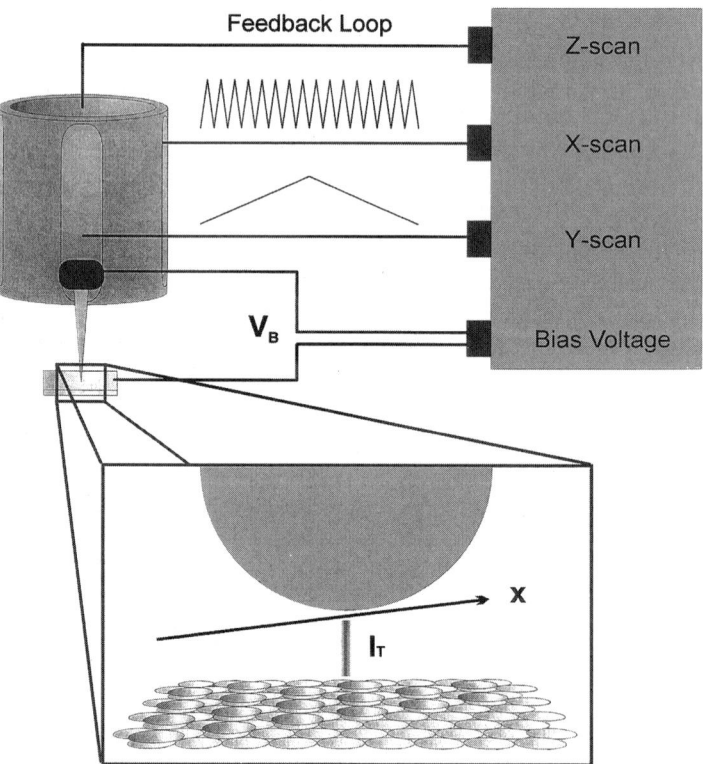

Figure 6.1. Schematic diagram of an STM. The X-scan and Y-scan electronic signals control piezoelectric actuator movement in the XY (i.e., surface plane) and the Z-electronic signal controls actuator movement along the surface normal.

6.2 STM CHARACTERIZATIONS OF ORGANIC THIN FILMS

face within a few angstroms of each other. At such separations, the electron wavefunctions of the two surfaces overlap, allowing electrons to tunnel through the insulating barrier between the two surfaces. Upon application of a bias voltage (V_t) across the tip and sample, which will be given herein as the voltage of the tip with respect to the sample, a net tunneling current flows. Since the wave function of a surface decays exponentially into the barrier, the dependence of the tunneling current exhibits a marked sensitivity to the tip-sample separation. This dependence can readily be appreciated from an approximate one-dimensional tunneling model [11]. At low values of V_t and at low temperature, the tunneling current (I_t) exhibits the following general dependence

$$I_t \approx 18 \frac{V_b k}{(10^4 \Omega) d} A_{\text{eff}} e^{-2kd} \tag{1}$$

where d (Å) is the separation distance between tip and sample, k (Å^{-1}) equals 0.513 $\Phi^{1/2}$, Φ (eV) is the mean barrier height between tip and sample, A_{eff} is the effective area of the tip ($A_{\text{eff}} = \pi(\frac{1}{2}L_{\text{eff}})^2$, L_{eff} is the effective lateral resolution ($L_{\text{eff}} \sim 2[(R_t + d)/k]^{1/2}$) of an image, and R_t is the tip radius. In general, values of Φ are based on the average work function for the two contacts, and the expression for L_{eff} is applicable only when d is smaller than R_t. Since the work functions for many metals are ~5 eV, I_t changes by an order of magnitude for a change in d of ~1 Å. It is also evident that if I_t is held at a constant value (e.g., within 2%), the variation in d is less than 0.01 Å. Thus, maps of I_t as a tip rasters across a surface are used to construct images that are viewed as contour plots of constant height above a sample surface.

There are two general STM imaging modes. Both are depicted in Figure 6.2. The first mode is known as the constant current mode, which maps the changes in d required via an electronic feedback loop to maintain a preset value of I_t as the tip moves across a sample surface. The second mode, termed the constant height mode, fixes the tip and sample Z-position, generating a map of I_t across the sample surface. In both cases, images of electronically homogeneous surfaces are often interpreted as maps of surface topography. The advantages and limitations of the two imaging modes, in terms of the type of information provided and type of sample, have been extensively discussed [16].

6.2.2 Spontaneously Adsorbed Monolayers

6.2.2.1 Introduction In recent years, there has been a virtual explosion in the number of studies that have explored the applications of monolayers formed by the chemisorption of alkanethiols (X(CH$_2$)$_n$SH) at gold electrodes [4–6]. This adsorbate-substrate combination is attractive to electrochemists as a route to creating interfaces that, in comparison with the compositional and morphological heterogeneity of polymer materials, have a well-defined composition, thickness, and spatial arrangement [7]. This system also offers the opportunity

Constant Current

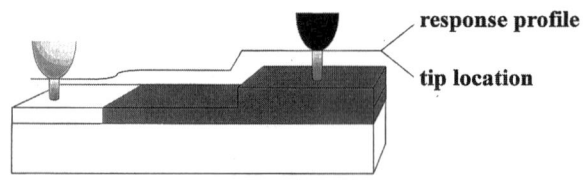

Constant Height

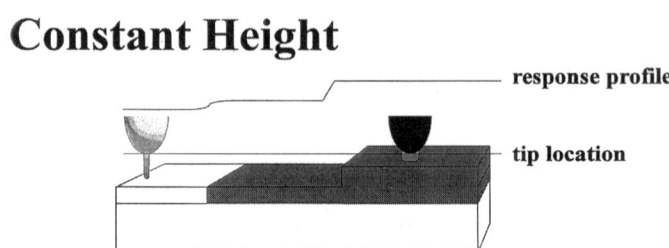

Figure 6.2. Schematic of STM modes of operation. In the constant height mode, the sample or probe is rastered in the XY plane (i.e., surface plane) without any change in the Z-position (i.e., surface normal) of either the sample or probe. In this mode the response is measured and plotted as a function of the XY coordinates. In the constant current mode, the sample-probe separation is varied to keep the measured property at a fixed setpoint value. Therefore, as the sample or probe is scanned in the XY plane, the Z-position is plotted as a function of the XY coordinates.

to tailor electrode surface structures by changing the precursor (i.e., variation of alkyl chain length and alteration of the end group) or the composition of the incubation solution (i.e., the combination of different precursors for formation of a multi-component monolayer). From a fundamental perspective, this approach has proven effective in constructing barrier films to electron transfer and ion transport [6,18,19] constituting important models for testing and extending theories of heterogeneous electron transfer and of the structure of the electrical double layer. This system has also been useful for the creation of interfaces for electroanalysis, electrocatalysis, and protein electrochemistry, and as a new route for the fabrication of microelectrode arrays [4,6,7,20].

Many of the early efforts to develop descriptions of the surface structures of thiol-derived monolayers employed macroscopic characterization techniques like infrared reflection spectroscopy and related approaches [7,8]. From such efforts, details concerning the electronic properties, surface free energies, spatial orientation, and structural imperfections of these layers are beginning to emerge. Importantly, the advent of STM and its offshoots, however, has opened the door for exploring issues related to the two-dimensional arrangement, domain size, formation mechanism, and the presence and location of structural defects at a microscopic level.

6.2 STM CHARACTERIZATIONS OF ORGANIC THIN FILMS

6.2.2.2 Thiols on Gold Early efforts aimed at characterizing this adsorbate-substrate system with STM focused on both molecular scale and topographic investigations. We first present the general findings of both types of investigations to set the stage for a more in-depth examination of the status of this research. In the case of molecular scale images, helium X-ray, and transmission electron diffraction microscopic precedents revealed that these layers existed predominately as a $(\sqrt{3} \times \sqrt{3})R30°$ adlayer superimposed on the surface of a Au(111) single crystal [21–24]. Figure 6.3 (top) depicts this two-dimensional arrangement, with respective nearest-neighbor and next-nearest-neighbor spacings of 0.50 nm and 0.87 nm. The arrangement shown places the sulfur head group in the three-fold hollows of the Au(111) lattice. Though not yet experimentally verified, this placement is based on the general preference of alkanethiolates for surface sites with higher coordination [25].

Images obtained using STM consistent with this general depiction are shown in Figure 6.3 (bottom) for a thiolate monolayer formed from ethanethiol. This image was obtained in a constant current mode with a V_b of -200 mV and an I_t of 2.0 nA. The 2.65 nm × 2.65 nm image exhibits the hexagonal periodicity and spacings expected for a $(\sqrt{3} \times \sqrt{3})$ adlayer, demonstrating the ability to probe the two-dimensional arrangement of this system using STM.

In addition to revealing the adsorbate spacing, STM has also been applied to assess the larger scale topologies of this adsorbate-substrate combination. These studies were motivated in large part by the need to probe the long-range topography of these systems to deduce correlations between the adlayer structure and their effectiveness as barriers to heterogeneous electron transfer. An example of a larger scale image is shown in Figure 6.4. This image exhibits a large number of apparent depressions embedded on the relatively flat surface of the underlying Au(111) substrate. Importantly, these depressions are not observed at uncoated gold or after exposure to the pure solvents used for sample preparation (e.g., ethanol and hexadecane), indicating that the depressions are a direct consequence of the formation of the adlayer structure. The next two subsections address issues related to larger scale topographies, which are of importance for correlating barrier properties and surface heterogeneity.

6.2.2.2.1 Barrier Properties and Structural Defects Because defects (e.g., pinholes, adsorbate, vacancies, and conformational kinks) represent weaknesses in barrier properties, some of the first explorations focused on the development of approaches for assessing the structural integrity of these adlayers. As discussed shortly, one approach is to image the real space structure at defects. An alternative strategy, however, utilizes what we refer to as "terrain altering" techniques. Since a structural defect represents a weakness in the barrier properties of the layer, solution-based processes that involve the deposition or removal of materials by reactions with the underlying gold surface can be used to decorate the location of defects. For example, a defect can be detected by creating a one-atom high terrace at gold by the underpotential deposition (UPD) of lead

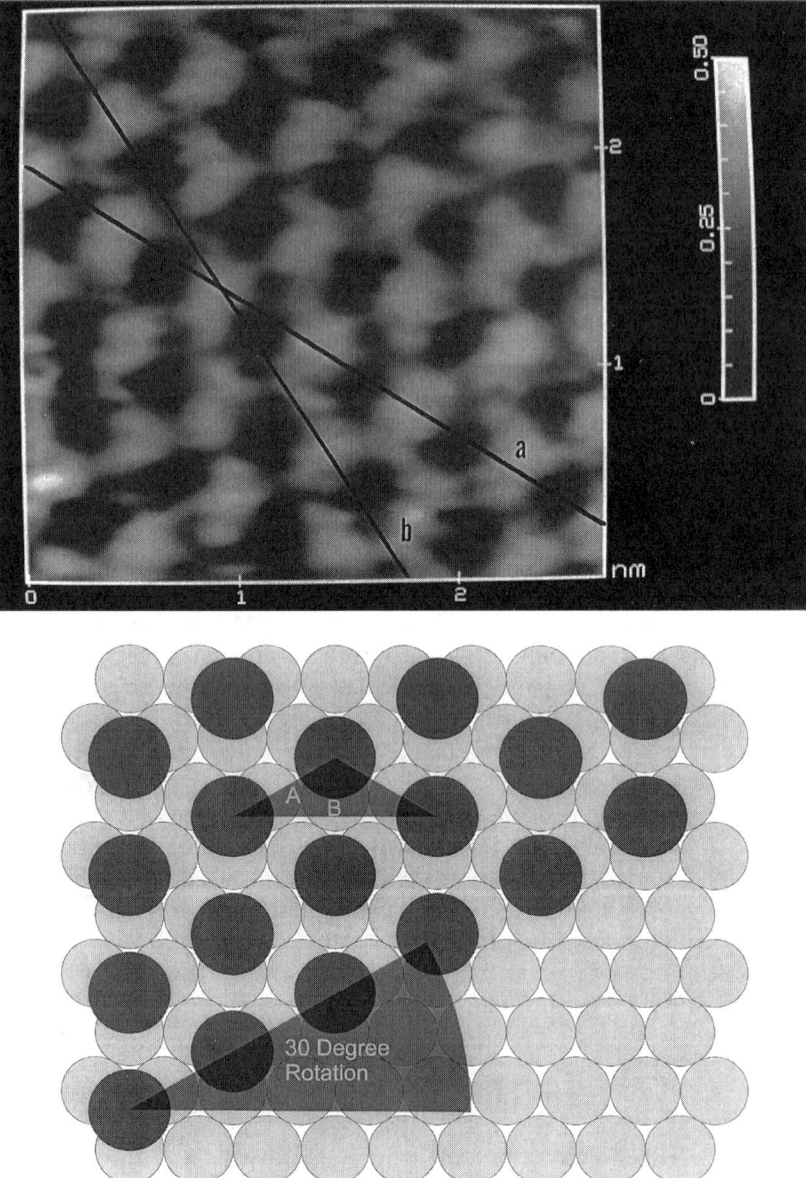

Figure 6.3. (Top) Scale drawing of the ($\sqrt{3} \times \sqrt{3}$)R30° adlayer on an underlying Au(111) lattice. The open circles represent the Au atoms, the shaded circles represent the adlayer, and the nearest-neighbor (A) and next-nearest neighbor (B) separations distances are 0.50 nm and 0.87 nm, respectively. (Bottom) Constant current STM image (2.65 nm × 2.65 nm) of a monolayer of ethanethiolate at Au(111): $V_b = -200$ mV; $I_t = 2.0$ nA. Line a highlights the nearest-neighbor separation distances (0.51 ± 0.02 nm), and line b highlights the next-nearest-neighbor separation distance (0.91 ± 0.04 nm). The z-scale is 0–0.50 nm. Reproduced with permission from Widrig et al. [28].

6.2 STM CHARACTERIZATIONS OF ORGANIC THIN FILMS

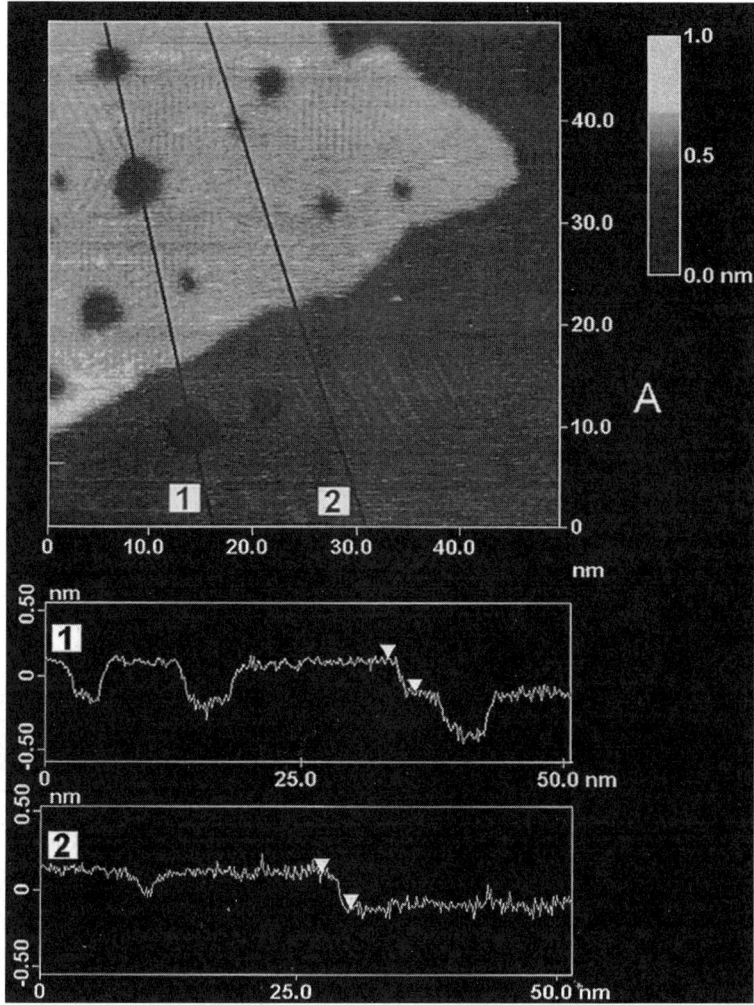

Figure 6.4. Constant current STM image (50 nm × 50 nm) of a monolayer of ethanethiolate at Au(111): $V_b = -200$ mV; $I_t = 2.0$ nA. The z-scale is 0–1.0 nm. The cursor height difference in line scan 1 is 0.25 nm and in line scan 2 is 0.24 nm. Reproduced with permission from McDermott et al. [36].

or copper [26]. On the other hand, a trend can be carved into the gold surface by a substrate specific etchant like cyanide [27] or triiodide [28].

One of the first uses of the terrain-altering strategy explored the UPD of copper to follow the film formation process as a function of immersion time [26]. In general, the extent of copper deposition diminished with increasing immersion time. At short immersion times (~10 min), the extent of copper deposition was attenuated by ~50% with respect to an untreated electrode. Longer immersion times resulted in a further decrease in copper deposition, with the much smaller levels of deposition evident after 1,000 min of immersion demonstrating the general viability of this type of characterization.

Subsequent studies have utilized a cyanide etchant to probe the size and density of structural defects [27,29]. Etching at octadecanethiolate layers at Au(111) in air-saturated KCN solutions resulted in the image in Figure 6.5 [27]. As is evident, etching results in the presence of large triangularly shaped depressions with a uniform spatial direction. These depressions are ~0.59 nm deep with sides ~65 nm in length. The similarities in orientation and the anisotropic shape of the depressions were interpreted via a defect etching mechanism whereby an adsorbate vacancy at a three-fold hollow binding site in the underlying gold acted as an initiation site for etching. Etching therefore initiates via the dissolution of three equivalent gold atoms that are exposed to attack by CN^-. The process continues with the desorption of the three surrounding thiolates, and is followed by the dissolution of 12 additional gold atoms. Although it is as yet unclear as to why etching at monolayer-covered samples proceeds anisotropically, continuation of this process is consistent with the appearance of the triangularly shaped patterns shown in Figure 6.5. By comparison, an etching process initiated at the on-top sites of a Au(111) substrate would yield

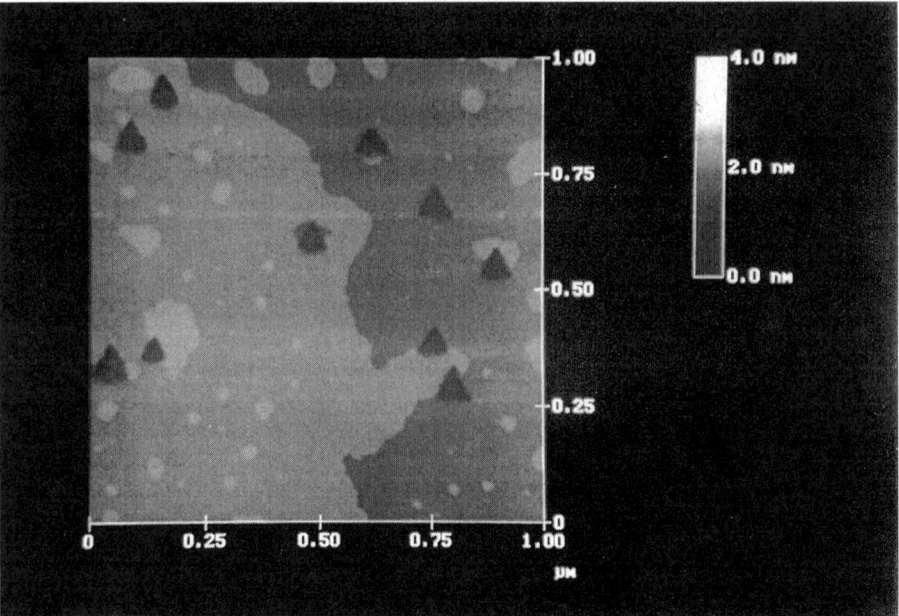

Figure 6.5. A 1 μm × 1 μm STM image of an octadecanethiolate monolayer on Au(111) after exposure to an etchant solution of CN^-. The side of the triangular etch pits are 65 nm ± 8 nm in length. The z-scale is 0–4.0 nm. Reproduced with permission from Sun and Crooks [27].

two sets of triangularly shaped etch pits, with one rotated 60° with respect to the other. This analysis therefore suggests that the thiolate adlayer is bound at the three-fold hollow sites of the underlying substrate, and, though indirect, is the only evidence to date for this system experimentally indicative of a binding site preference.

The CN^--etching procedure was also utilized to assess the chain length dependence ($n = 5$, 11, and 15) of the relative barrier strengths of alkanethiolate monolayers formed from vapor and liquid-phase depositions [27]. Both liquid and vapor-deposited monolayers were resistant to etching, with the resistance greater the longer the chain length. However, the layers from intermediate length precursors ($n = 11$) were more resistant to etching if formed by vapor deposition, whereas those with long chains ($n = 15$) were more resistant if formed by solution deposition. The difference in the barrier properties for the layers formed by the two preparations suggests that the solvent may play an important role in the formation of these systems. That is, layers formed from short chain length precursors are affected by the possible incorporation of solvent, which results in a more defective structure than the vapor deposited analogs. The formation of layers from longer chain length precursors is, however, assisted by the solvent by increasing the fluidity of the chain structure during formation, and yields a more effective barrier film.

Another study showed that the structural integrity of an octadecanethiolate layer could be manipulated by changes in the tunneling conditions. Both a tip-induced etching and a tip-induced deposition of material could be achieved [30], presenting an opportunity for pursuing nanolithographic patterning of these interfaces. More recent studies, as in the following discussion, revealed that these layers can be easily damaged while imaging.

In general, these studies have shown that as-formed monolayers from alkanethiols of intermediate to long chain lengths are fairly resistant to etchants at a microscopic scale and are thus free of defect sites large enough to permit contact between an etchant and the underlying gold surface. These data, however, do not exclude the possibility of smaller defects in the chain structure that preclude contact between etchant and substrate. Such small defects, which include chain conformers and domain boundaries from mismatches in registry between sulfur with gold, are likely to be smaller than that of an adsorbate vacancy, but still represent a weakness in the effectiveness of the layer as a barrier to electron transfer.

6.2.2.2.2 Morphology and Molecular-Scale Features

6.2.2.2.2.1 DEPRESSIONS As evident in Figure 6.4, the STM-determined morphology of alkanethiolate adlayers at Au(111) surfaces consists of features that are attributable to the gold substrate (e.g., 0.24 nm steps between atomically smooth terraces) as well as those that do not appear at Au(111) prior to exposure to a thiol-containing solution. In the latter case, the features appear as 2–5 nm diameter depressions and are thus associated with the formation of the monolayer. There have been a number of STM investigations of monolayer morphology addressing the structural origins of these depressions [21,26,29–41]. The principle findings of these studies elucidate whether the depressions originate as defects at the gold substrate [21,26,29–41] or as topographic [30,38] or electronic [21,39,40] perturbations within the chain structure of the alkanethiolate layer.

Some of the early evidence indicated that the depressions originate from single-layer-deep pits in the gold surface. Several observations support this assertion. First, depression depths were determined to be ~0.25 nm, a value similar to the height expected for a step on a Au(111) surface, and less than that expected for a defect in the chain structure in a monolayer [29,31–35]. Second, variations in imaging conditions such as the tunneling gap resistance (i.e., R_Ω, which equals V_t/I_t) did not affect the measured depths as might be expected for defects in the chain structure of the monolayer [31]. Third, depressions have been observed to move across a terrace to a step edge, leaving a corrugated step edge with no discernible height difference from an adjacent, lower level terrace [29,31,33–35]. Finally, these features are impervious to the UPD of Cu and unaffected by CN^- etchants [26,29,41]. All of these observations are inconsistent with the depressions originating from a defect in the chain structure in the adlayer.

Other observations support the origination of the depressions within the chain structure of the monolayer [21,30,37–40]. In particular, depression depth measurements of ~5–26 Å [30,37,38] are more consistent with defects that span a portion of the chain structure of the monolayer. One study speculated that some of the STM-measured depths may be erroneously low because of the onset of interactions between a tip and the opposite side of a depression before an interaction between a tip and the bottom of a depression [37]. Reports utilizing STM coupled with SFM concluded, on the other hand, that the depressions arise from an electronic effect [21,39,40]. These studies proposed that local inhomogeneities in the chain structure decrease the conductance of the tunneling barrier of the adlayer, resulting in a tip-induced compression at such inhomogeneities while imaging. This model therefore argues that the depressions are not indigenous topographic features of the adlayer, but rather are induced by the imaging process.

Subsequent reports have offered conclusive molecular-scale evidence for placement of the depressions at the gold surface. In these studies, the $\sqrt{3} \times \sqrt{3}$ periodicity of the adsorbate, along with the other characteristic two-dimensional arrangements of the adlayer discussed in the next subsection, extend into the depressions. Such images, as presented in Figure 6.6, show that the depressions are present even when filled with an ordered adsorbate layer and are therefore not a consequence of defects in the monolayer structure [33,36]. Recent results using SFM in a frictional force mode detected depression densities similar to those found in the STM efforts, further supporting this conclusion [36].

Given this evidence, the mechanistic pathway that leads to the formation of the depressions becomes an issue. Two possible routes have been put forth. One

6.2 STM CHARACTERIZATIONS OF ORGANIC THIN FILMS 261

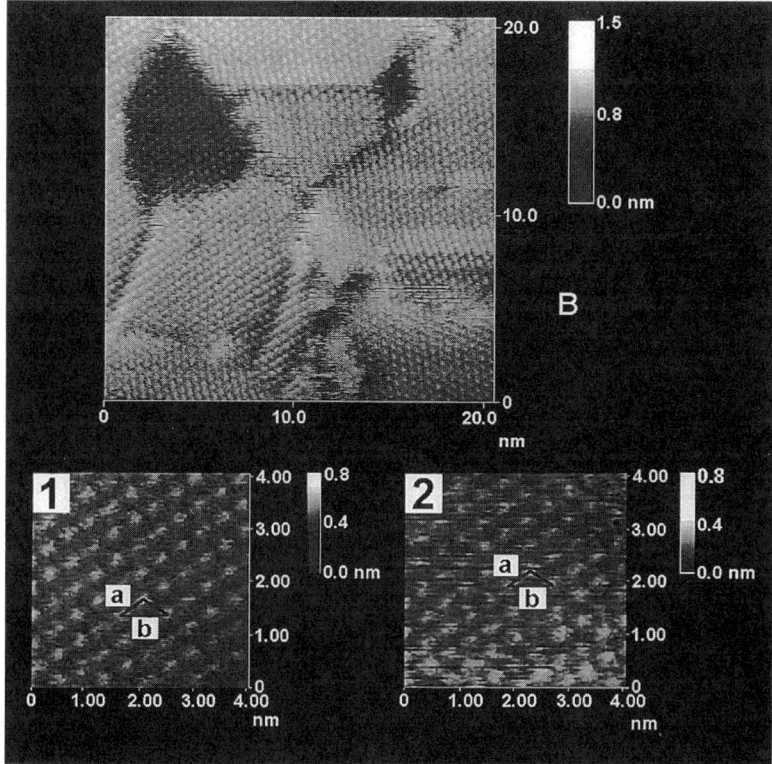

Figure 6.6. Constant current STM image (20 nm × 20 nm) of a decanethiolate monolayer at Au(111): V_b = 938 mV; I_t = 54 pA. The z-scale is 0–1.5 nm. The insets are 4 nm × 4 nm images of the adlayer spacings (1) outside the depression (a = 0.52 nm and b = 0.91 nm) and (2) inside the depression (a = 0.51 nm and b = 0.88 nm). Reproduced with permission from McDermott et al. [36].

poses an adsorbate-induced etching of the gold substrate during adlayer formation, and is supported by the detection of gold via atomic absorption spectroscopy (AAS) in the deposition solutions after layer formation [31,35]. These data show that increasing the thiol concentration, and to a lesser extent, decreasing the length of the alkyl chain, results in increased amounts of dissolved gold as expected for an adsorbate-induced etching process [35]. Support for this proposal also arises from the observed dissolution of gold in sulfide and thiocyanate solutions [42]. There are, however, some discrepancies with this model. That is, the amount of gold detected via AAS is in many cases much lower than that expected from the concentrations predicted using the size and densities of the depressions. This discrepancy suggests that an alternative pathway may be of greater significance. The recent observation of depressions of similar sizes and densities at monolayers prepared by a vapor deposition process with those from the conventional solution-based process also mitigates against an etching process as the major mechanism [29]. As an alternative, an adsorbate-induced reconstruction mechanism of the underlying gold surface has been offered as a parallel process for depression formation, relying on preliminary descriptions of the nucleation and growth of the adlayer (36). Though it is premature to determine the relative importance of the two processes, ongoing experiments in several laboratories should provide a clearer picture of the situation in the near future.

6.2.2.2.2.2 MOLECULAR-SCALE STRUCTURES As noted previously, the early STM investigations of alkanethiolate monolayers characterized the structures formed at gold from methyl-terminated systems on a nanostructural level. These images revealed a hexagonal adsorbate lattice with a nearest neighbor spacing of 0.50 nm and a next-nearest neighbor spacing of 0.87 nm [28]. Importantly, these spacings are in agreement with the $(\sqrt{3} \times \sqrt{3})R30°$ structure deduced from transmission electron, X-ray, and He-diffraction experiments [21–24] for the same types of systems on Au(111) (n = 9–21).

Another study (43) investigated the structure of thiolate films at Au(111) that had been derivatized with end groups of various average molecular diameters (d). A goal of this effort was to determine the effect of the size of the end group on the packing density of the layer, and the interplay between the spacing density preferences of the polymethylene spacer groups and the end group. The precursors included 4-aminothiophenol (d = 4.3 Å), Ru(bipyridine)$_2$(4-methyl-4'-(12-mercaptododecyl)-2,2'-bipyridine) (PF$_6$)$_2$ (d = 13 Å), (C$_5$H$_5$)Fe(C$_5$H$_4$CO$_2$(CH$_2$)$_{16}$SH) (d = 6.5 Å), and [2]staffene-3,3'-dithiol pentaamine-ruthenium (II) hexafluorophosphate ($d \sim$ 7 Å). Surprisingly, a $\sqrt{3} \times \sqrt{3}$ adlayer spacing was observed in all cases (see Fig. 6.7) even for precursors with end groups too large to access the $\sqrt{3} \times \sqrt{3}$ packing density dictated by the polymethylene chains. Surface coverages based on the charge passed to electrolyze the end groups, however, indicated that the packing densities were closer to predictions based on limitations of the sizes of the end groups and were significantly greater than those expected from the adlayer spacings in the images. To account for these discrepancies, the molecular structure detected in these images was attributed to an adsorbate-induced perturbation in the electron density at the underlying Au(111) surface and not the molecular structure of the adlayer.

As is evident, the results of the first two molecular resolution studies are in direct conflict with respect to a mechanistic interpretation of the imaging process as well as to the viability of the structural insights from such images. The first of the two reports proposed that electrons tunneled primarily between the tip and gold-bound sulfur [28], indicating that the images reflected the real two-dimensional spacings of the adlayer. The latter of the two reports, however, suggested that such images resulted from an adsorbate-induced perturbation of the electronic structure of the underlying gold substrate [43], questioning the interpretation of the images as true two-dimensional representations

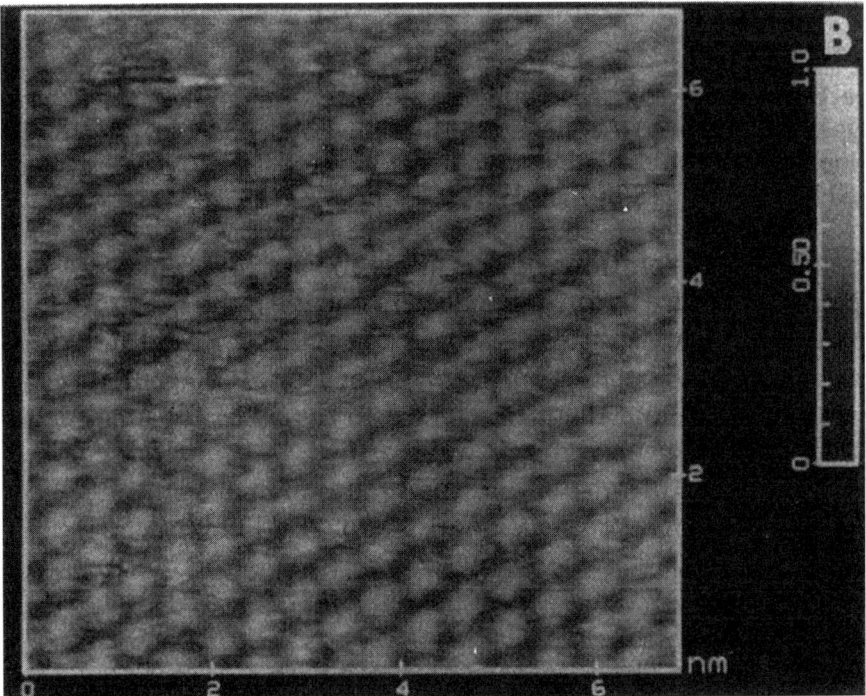

Figure 6.7. Constant height images of a monolayer prepared from [Ru(bipyridine)$_2$(4-methyl-4'-(12-mercaptododecyl)-2,2'-bipyridine)] at Au(111): V_b = 200 mV; I_t = 0.5 nA. The nearest-neighbor separation distance is 0.51 ± 0.03 nm and the next-nearest-neighbor separation distance is 0.87 ± 0.05 nm. The z-scale is 0–1.0 nm. Reproduced with permission from Kim et al. [43].

of localized structure. We believe, however, that the ability to detect highly localized structural defects (i.e., nanometer-wide boundaries between ordered adlayer domains), coupled with the observation of the $\sqrt{3} \times \sqrt{3}$ adlayer spacings under conditions where the tip resides outside the alkyl chain structure [21,39,40] is diagnostic of the real two-dimensional spacings of these systems. Nevertheless, it is clear that more advanced theoretical treatments of the imaging mechanism in these systems are needed in order to resolve this issue.

Subsequent to the preceding studies, there were few reports on the molecular-scale imaging of alkanethiolate monolayers until recently. We suspect that the difficulties in obtaining molecular resolution at these surfaces results from the extent of the interaction of the STM tip with the sample. Such interactions appear especially problematic in the case of the longer chain length monolayers at high tunneling currents, where images often differ from the $\sqrt{3} \times \sqrt{3}$ periodicity. In one interesting case, images at large values of I_t displayed the periodic square-shaped pattern shown in Figure 6.8 [30,42]. As is evident, the pattern is composed of three bright spots along each side, with separations between spots

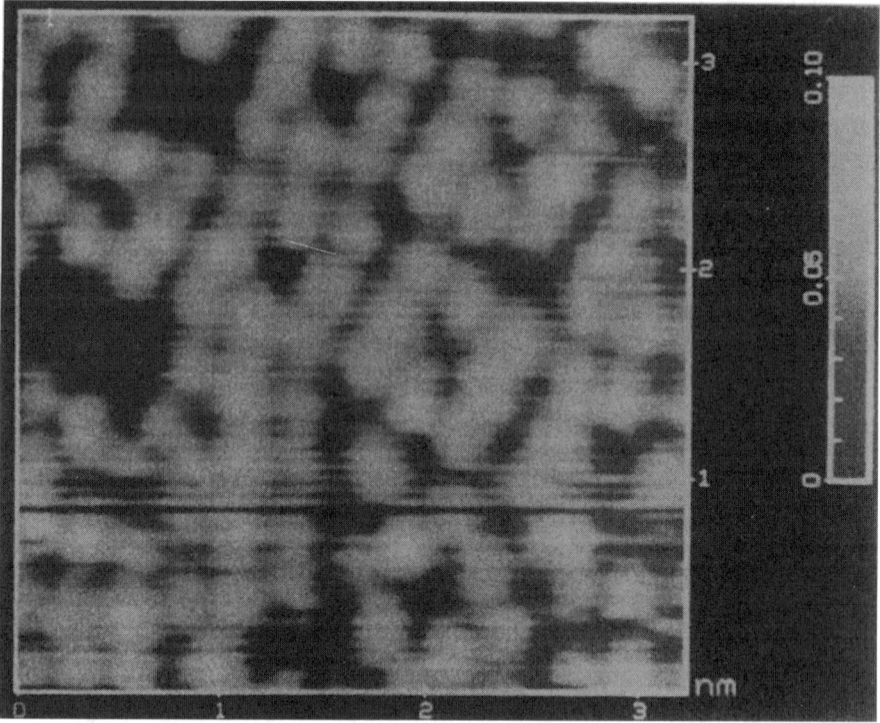

Figure 6.8. Constant height STM image of a 3.2 nm × 3.2 nm etched area at the surface of an octadecanethiolate monolayer on Au(111): V_b = 6.6 mV; I_t = 25 nA. The z-scale is 0–0.10 nm. Reproduced with permission from Kim et al. [43].

of ~0.27 nm. The same patterns were observed, at Au(111) samples exposed to sulfide and thiocyanate solutions [42]. The origin of these patterns has been ascribed to a reconstruction of the gold surface that is induced by tip effects or by thiocyanate or sulfide dissolution [30], with the patterns reflecting the reconstructed surface.

There is, however, a second possible origin of these patterns. This possibility is based on a recent in situ STM investigation of the electro-oxidation of sulfide on Au(111) [44]; square patterns virtually identical to those above were observed for the oxidation of sulfide. Thus, an image for adsorbed sulfide could be reversibly transformed to one exhibiting the square pattern. Interestingly, the spacings between the eight spots of each square-shaped pattern is similar to that of elemental sulfur, i.e., S_8. The square-structured pattern may then arise from a tip-induced oxidation of the thiolate adlayer. We note that this possibility is supported by a recent demonstration of the propensity of short chain thiolates to undergo S—C bond cleavage to yield an adsorbed sulfide-type film [45,46]. The results of ongoing experiments in several laboratories should shed more insight into this issue.

More recent efforts have successfully imaged these systems at lower values of $I_t \approx 2$ pA, and higher values of $V_t \approx 1$ V. In imaging dodecanethiolate monolayers, it was found that high values of R_Ω ($R_\Omega > 100$ GΩ) were required to avoid mechanical interactions between the tip and sample [33]. Interestingly, images at higher values of R_Ω (i.e., $R_\Omega = 670$ GΩ) revealed the pattern of the adlayer structure, whereas those at lower values of R_Ω (e.g., $R_\Omega = 33$ GΩ) exhibited the pattern of the underlying gold substrate [33,35]. The likelihood of tip interactions with the adlayer structure is also supported by a study of the layer formed at Au(111) from mercaptohexadecanol [32]. In this case, force gradient measurements indicated that low values of R_Ω ($R_\Omega = 0.1–1$ GΩ) resulted in an elastic compression of the chain structure. Recent studies have also detected the movement of features associated with the adlayer structure that have been correlated to imaging effects [31–33,35,43]. As a consequence, the most recent efforts have employed large tunneling resistances to minimize the possibility of tip-induced effects [47–52].

The previous findings indicate that imaging at low values of R_Ω may induce a perturbation of the adlayer through a tip-induced compression of the adlayer structure near the chain terminus [31,33]. This interpretation is supported by two related results. First, the success rate in imaging the longer chain adlayers (e.g., octadecanethiolate at Au(111)) at low gap resistances was notably lower than for the shorter chain adlayers (e.g., ethanethiolate at Au(111)) [28]. This correlation is consistent with the infrequent fabrication of a tip with an unusually sharp asperity, which could minimize a tip-induced disruption of the structure of the adlayer while imaging at a smaller tip-sample separations. Second, a recent molecular dynamics study demonstrated that unusually sharp asperities can penetrate into the outermost portion of the adlayer without fully disrupting the underlying chain structure [53]. These dynamics simulations, when coupled with the noted infrequency in the ability to image longer chain adlayers, further support the earlier arguments related to the mode in which electrons tunnel between the tip and sample at low gap resistance.

Interestingly, a variety of new molecular structures, in addition to the $\sqrt{3} \times \sqrt{3}$ adlayer, have been detected when imaging at high tunneling resistances. As shown for a dodecanethiolate monolayer at Au(111) in Figure 6.9, "missing row" structures have also been observed [33,51]. Within each row, the adlayer exhibits a $\sqrt{3}$-nearest-neighbor separation distance; however, the voids between rows are inconsistent with the presence of a fully packed adlayer. A detailed examination of such samples concluded that the most common missing row structure is comprised of a missing row alternating regularly with four rows of chemisorbed molecules in a $5\sqrt{3} \times \sqrt{3}$ adlayer arrangement [51].

Characterizations of monolayers from various alkanethiols ($n = 3, 5, 7, 9, 11$) have also found a fundamentally different adlayer structure not observed in previous STM experiments. This structure, revealed by the image in Figure 6.10a and cross-sectional profiles in Figure 6.10b–d, is a repeating rectangular unit mesh containing four thiolate moieties with a $3a \times 2\sqrt{3}a$ nearest-neighbor

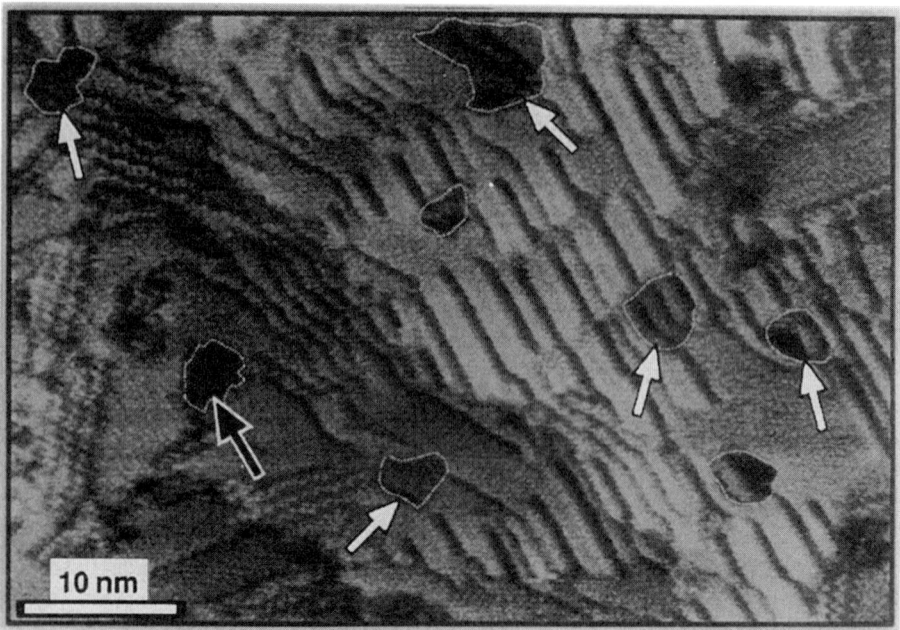

Figure 6.9. Constant current STM image (55 nm × 40 nm) of dodecanethiolate at Au(111) obtained at R_Ω of 670 GΩ. Reproduced with permission from Schonenberger et al. [33].

spacing, where a equals the 0.29 nm interatomic spacing of Au(111) [47,48]. This structure, described as a c(4×2) superlattice with respect to the hexagonal thiolate lattice, was previously observed in helium and X-ray diffraction experiments [54,55]. Similar types of structures have also been proposed in earlier low temperature infrared reflection spectroscopic studies and in low energy electron diffraction experiments [56,57]. Structurally, the superlattice (see Fig. 6.10e) [47,48,54,58] reflects a twisting of the polymethylene chains about their molecular axes, a situation speculated to give rise to the height modulations observed in the images presented in Figure 6.10b–d. An alternative suggestion is a variation in the gold-sulfur interaction (i.e., hybridization of the S—C bond) [25].

In addition to the superlattice, several other structures have been observed at monolayers from short chain alkanethiols. In a study characterizing the formation process at these monolayers (n = 3, 5), UHV-STM images show that these samples initially exhibit regions of disorder interspersed between the depression features discussed earlier [47,48]. The disordered regions are attributed to a two-dimensional liquid-phase monolayer. Images acquired at later times in vacuum reveal that the area for ordered regions increased at the expense of the area for the disordered regions, and the number of depressions decreased. The ordered regions were further characterized by lower coverage structures that have a $p \times \sqrt{3}$ spacing, where p usually varies from 8 to 10. The association

6.2 STM CHARACTERIZATIONS OF ORGANIC THIN FILMS

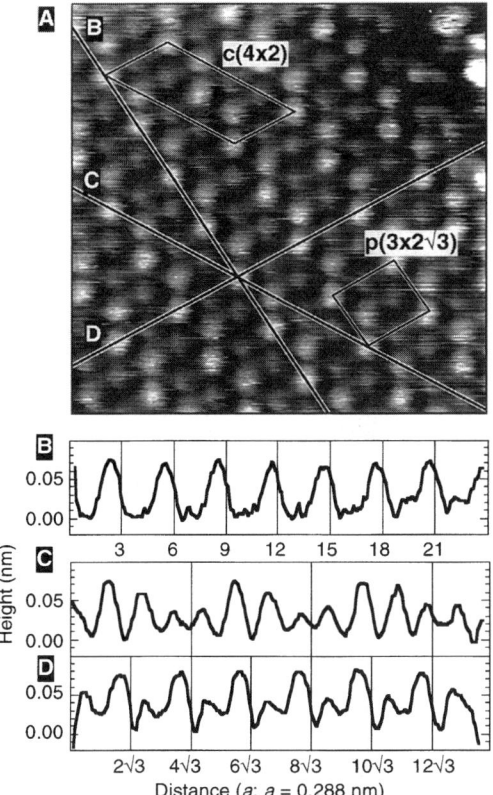

Figure 6.10. (a) Constant current image (6.0 nm × 6.0 nm) STM image of an octanethiolate monolayer on Au(111). The $p(3\times2\sqrt{3})$ unit mesh and the $c(2\times4)$ superlattice unit cell are outlined. (b) Plot of cross-section B in (a) running in the gold nearest-neighbor direction. Plots of cross sections (c) and (d) running in the two gold next-nearest-neighbor directions. (b)-(d) reproduced with permission from Poirier et al. [47]. (e) Sketch of the herring-bone packing structure of an alkanethiolate monolayer on Au(111). The sulfur atoms are bound at the three-fold hollow sites of the gold surface and form a $(3\sqrt{3}\times\sqrt{3})R30°$ lattice. The vectors represent the alkyl chain backbone planes and point in the direction of the S—C bond (after Poirier et al. [47]).

between the $p\times\sqrt{3}$ structure and the lower packing density phase was deduced by observing a conversion of a superlattice to $p\times\sqrt{3}$ structure for longer chain monolayers after inducing desorption by elevated temperature treatments. As such, an ordering process was envisioned in which the slow desorption of short chain thiolates proceeded concurrently with the nucleation of crystalline domains [47]. Investigations are needed to determine whether the conclusions from the nucleation and growth observations found in these vacuum studies can be translated to the events that transpire in solution-based nucleation and growth processes.

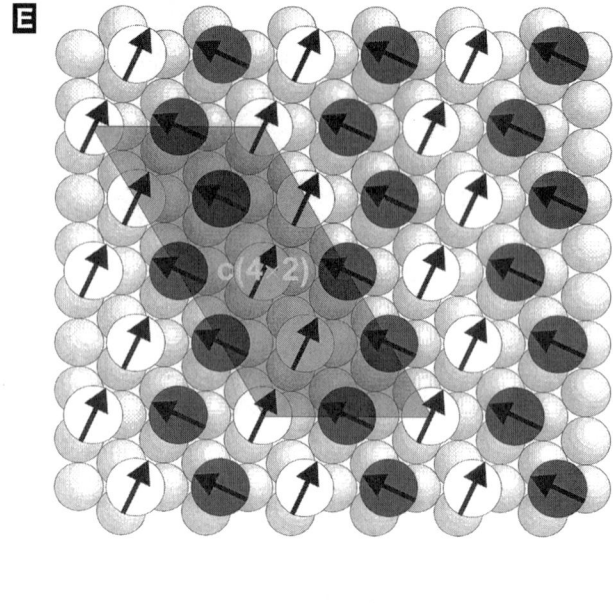

Figure 6.10. (*Continued.*)

6.2.2.2.2.3 MOLECULAR MOTION AND SURFACE EVOLUTION Imaging by STM has also provided evidence for the evolution of the surface structure through the detection of adsorbate movement and measurements of surface diffusion rates [59]. In these cases, the movement of the depression features was followed by raising the temperature of a sample after removal of the sample from the incubation solution. Images showed that the depressions migrated into the steps defined by the lower, neighboring terraces [29,34]. As a consequence, the voids in a terrace from the depressions transformed into a change in the shape of a step edge. This observation suggests that the diffusion of a depression is due either to a highly labile Au-S bond that allows the underlying gold atoms to migrate across the surface or a weakening of the underlying Au-Au bonding, which facilitates the diffusion of an adsorbate-substrate complex.

At high values of R_Ω (i.e., imaging conditions where tip-sample interactions are minimized), annealing in air at ~100°C resulted in an increase in the size of the depressions and a decrease in the number of the depressions [50,51]. An example of such an image is presented in Figure 6.11. These and other reports have found that the adlayers assemble into domains ~5–15 nm in size [47–52,60]. These domains are bounded by both depressions and features associated domain boundaries from sulfur-gold and chain tilt mismatches. Upon annealing in air, the domain boundaries arranged into a striped pattern, which may reflect a structural change because of thiolate desorption. Prolonged

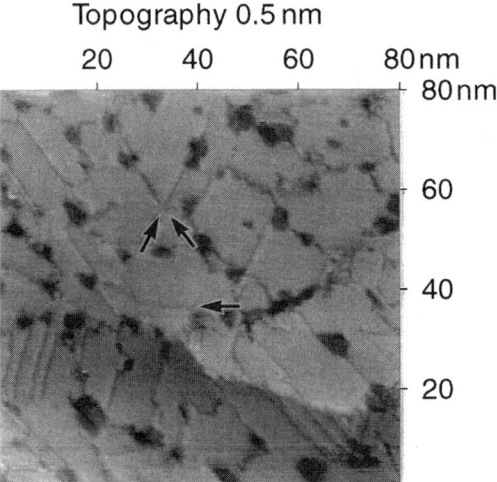

Figure 6.11. Constant current STM image (80 nm × 80 nm) for a monolayer of dodecanethiolate at Au(111): $V_b = 927$ mV; $I_t = 9$ pA. The domains are evident by straight and well-oriented boundaries as marked by arrows and are in the nearest-neighbor direction of the sulfur lattice. The z-scale is 0–0.50 nm. Reproduced with permission from Sprik et al. [60].

annealing resulted in the disappearance of the depressions and a decrease in the distance separating the lines, due again to increased local ordering [50].

Studies where the monolayer is formed at elevated temperatures yielded surfaces that displayed an increase in domain size and an alignment of domain boundaries along the next-nearest-neighbor direction of the adlayer lattice [48]. The increase in domain size is attributed to a decrease in the number of sulfur-gold lattice mismatches.

The consistent ability to image at the molecular scale has similarly yielded observations of the evolution of the two-dimensional architecture over time. An ex situ STM study of dodecanethiolate adlayer formation describes the presence of low density molecular structures at partial-coverage monolayers [51]. A study at butanethiolate adlayers shows that the surface structure evolves from the liquid-like state already discussed [49,52] into crystalline-like domains. Although real-time monolayer formation has not yet been observed, these studies promise to further our insight into the formation processes and structure of this important class of monolayers.

As noted earlier, one of the important aspects in interpreting the reactivity of modified electrode surfaces is a description of the two-dimensional arrangement of the immobilized redox moieties. Thiols on gold systems offer several routes to the design of structures with a controlled spacing between redox species [5,6]. Although there are several reports of large-scale STM imaging of end-group derivatized monolayers [21,32,39,40,43,61], there are few studies describing molecular resolution. Apparently, molecular resolution is obtained with significantly more difficulty at these types of samples versus the methyl-terminated systems. Though not well understood, part of this difficulty may arise from the affinity of polar end-groups for impurities and coadsorbates such as water, the solvent used in film formation (e.g., ethanol), or the presence of a partial bilayer of adsorbates [60], all of which may disrupt the structural order near the chain terminus. Experiments under more carefully controlled preparation conditions and imaging environments (e.g., ultra high vacuum imaging) are needed to test this assertion.

Lastly, a recent study examining azobenzene-terminated alkanethiols at gold surfaces observed that the adlayer formed two stable rectangular arrangements [62] each with lattice constants of 0.61 and 0.79 nm. These lattices were incommensurate with the Au(111) substrate and extended over terrace edges without detectable offset in the pattern, arguing that the observed ordering in this particular system is dominated by end-group interactions.

6.2.3 Polymer Films

6.2.3.1 Introduction If the many types of films and coatings in existence today were rated on technological importance, polymer films would almost certainly sit at the top of the list. For electrochemists, interest has focused largely on conductive polymers because of ease of preparation and stability in addition to possible applications to optoelectronic devices, chemical sensors, rechargeable batteries, and related technologies [63]. STM was the first of the scanning probe techniques applied to electrochemically deposited conducting polymer films at electrodes. The ability of STM to resolve surface structure at high resolution as well as probe surface electronic states has been exploited to examine conductive films with regards to morphology, nucleation and growth mechanisms, and modes of conduction.

Before continuing, however, the difficulty in interpreting the STM images of polymer films should be noted. Although the polymer films discussed are conductive over macroscopic length scales, the localized conductivity is dependent on morphology and doping homogeneity. Thus, variations in conductivity across a surface can have a potentially significant influence on the observed image. Features interpreted as changes in topography may in fact reflect differences in local conductivity. This situation is further complicated because many of the images of conducting polymer films, as described in the following sections, are not as well defined as those discussed in the earlier section on organic monolayer films. In many cases, the difficulty in obtaining readily interpretable images of these systems likely stems from the interaction of the STM tip with the soft, more compliant polymeric materials and the subsequent perturbation of the sample surface. This complication is, however, likely to improve as more studies begin to characterize samples at higher values of R_Ω.

The following sections are broken down on the basis of the nature of the initial monomer. In addition, STM has also found utility in unraveling details of passivating polymer films in the context of electrode fouling, which is discussed in a subsequent section.

6.2.3.2 Polypyrrole (PPy) A significant amount of research has been directed at thin films of PPy on electrode surfaces in part because of ease of formation and relative stability. Early STM studies of PPy films focused largely on nucleation and growth mechanisms [64–69] and the structural variations induced by the dopant anion [65,67] and by the electrochemical deposition conditions [66,70,71]. In these investigations, several different electrode materials have been used to initiate film formation including highly oriented pyrolytic graphite (HOPG) [64,65,67–70], platinum [66,72], and gold [71,73,74].

Images of the initial stages of PPy grown galvanostatically at HOPG reveal a discontinuous structure consisting of semicrystalline islands adjacent to the electrode surface at low coverages that grow and coalesce into an amorphous, nodular structure at higher coverages [64,65,67–69]. This structural evolution is detailed in the images in Figure 6.12 [67]. Figure 6.12a is an image of a PPy island formed under low coverage conditions at HOPG. As shown, these islands are composed of interconnected polymer strands. The images in Figure 6.12b, c correspond to fibrilular transition structures that are produced by the coalescence of the islands as the coverage of PPy increases. The amorphous, modular structure that is characteristic of a thick (>900 nm) PPy film is presented in Figure 6.12d. Interestingly, Figure 6.12c reveals the boundary between the intermediate (bottom) and final (top) nodular structure.

A subsequent study [69] built on these earlier findings, showing that the nucleation of a PPy film takes place at a step edge defect where monomer can first physically adsorb onto the HOPG surface. Polymerization is then initiated via the formation of semi-crystalline islands that extend from monomer adsorbed at these sites. We also note that the observation of islands dispersed on the HOPG surface in the early stages of growth indicate that at lower coverages, the PPy film thickness calculated by the number of coulombs passed represents an average value where the thickness of the individual islands is actually greater than this average.

The initial islands observed at HOPG are typically connected by polymer strands with diameter of 1.5 to 2.0 nm that extend hundreds of nanometers across the substrate [65,67]. The detailed structure of these strands has been shown to depend on the nature of the dopant anion. For example, PPy films doped with *p*-toluenesulfonate form long, linear helical strands while films doped with tetrafluoroborate exhibit randomly oriented structures [65]. A similar island nucleation and growth mechanism has been detected during an in situ STM study of PPy deposition on Pt [66].

The two-dimensional structure of PPy formed at electrode surfaces has also been shown to depend on the rate of deposition. Films deposited at HOPG with current densities of 0.03 mA/cm^2 appear smooth and oriented relative to those deposited at higher current densities (1 mA/cm^2) [70]. A similar relationship between current density (i.e., deposition rate) and roughness has been found

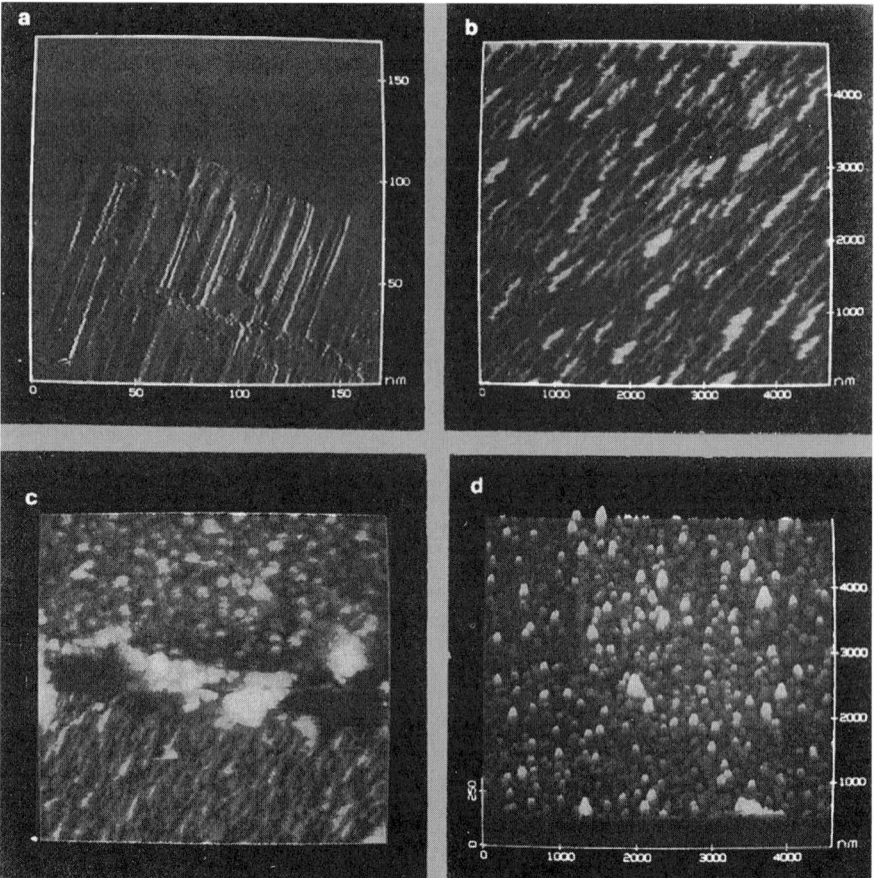

Figure 6.12. STM images of the evolution of the morphology of a PPy film as a function of coverage. (a) Initial island formation at low coverage. (b) Fibrilular transition structure formed at intermediate coverages. (c) Boundary between transition structure and fibrilular structure. (d) Thick polymer film with a nodular structure. The coverage of the PPy film was controlled by varying the charge passed during the deposition from 30 to 200 mC. Reproduced with permission from Yang et al. [67].

for PPy deposited at gold [71]. In this study, the rougher films exhibited an enhanced rate of O_2 uptake, as monitored by electron spin resonance, indicating a highly porous structure. Additionally, electrochemical STM images have revealed that the roughness of PPy at gold was dependent on the deposition rate as manipulated potentiostatically [71].

Two recent reports are concerned with the creation of nanometer-sized PPy structures and the electropolymerization of surface-confined pyrrole. Modulation of the potential between the STM tip and the surface in a pyrrole-containing solution has been used to create nanoscale PPy patterns at gold [73]. This effort demonstrated that nanometer-sized dots of PPy could be written onto, read, and erased from a gold surface. This results represents a potential advance in nanolithography. To build on this development, issues related to the drift and hysteresis of the piezoelectric scanner tubes as well as the long time scales requisite for construction these patterns need to be significantly improved. We also note that STM images of a PPy film formed from the electropolymerization of an ordered monolayer of pyrrole tethered to gold via an alkanethiolate linkage have recently appeared [74].

6.2.3.3 Polyaniline (PAN) Since the demonstration of an insulator-to-metal transition upon protonation or oxidative doping of emeraldine base [75], the PAN family of polymers have undergone extensive experimental scrutiny in terms of its two-dimensional structure [76–84], and the spatial distribution of its conductivity [76,79,84].

In structural studies, the galvanostatic (70 μA/cm^2) nucleation and growth mechanism of PAN at gold was shown to proceed via an island coalescence process analogous to PPy film formation [76]. Driven by the possibility of gaining enhanced control over interfacial properties, a series of studies have compared functionalized PAN films at Pt formed by both electrochemical and chemical means [77,78]. Borate-substituted PAN films prepared galvanostatically (30 μA/cm^2) exhibited a uniform conductivity [77]. The surface structure of this interface consisted of amorphous islands and coiled strands similar to PPy films [65,67,68]. The chemically formed films proved to be more difficult to image which, when combined with X-ray photoelectron spectroscopy results indicative of a lower uptake of counterions, pointed to a lower average conductivity. Both chemically and electrochemically prepared films of polyhydroxy-aniline exhibited amorphous islands ~1 μm in diameter with extended helical strands at the edges of the islands [78]. Differences in surface structure have also been noted between PAN and poly-hydroxy-aniline films deposited by cyclic voltammetry [79]. The PAN film exhibited a terraced structure diagnostic of a layer-by-layer growth mechanism, whereas the surface of the hydroxylated film was comprised of bundle-like structures more consistent with an island-based growth.

Detailed electrochemical and STM interrogations of PAN films formed by potential cycling have illustrated the effects of the scan rate [81,82] and the number of cycles [82]. The former studies have provided insights on electropolymerization kinetics. Polymerization at 50 mV/sec results in a poorly resolved amorphous surface, while deposition at 20 V/sec yields a surface structure consisting of an array of uniform particles [81]. These differences in structure were dissected by kinetic arguments. That is, scans at slower rates undergo a greater number of tail-to-tail and head-to-head dimerizations as well as head-to-tail polymerizations, resulting in a wider distribution of products. Higher scan rates, in contrast, favor the more rapid head-to-tail reaction, leading to a more uniform polymer structure.

The electronic structure of PAN films has also been probed with scanning tunneling spectroscopy (STS) in an attempt to gain further insight into the conduction mechanisms of this type of polymer [76,79,84]. All of these studies reported a π-π^* bandgap for PAN of ~4 eV, which agrees well with values from optical measurements. Importantly, the measured STS spectra were found to vary across the surface of protonated PAN films, an observation that points to an inhomogeneity in both morphology and doping [84]. This latter result illustrates the importance in considering the possible contributions of spatial variations in the surface electronic structure of these materials when interpreting the observed images.

6.2.3.4 Polythiophene (PT) STM studies on PT films have probed both film structure [85,86] and nucleation/growth mechanisms [87]. Early STM images of PT and functionalized analogs revealed a range of film structures that were dependent on the structure of the monomer [85,86]. The morphology of PT films deposited galvanostatically on Pt was characterized by helical and chain-like structures. Films produced from 3-methylthiophene exhibited an ordered zigzag structure, while those from 3-bromothiophene revealed a kinked-linear morphology. A more detailed investigation of the helical structure of PT has also been performed [67], driven in part by predictions based on X-ray analysis [88]. Fully doped films were compared to as-prepared films that were de-doped electrochemically. The periodicity or pitch observed in the helical strands of the fully doped film was remarkably uniform (i.e., 0.8 nm), whereas a partially de-doped film displayed two pitch values, 0.8 nm and 0.5 nm. The decrease in pitch was attributed to the release of dopant counterions during the electrochemical reduction.

To understand further the details of PT film nucleation and growth, a comparison of deposition on uncoated indium-tin oxide (ITO) and chemically modified ITO was conducted [87]. Images of the early stages of polymerization of 3-methylthiophene on uncoated ITO revealed large spatially separated islands while deposition on ITO modified with $SiCl_4$-bithiophene yielded a gradual, uniform film growth. These observations pointed to two different nucleation mechanisms. On unmodified ITO, nucleation proceeds by oligomer formation in solution followed by subsequent deposition of large deposits. On the chemically modified substrate, the bound bithiophene can interact with monomeric species to form surface nucleation sites. These results clearly demonstrated the value of chemically tailored surfaces for controlling the deposited film structure.

6.2.3.5 Miscellaneous Conducting Polymers Several other types of conducting polymer films at electrode surfaces have been interrogated with STM. Examples include poly[Re(CO)$_3$(4-vinyl,4'-methyl-2,2'bipyridine)Cl] [89], poly-1-aminoanthracene [90], poly(o-toluidine) [91], and the co-polymer poly[Fe(4-vinyl-4'-methyl-2,2'-bipyridyl)$_2$-(CN)$_2$]-poly(4-vinyl-4'-methyl-2,2'-bipyridyl) [92]. As with these investigations, STM has been applied to probe issues related to structural morphology and film nucleation and growth.

6.2.3.6 Poly(phenylene oxide) (PPO) Applications of electrochemical sensors are in many instances complicated by electrode fouling and subsequent deactivation. For example, the electro-oxidation of phenol produces a passivating, polymeric film of PPO on electrode surfaces. Because of the strong interest in the electrochemical determination of such compounds in environmental, industrial, food, or clinical matrices, the structure and formation mechanism of PPO films at electrodes have been examined by STM [93,94]. An early investigation was concerned with the effect of deposition procedure and correlated STM images with electrode activity as represented by the voltammetric responses of marker redox couples at glassy carbon (GC) electrodes [93]. STM images of PPO on GC formed by voltage cycling at low scan rates (5 mV/sec) showed a rough, granular structure as compared to a smooth fibrous morphology for films deposited at 50 mV/sec. The magnitude of the current flow in the cyclic voltammetry of ferricyanide also exhibited a notable dependence on deposition scan rate. Films produced at 50 mV/sec pacified the oxidation of ferricyanide after a single cycle while four cycles were required to achieve the same level of inactivity at 5 mV/sec. Similar results were reported for films deposited at constant potential for various times. The correlation of STM observations with the voltammetry clearly demonstrates the utility of this integrated approach to gain insights into relationships between electrode surface structure and electrochemical reactivity.

A more recent in situ STM examination has provided a molecular scale perspective of phenol-phenolate adsorption on Au(111) electrodes [94]. This investigation combined electrochemical, STM, and spectroscopic descriptions for the development of both micro- and macroscopic descriptions of the adlayer. Figure 6.13a shows a STM image of a Au(111) surface emmersed in a 0.1 mM solution of phenol at a pH of 9.3. At this pH, ~0.1% of the aromatic compound is in the phenoxide form. The image shows molecular scale features arranged

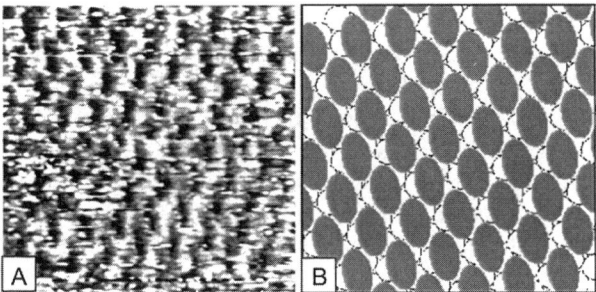

Figure 6.13. (a) In situ STM image of adsorbed phenoxide ion at Au(111) at 50 mV vs NHE. Nearest-neighbor spacing is 0.50 ± 0.02 nm and corresponds to a ($\sqrt{3} \times \sqrt{3}$)R30° adlayer: V_b = 28 mV; I_t = 2 nA. (b) Model of the phenoxide overlayer showing the underlying Au(111) lattice. The open circles represent Au, and the filled ovals represent phenoxide. Reproduced with permission from Richard and Gewirth [94].

in a hexagonal pattern, indicative of a $(\sqrt{3} \times \sqrt{3})R30°$ overlayer structure. Figure 6.13b gives one possible arrangement of this structure. Based on van der Waals radii as well as descriptions derived from in situ infrared reflection spectroscopy, the phenol molecules were found to adsorb "edge-on," which would facilitate binding through the O atom. After the initial electro-oxidation of this ordered layer, a disordered architecture comprised of close-packed oblong-shaped disks resulted. These images were interpreted as arising from phenoxide oligomers lying with their aromatic rings parallel to the electrode surface. Thus, the adsorbed monomer must change its orientation for polymerization. These results illustrate the ability of in situ STM to provide detailed molecular scale descriptions of organic films at electrode surfaces.

6.2.4 Electrochemically Addressable Biomolecular Films

The possibility of visualizing directly the structure and orientation of biologically important species such as DNA and proteins with SPM techniques has prompted many research groups to focus their efforts on this front. Some of this work has been directed at STM studies of amperometric biosensors [68,95] as well as of biomolecules adsorbed on electrodes [96–105]. The usage of STM in biosensor applications has been mainly concerned with correlating surface structure to sensor response. For example, for a PPy/glucose oxidase (GOD) electrode, it was found that both the surface morphology and amperometric response to glucose depends on the deposition procedure [95]. Because of the lower conductivity of GOD relative to PPy, the entrapped enzyme was visualized in the PPy structure as "holes" in STM images.

One of the initial difficulties in obtaining STM images of adsorbed biomolecules was the elimination of a tip-induced motion of the adsorbate. Thus, some of the early work in this area was concerned with developing schemes to attach biomolecules to electrode surfaces [96,97]. Pathways for electrochemically depositing enzymes [97] and nucleic acids [96] at gold were devised such that the adsorbed layer could withstand the tip-surface interaction forces and be reproducibly imaged. The two-dimensional structure of films of DNA bases deposited under electrochemical control has also been examined with STM [98–100]. The molecularly resolved images of these systems suggest the possibility of resolving single bases of adsorbed DNA [98]. The ability of STM to image electrodes under potential control has also enabled the observation of the electrochemical oxidation of a guanine monolayer [100].

Due to their importance in biological oxygen transport and in their wide usage as model systems for electrocatalysis, adsorbed porphyrin films have recently been studied [101,102]. In one study, electrochemical and STM results showed that a thin film of protoporphyrinato iron(III) chloride on HOPG is comprised of 3.0 nm diameter aggregates, while a thicker film produced by electropolymerization exhibits a more dense morphology [101]. No evidence was found indicative of a monomolecular layer adsorbed with the porphyrin ring parallel to the electrode surface as was proposed based on electrochemical

6.2 STM CHARACTERIZATIONS OF ORGANIC THIN FILMS

and spectroscopic descriptions of this system [106]. However, in another study, nonmetallized protoporphyrin along with iron and zinc protoporphyrin were molecularly resolved on the basal plane of HOPG. These molecular resolution images show that all three types of adsorbates pack in similar two-dimensional arrays in which the plane of the porphyrin ring is oriented parallel to the surface [102]. The images also revealed significantly different internal structures for the three macrocycles presumably due to the variation in the complexed metal ion. Molecular resolution in situ STM has also revealed an ordered two-dimensional structure for porphyrins adsorbed to iodine modified gold electrodes [103].

Many researchers approach the complexities of examining biomolecular films with STM (e.g., sample preparation and image interpretation) by studying simpler organic monolayers on electrode surface [104,105]. The behavior of these model systems can then be extended to more complicated systems. In a recent report, electrochemical STM was utilized to observe potential induced phase transitions in a monolayer of 2,2′-bipyridine (22BPY) on Au(111) at molecular resolution [104]. These molecules adsorb "on edge" from an aqueous solution of 0.1 M $NaClO_4$, with the pyridyl nitrogens associated with the gold surface, and aggregate in a face-to-face configuration to form polymeric linear chains of various lengths. Figure 6.14 shows the evolution of the monolayer structure as the

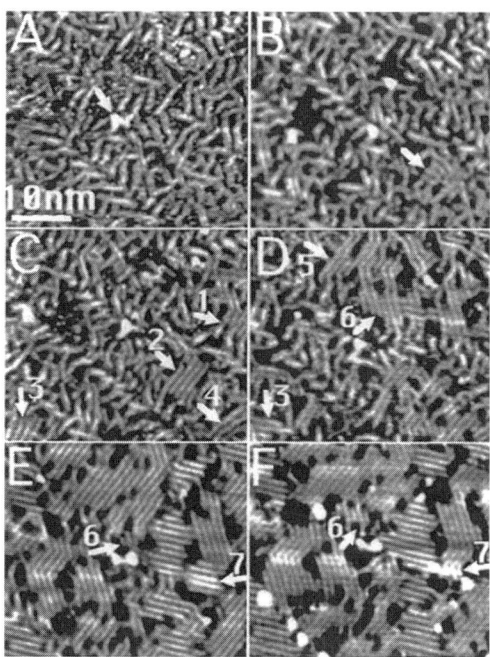

Figure 6.14. In situ STM images of the surface charge induced transition from disordered to ordered phases for 2,2′ bipyridine at Au(111): (a) −0.10 V, (b) +0.14 V, (c) +0.18 V, (d) +0.22 V, (e) +0.25 V, and (f) +0.40 V (vs SCE). Reproduced with permission from Chuna and Tao [104].

electrode potential (i.e., surface charge density) is increased. At a potential of −0.1 V (vs. the saturated calomel electrode), the chains are arranged in random orientations with respect to each other (Figure 6.14a). As the potential of the electrode is sequentially increased to 0.4 V (Figure 6.14b–f), the monolayer rearranges to form a collection of ordered domains. The orientation of the domains are aligned along three directions, each oriented 120° with respect to the other. It was therefore concluded that the domains of the adlayer were preferentially oriented along the hexagonal Au(111) lattice, although the underlying gold surface was not concurrently imaged. In addition, the phase transition induced by changes in the surface charge were shown to be fully reversible, reflecting differences in the concentration of adsorbed 22BPY. This explanation is based on electrostatic arguments in that a more positive electrode potential will increase the adsorption energy of the negatively charged nitrogen atoms of 22BPY. This study clearly demonstrated the utility of STM in providing detailed information concerning electrochemical transformations of organic adsorbates.

6.3 SFM CHARACTERIZATIONS OF ORGANIC THIN FILMS AT ELECTRODE SURFACES

The goal of this section is to introduce the fundamental principles of SFM and to present a brief review of its successful applications to the characterization of organically modified electrodes. Following a brief introduction of operational principles and scope of applicability, examples that demonstrate where SFM characterizations have produced insight into the structure and reactivity at electrified interfaces that have been modified with monomolecular and polymeric films are described.

6.3.1 Instrumentation and Theory

6.3.1.1 Introduction A number of excellent reviews of SFM are available, ranging from broad introductions [13,14,107] to detailed examinations of fundamental principles [108,109] and instrumental design issues [110]. As shown in Figure 6.15, the SFM is similar in several respects to the STM—a probe is rastered over a surface while detecting variations in interfacial interactions. With SFM, however, the deflection of a cantilever-mounted tip, which is affected by changes in the physical and chemical forces between the tip and sample surface, is monitored while scanning. As such, insulating as well as conductive samples can be imaged using SFM. Furthermore, the lack of the use of current in the imaging mechanism facilitates applications of SFM to in situ electrochemical explorations. The next three sections focus on (1) SFM imaging modes, (2) accessible experimental ranges, and (3) spatial resolution.

6.3.1.2 Modes of Operation There are a larger number of SFM imaging modes [11,13,14,17], several of which are summarized in Figure 6.16. In the

6.3 SFM CHARACTERIZATIONS OF ORGANIC THIN FILMS

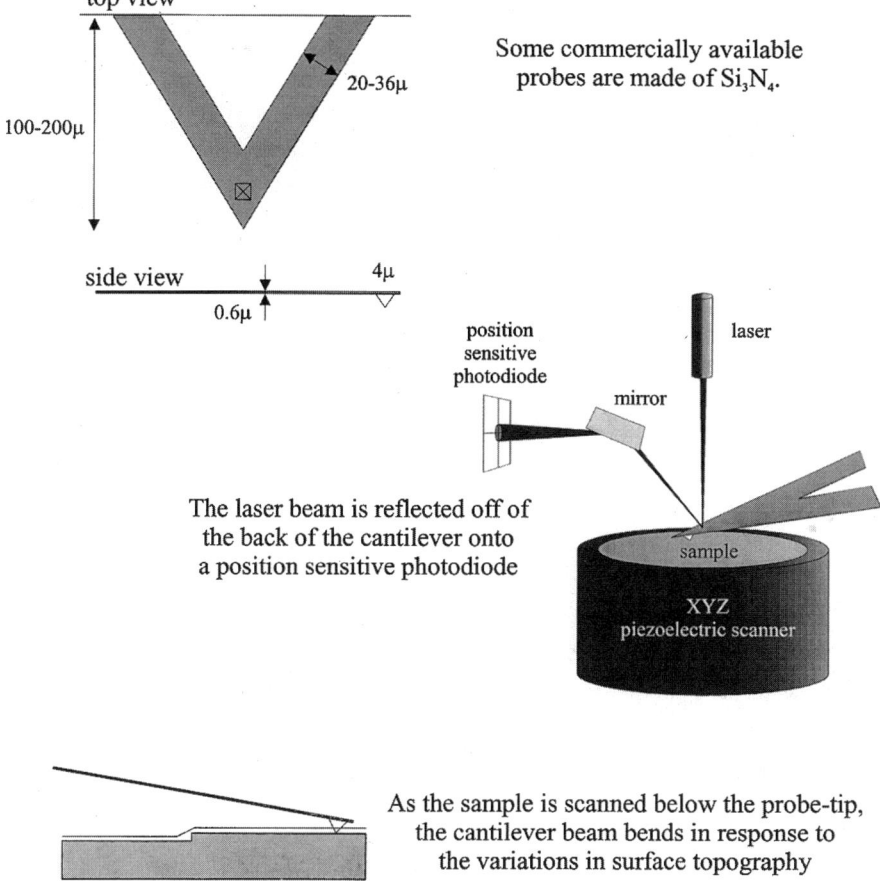

Figure 6.15. Schematic of the principle components for an optical lever type AFM.

most used mode, often referred to as the contact mode [111,112], a cantilevered tip is brought into direct physical contact with a sample surface. As the tip is rastered across the surface, a process accomplished with SFM by movement of the sample by a set of piezoelectric actuators, the deflection of the tip is affected by changes in the extent of its interaction with the sample surface. Maps of tip deflection are then used to develop topographic profiles of the sample surface (see Fig. 6.16a).

Topographical information obtained in a contact mode, however, can be compromised in instances where surface properties such as elasticity, adhesion, or friction vary at a microscopic level [113–117]. Interestingly, these potential complications have been exploited to devise new SFM imaging modes. Elasticity, for example, may be determined by a mode known as force modulation [118–120]. With force modulation, the vertical position of the sample is mod-

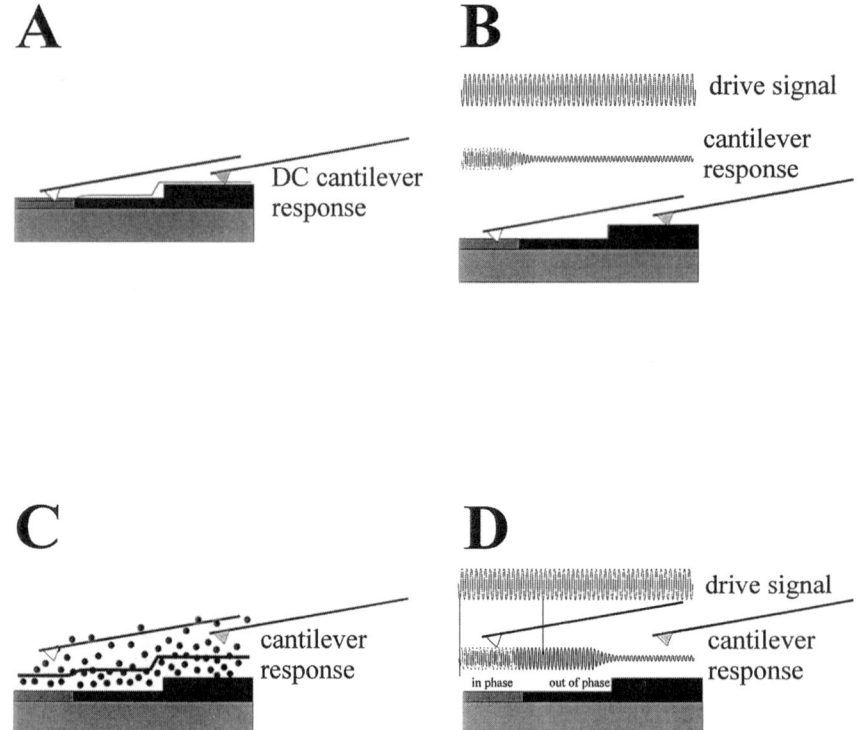

Figure 6.16. Measurable properties and modes of operation for SFM: (a) the classical contact mode measures the topography; (b) an AC contact mode with force modulation measures the viscoelastic response of the tip-sample contact; (c) a DC noncontact mode has been used for initial work on imaging long-range interactions; (d) an AC noncontact mode and is currently the most popular mode for imaging soft samples.

ulated while the tip remains in contact with the sample, and the resulting cantilever deflection correlated with the elasticity of the tip-sample contact region. There are two general time-domains for the determination of elasticity: the high frequency domain and the low frequency domain. In the high frequency domain, as presented in Figure 6.16b [118], the position of the cantilever is modulated vertically by applying a sinusoidal waveform to the Z-positioning piezoelectric actuator at frequencies in the kilohertz range. By monitoring the amplitude and phase of the cantilever response, elasticity becomes the imaged response. In contrast, the low frequency (1–10 Hz) domain is studied by the application of a large amplitude triangular waveform to the z-positioning actuator. The slope of the correlation between the cantilever deflection and the vertical displacement of the sample, plots of which are often referred to as deflection-distance curves (see Fig. 6.17), force-distance curves, or simply force curves, can then be used to determine the elasticity.

6.3 SFM CHARACTERIZATIONS OF ORGANIC THIN FILMS

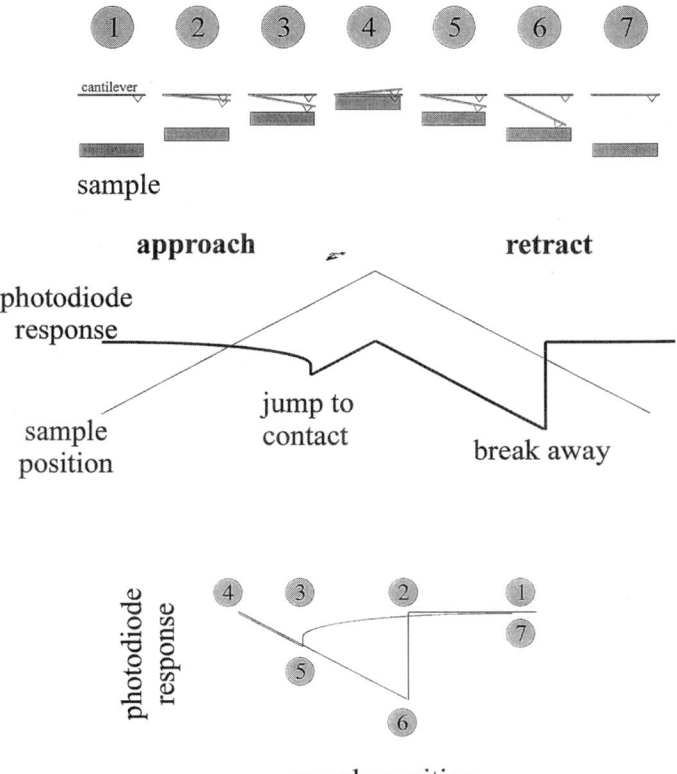

Figure 6.17. Normal force measurements with SFM: force curves: (1) the probe-sample separation, z, is large and they are not interacting; (2) z is decreased and the probe and sample begin to interact, but the strength is less than the restoring force of the cantilever; (3) at this point, the interactions become greater than the cantilever restoring force, and there is an instability as the probe "jumps to contact" with the sample; (4) at this point z is decreased; (5) z is still being decreased; (6) at this point, the cantilever force exceeds the adhesive forces and the probe and contact is broken; (7) and the probe and sample are returned to the noninteracting large separation.

Force curves can also be used to determine the adhesive force (F_{ad}) actively holding the tip in contact with the sample. Since the tip is mounted at the end of a "Hooke-like" spring, the value of F_{ad} required to separate the tip from the surface can be determined from the deflection of the cantilever required to break contact with the sample surface via eq. (2) [121–129].

$$F_{ad} = k_N \Delta z \qquad (2)$$

In eq. (2), k_N represents the force constant for the normal bending mode of the cantilever. Detailed discussions about the intricacies of the measurements of F_{ad}

with SFM are covered by Burnham [125,126]; this measurement is discussed further in a subsequent example.

To this point, we have focused on imaging mechanisms that have centered about deflections of the cantilevered tip normal to the surface. However, the force generated as the tip is moved laterally across the sample surface can also be used as an imaging mechanism. Such forces induce a torsional deflection of a cantilever about its major axis. The top of Figure 6.18 represents how an optical beam deflection system can detect these torsional deflections of the cantilever. This approach results in a plot often denoted as a friction loop, which is a plot of the torsional response of the cantilever with respect to the movement of the sample in a given direction, as shown in the lower portion of Figure 6.18. Thus, the slope of the static portion of a friction loop (points b-e) can be used

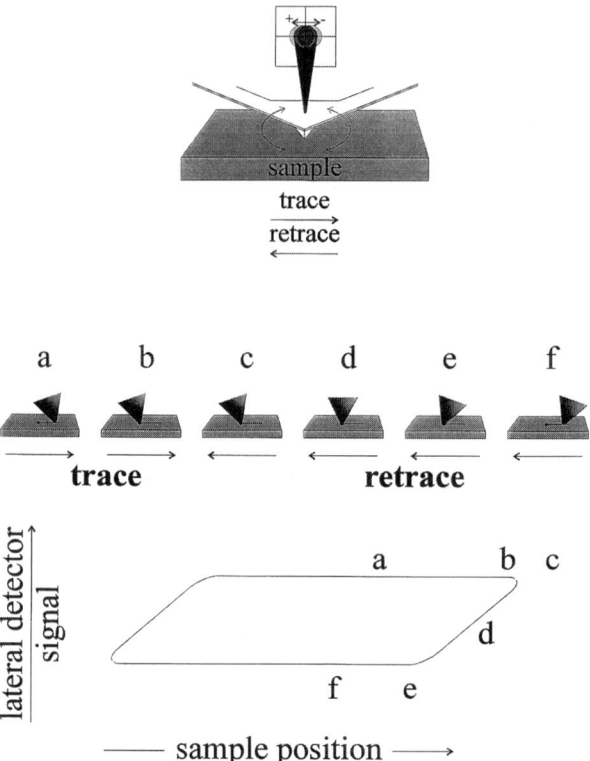

Figure 6.18. Lateral force measurements with SFM friction loops: (a) as the sample is scanned to the right, the cantilever experiences a torsion that deflects the laser beam to the left; (b) eventually, the scan reaches the right limit; (c) the scan direction reverses and initially there is no change in the torsion of the cantilever; (d) however, as the scan progresses to the left the cantilever rotates about a fixed tip-sample contact point; (e) eventually, cantilever restoring force (torsional) exceeds the static friction; (f) and the probe begins sliding across the surface.

to calibrate the detector response, which when coupled with the torsional force constant of the cantilever (k_t), can be correlated with the frictional forces at work in the contact zone [130–133]. Recent efforts using this approach to map surface composition at nanometer length scales will be discussed shortly.

There are also a host of SFM imaging modes that rely on tip deflections through long range forces (e.g., double layer forces) [122,134], van der Waals forces [135,136], electric and magnetic forces [136–139], and a few others [135,140,141]. Typically, these forces are determined by an AC technique where the SFM probe is oscillated near resonance and changes in the amplitude, frequency, and phase of the oscillation are monitored as a function of probe-sample separation [112,137]. These three measurables are influenced by the interaction force and its gradient between the probe and the sample. Access to these interactional and material properties of surfaces at these length scales offers the surface chemist an unprecedented view of the surface before, during, and following modification with organic films.

6.3.1.3 Resolution Issues While STM has produced molecular *resolution* images of surface structures, the corresponding studies with SFM have typically produced images with molecular *periodicity*. The distinction arises from the mechanistic differences between the measurements made by the two instruments. With STM, each pixel in an image represents the exponential dependence of the current that flows between an "atomically sharp" filament [16] and sample surface. In contrast, each pixel in a SFM image reflects the deflection of a generally larger area probe tip that is induced by the collective interactions of tens or hundreds of molecules via continuum mechanics as described by the Hertz and JKR theories [9,14]. In other words, the latter situation is the result of a large tip-sample contact area, the dependence of the contact area on the normal force between tip and sample, the size and shape of the contacting area of the probe, and the elasticity of the sample. Organic adlayers are particularly susceptible to large contact areas, since they are comparatively more compliant, i.e., have lower Young's moduli. Thus, to obtain molecular resolution of organic layers with contact SFM, sharp tips, low forces, and relatively rigid layers are required [116,117,142–147]. Although a recent study has demonstrated that under appropriate conditions, a noncontact technique can produce an image at molecular resolution [148], the relatively high compliance of most organic films precludes this level of scrutiny.

6.3.2 Spontaneously Adsorbed Monolayers

6.3.2.1 Introduction As described in the STM section, the application of spontaneously adsorbed monolayers to the modification of metal surfaces has captured the imagination of electrochemists on a global scale. This section describes SFM investigations of such interfaces, which may be conveniently divided in two categories: (1) structural morphology and (2) changes in tribological properties. Examples of studies in each category are discussed in turn.

6.3.2.2 Morphology

Initial studies of these types of monolayers with SFM focused primarily on structural characterization. The first molecular level images of alkanethiolate monolayers produced by SFM verified the $\sqrt{3} \times \sqrt{3}$ packing at Au(111) [149], and were subsequently extended to adlayers prepared from perfluorinated alkanethiols [150]. Importantly, the images for the perfluorinated system exhibited the packing expected from the increase in the molecular size of the perfluorocarbon chain (see Fig. 6.19). Because of the limited resolution of SFM, however, many of the structural subtleties measured in

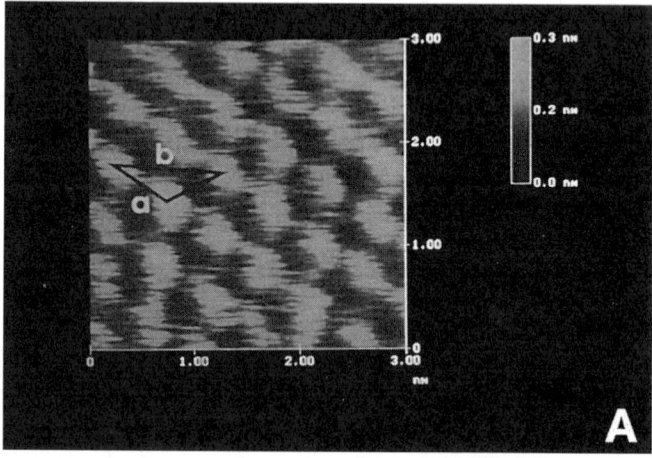

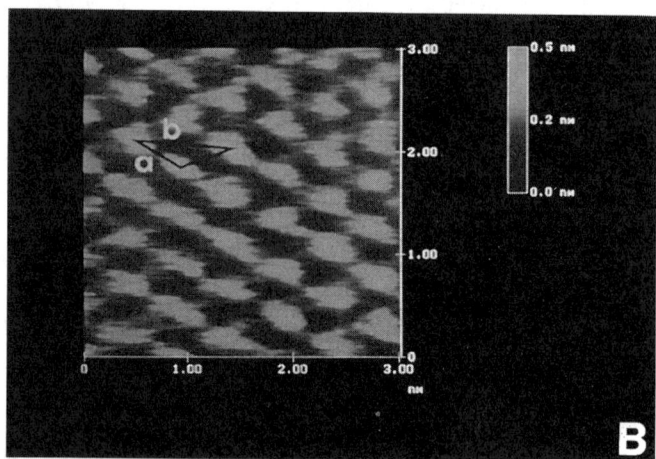

Figure 6.19. AFM images (3 nm × 3 nm) of monolayers chemisorbed at Au(111) from (a) $CF_3(CF_2)_7(CH_2)_2SH$ and (b) $CH_3(CH_2)_9SH$. For (a), $a = 0.58$ nm and $b = 1.01$ nm, and for (b), $a = 0.50$ nm and $b = 0.91$ nm. The z-scale is 0–0.50 nm. Reproduced with permission from Alves and Porter [150].

6.3 SFM CHARACTERIZATIONS OF ORGANIC THIN FILMS

STM are rarely detected. For example, STM images reveal a host of complex molecular superstructures that are currently unobserved via SFM. We suspect that this situation is a combination of the relatively small domain size of these superstructures with the moderately larger contact area of the SFM tip. Indeed, only recently have SFM images revealed the ubiquitous depressions common to thiolate monolayers [32,36,151].

These studies have recently been extended to the electrochemical arena [152]. The results are summarized in the images in Figure 6.20. Beginning with a preformed $\sqrt{3} \times \sqrt{3}$ adlayer of an alkanethiolate, the potential applied to a Au(111) electrode was scanned to large cathodic values. At extremes in cathodic potentials, the adlayer is reductively desorbed [153] and the $\sqrt{3} \times \sqrt{3}$ periodicity was replaced by that of the underlying Au(111) electrode. Importantly, the angular displacement of the patterns in the two separate images confirmed the 30° rotation of the adlayer structure with respect to the underlying lattice. This result clearly demonstrates the value of SFM to electrochemical studies.

6.3.2.3 Mechanical and Chemical Properties This section examines the use of friction, adhesion, and elasticity as alternatives to imaging electrochemical interfaces via SFM. The majority of these investigations have focused on proof-of-concept explorations, seeking to define the range and scope of the application of friction- and adhesion-based measurements to advancing the chemical content of SFM imaging. This section first covers details concerning applications that define the state-of-the-art in applications to compositional mapping and then the few examples of such characterizations in electrochemical environments.

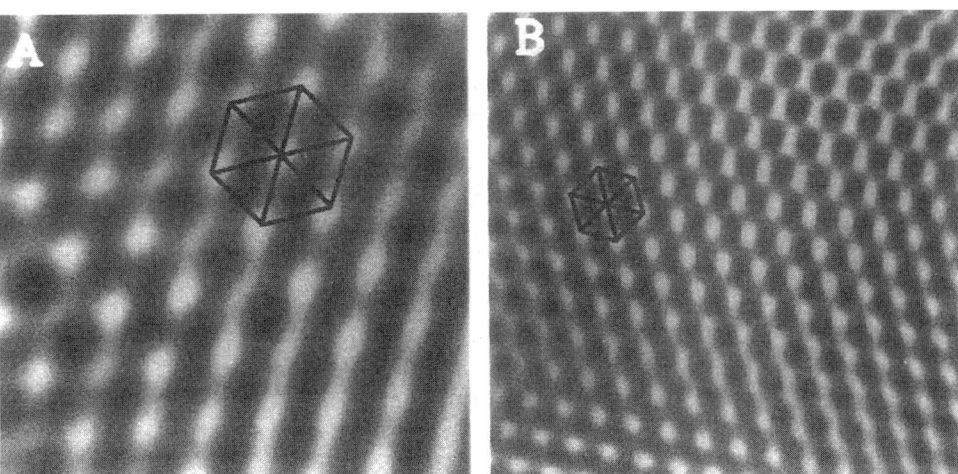

Figure 6.20. In situ SFM image (3 nm × 4 nm) of (a) an octadecanethiolate monolayer at Au(111) in 100 mM NaClO$_4$ solution at an applied potential of +290 mV vs a Ag/AgCl electrode and (b) after desorption of the monolayer by application of a large negative potential. Reproduced with permission from Pan et al. [152].

The underlying basis of strategies that utilize adhesion and friction as compositional characterization tools is the dependence of such interactions on the chemical functional groups at the outermost few angstroms of the two surfaces that form a microcontact. Evaluations of the depth sensitivity of wetting, which demonstrated that the contact angle for a variety of probe liquids was largely controlled by the functional groups at the surface of different types of spontaneously adsorbed monolayer films [154,155], serve as a foundation for this general concept. Existing theories, however, describe the dependence of the interactions at two contacting surfaces to a complex mixing of contact area, elasticity, load, and surface free energy [9]. In exploring such possibilities, the first report in this area used two-component, phase-separated films prepared by Langmuir-Blodgett transfer techniques from fluorocarbon and hydrocarbon precursors [156]. The results indicated that the friction at the fluorocarbon domains were larger than that at the hydrocarbon domains. Importantly, measurements of the elastic compliance at each type of domain revealed that the modulus of the fluorocarbon domain was lower than that of the hydrocarbon domain, demonstrating that differences in elasticity could be exploited as a means to achieve compositional mapping via SFM.

More recent studies have built on this precedent using alkanethiolate monolayers with different end groups chemisorbed at gold [157–161] to test the viability of both friction and adhesion measurements as mapping techniques. In these efforts, the adlayer architectures were designed to minimize contributions of elasticity to the friction and adhesion measurements by using long polymethylene chains and end groups with limiting packing densities similar to that of the underlying polymethylene chain. The former yielded mechanically rigid structures through enhanced cohesive interactions, whereas the latter minimized disruptions in the structural order of the underlying polymethylene chains. These studies also explored the use of chemically modified probe tips to gain enhanced control over the composition of the tip-sample micro-contact. Armed with these structural approaches, studies have probed whether differences in surface free energies can be exploited to attain *chemical* discrimination in instances where the *mechanical* properties of the film (e.g., elasticity) have been effectively decoupled from the measurement.

A series of results that exploited the structural attributes of the adlayer noted earlier for assessing the mapping capability of SFM-based friction measurements are presented in Figures 6.21, 6.22, and 6.23 and Table 6.1 [159]. These data were collected using a set of long chain alkanethiolate monolayers with different end groups (e.g., —CH_3, —CH_2Br, —CO_2CH_3, —CH_2OH, and —CO_2H) and the same unmodified probe tip. The normal forces (F_N) between the probe tip and sample surface ranged up to ~100 nN. As evident in Figure 6.21, the friction (f) at each of the microscopic contacts exhibits a linear dependence with F_N that can be described as

$$f = f_0 + \alpha F_N \tag{3}$$

6.3 SFM CHARACTERIZATIONS OF ORGANIC THIN FILMS

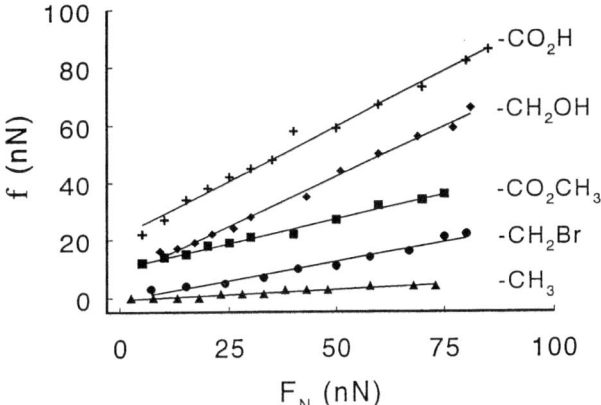

Figure 6.21. Plots of f versus F_N for a single unmodified probe tip and a series of end-group modified alkanethiolates at Au(111). The end groups are denoted on each plot. The lines through the data are linear least-squares fits with the slopes (α) and intercepts (f_0) listed in Table 6.1. The monolayer precursors were octadecanethiol, 1-bromo-21-mercaptoheneicosane, 16-mercaptohexadecanoic acid, 16-mercaptohexadecan-1-ol, and methyl 16-mercaptohexadecanoate. Reproduced with permission from Green et al. [159].

where the constants α and f_0 can be correlated with the chemical composition of the interface, where α is the effective coefficient of friction, and f_0 is the friction at zero load (a result of the nonzero contact area present when the probe-sample contact is broken). More importantly, the dependencies in Figure 6.21 follow the trend expected based on surface free energy considerations. That

TABLE 6.1. Frictional Parameters for the Contacts Formed Between an Uncoated Si_3N_4 AFM Tip and Several Different End-Group Terminated Alkanethiolate Monolayers (Au-S$(CH_2)_n$-X Chemisorbed at Au(111)[a]

End-group[b]	$\cos \theta_a$[c]	a	f_0 (nN)	f (nN) for F_N = 20 nM
—CO_2H (n = 15)	~1	0.76	22	42 ± 7
—CH_2OH (n = 15)	~1	0.70	8	40 ± 6
—CO_2CH_3 (n = 15)	0.39	0.34	10	20 ± 5
—CH_2Br (n = 20)	0.12	0.26	~0	9.4 ± 2
—CH_3 (n = 17)	−0.37	0.070	~0	1.4 ± 1

[a] Values of a and f_0 are the respective slopes and y-intercepts obtained from linear least-squares fits of the data in Figure 6.21.
[b] n equals the number of methylene groups in the polymethylene chains.
[c] This column correlates the surface free energies of the different monolayers with the advancing contact angle (θ_a) of water as a probe liquid (154). Thus, through the Young relation [7], the larger $\cos \theta_a$, the greater is the surface free energy.

is, the higher the free energy of the sample surface, the higher is the friction. This conclusion is further delineated in Table 6.1, which presents the slope and intercepts of the dependencies of the data in Figure 6.21, the values of f at a F_N of 20 nN, and the cosine of the advancing contact angles at each surface with water as the probe liquid. Together, these data clearly demonstrate a correlation between the SPM-measured friction and the surface free energy of the contacting tip-sample interface, providing a basis using this technique for compositional mapping based on the chemical functional groups at a sample surface.

The ability to map surface composition at a nanometer-length scale is demonstrated by using the partially formed, spatially segregated bilayer structure that is shown in Figure 6.22, where X is —COOH. The results of characterizing this structure in both topographic and frictional modes are presented by the images in Figure 6.23 [159]. With this bilayer structure, the uncoated tip is in contact with —CH_3 end groups when located at the top of the second layer of the bilayer and is in contact with —COOH end groups when located at the top of the first layer of the bilayer. Thus, the topographic image defines the tip location and can be used for a correlation of the SFM-measured friction and end group composition. On the basis of the direct and expected correspondence with the topographic profile, the differences in the frictional image reflect the contact of the probe tip with small domains of two distinctly different functional groups. Furthermore, an analysis of the resolution of the compositional mapping from these images, given as the minimum distance clearly separating two chemically distinct domains, defines the limit for these samples at ~10 nm.

In an exciting alternative to the preceding imaging strategies, experiments where the potential difference between sample and probe tip was used to assess

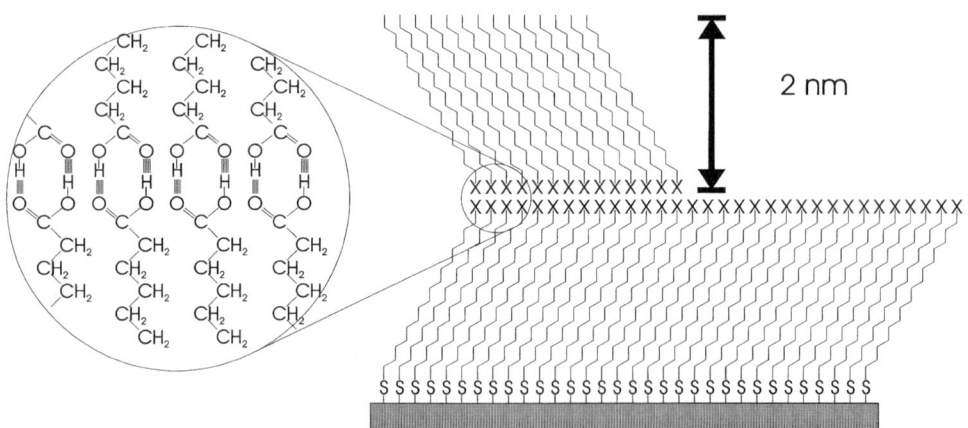

Figure 6.22. Idealized depiction of a partially formed bilayer comprised of mercaptohexadecanoic acid (bottom layer) and stearic acid (top layer).

6.3 SFM CHARACTERIZATIONS OF ORGANIC THIN FILMS

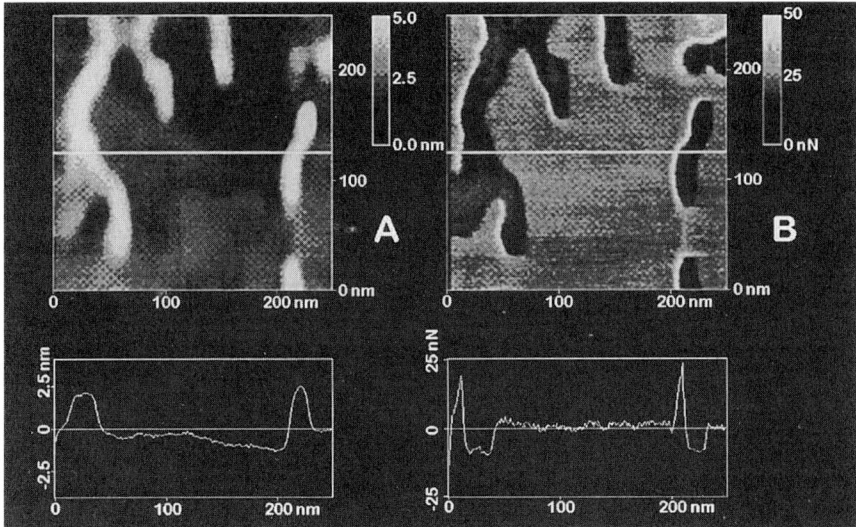

Figure 6.23. AFM (250 nm × 250 nm) topographic (a) and friction force (b) images of the partially formed bilayer comprised of mercaptohexadecanoic acid (bottom layer) and stearic acid (top layer) as depicted in Figure 6.22. Height and friction data were collected concurrently. The z-scale is 0–5.0 nm. The scan directions are from left-to-right at a rate of 10 Hz, using an unmodified Si_3N_4 tip. The normal force (F_N) between the tip and sample was ~10 nN. The topographic features in the cross section of (a) are ~2 nm in height, and the frictional variation defined by the cursors in the cross section in (b) are ~15 nN. The lateral distance required to complete the transition from low friction to high friction, as illustrated by the horizontal separation of the cursors in the cross section of (b) is ~10 nm. Reproduced with permission from Green et al. [159].

how the screening by the dipole moment of the end group altered F_{ad} [162,163]. This measurement can be viewed as a microscopic version of a conventional surface potential or Kelvin probe measurement [7]. The measurement is performed by fixing the tip-sample separation distance and following the change in the interaction force as the potential difference between the tip and sample is varied. The data, which was obtained using samples and probe tips modified with —CF_3, —CH_3, —CH_2OH, or —CO_2H end groups, exhibited a parabolic potential dependence, with attractive forces at both positive and negative potential differences. Most notably, the minimum in these curves, defined as contact potential differences, correlated with differences in the dipole moments of the end groups. Studies using two-component, phase-separated films prepared by Langmuir-Blodgett transfer techniques from fluorocarbon and hydrocarbon precursors have also demonstrated the sensitivity of this type of measurements to the lateral distribution of surface dipoles [162,163].

Adhesion-based measurements via SFM have recently been applied to the real-time monitoring of the electrochemical transformation of a surface-bound redox group [161]. These measurements were conducted using monolayers prepared from a ferrocene-terminated alkanethiol (11-mercaptoundecyl ferrocenycarboxylate (FcT)), chemisorbed at a Au(111) substrate, and a probe tip modified with a methyl-terminated alkanethiolate monolayer (i.e., octadecanethiolate) as the tip-sample combination. The results are summarized in Figures 6.24, 6.25, and 6.26. Figure 6.24 presents the cyclic voltammetric current-potential (i-E) curve for the oxidation and reduction of the gold-bound FcT monolayer mounted in the electrochemical cell of the SFM. As observed in earlier electrochemical studies of this adlayer [164], the i-E curves have asymmetric shapes for both the anodic and cathodic scans. A set of the force curves obtained during the acquisition of the i-E curves in Figure 6.24 is shown in Figure 6.25. The force curves on the left side of Figure 6.25 were collected at −0.47, −0.37, −0.13, −0.09, −0.04, +0.11, and +0.20 V, and those on the right side were collected at the same applied voltages during the subsequent cathodic scan. At the early stages of the anodic scan, the force curves remain unchanged, consistent with the lack of a detectable oxidative conversion of the ferrocenyl group at these values of applied potential. As the anodic scan continues, however, a notable decrease in the translation distance required to rupture the contact between the probe tip and sample is observed. The force curves obtained for the subsequent cathodic scan reveal that the changes in F_{ad} are qualitatively reversible.

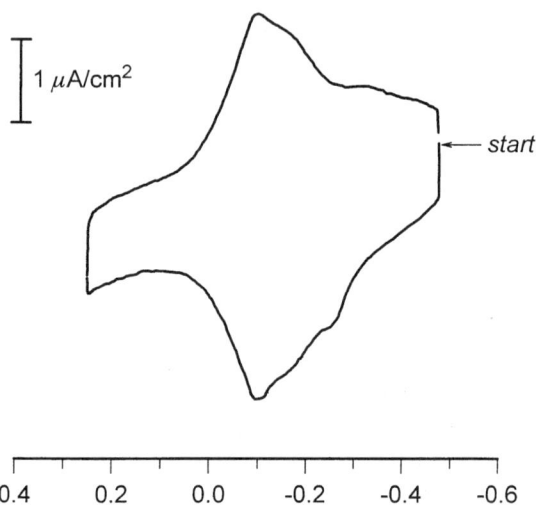

Figure 6.24. Cyclic voltammetric current-potential curve for a FcT monolayer chemisorbed at a Au(111) electrode in an AFM electrochemical cell (scan rate: 10 mV/s; supporting electrolyte: 1.0 M $HClO_4$). A Pt wire, which was ~+0.6 V with respect to a saturated calomel electrode, was used as a pseudoreference electrode; the x-axis is given with respect to the Pt electrode. Reproduced with permission from Green et al. [161].

6.3 SFM CHARACTERIZATIONS OF ORGANIC THIN FILMS

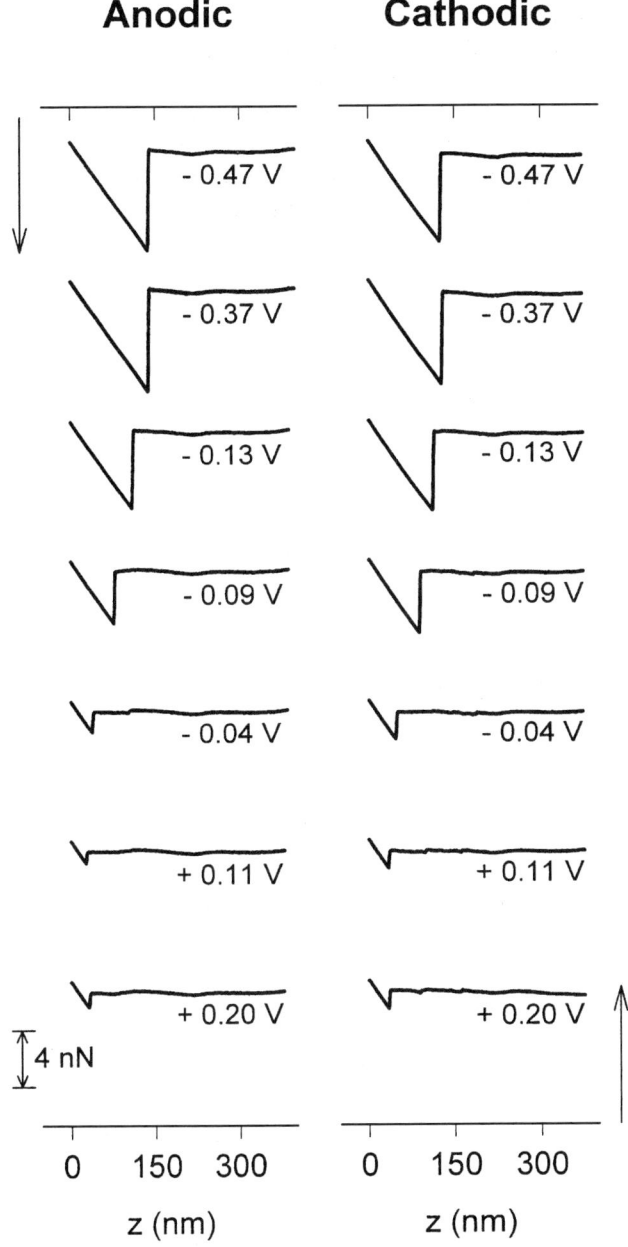

Figure 6.25. Adhesion force curves (F_{ad}) obtained during the cyclic voltammetric scan in Figure 6.23. The approach curves have been omitted for clarity. The probe tip is a gold-coated silicon nitride cantilever modified with a monolayer of octadecanethiolate. The x-axis represents the vertical displacement of the sample stage. Reproduced with permission from Green et al. [161].

A more detailed summary of the dependence of F_{ad} on the redox conversion of FcT is more firmly established by the plots in Figure 6.26. Figure 6.26a summarizes all of the force curves taken over the voltammetric scan, and Figure 6.26b presents the fractional surface concentration (φ) of the reduced form of the ferrocenyl group as a function of applied potential for both the anodic and cathodic scans. The correspondence of the changes in F_{ad} and the transformation of the ferrocenyl group is firmly established by the strong similarity of the dependencies.

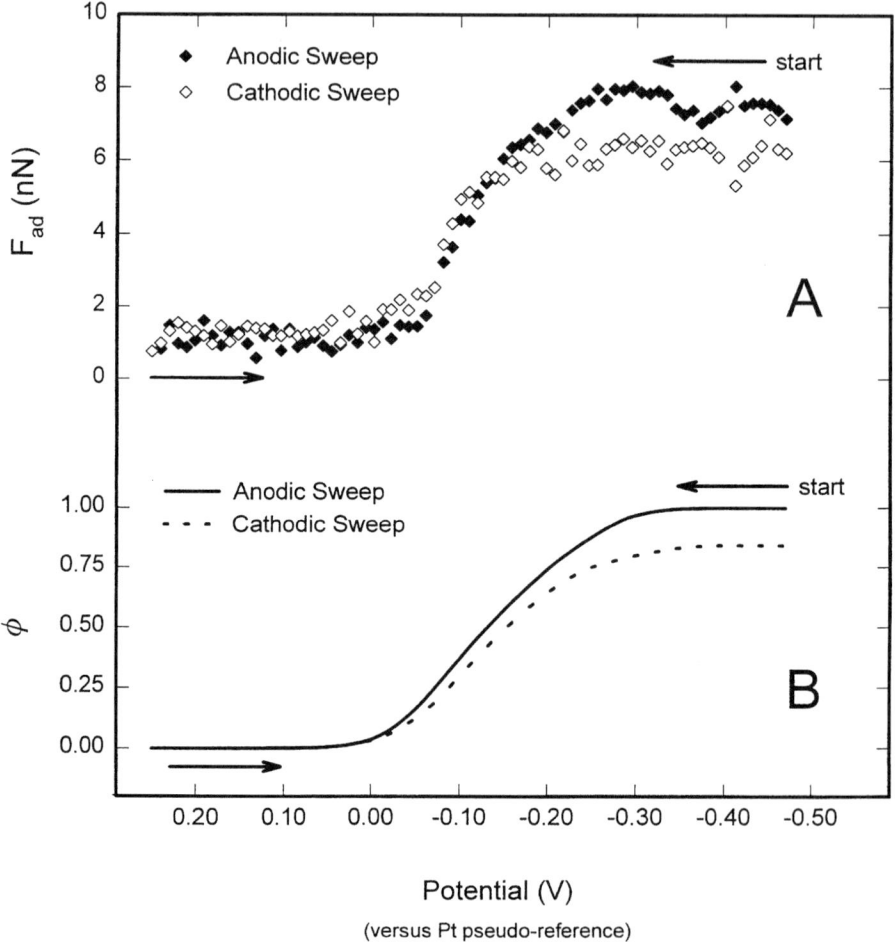

Figure 6.26. (a) Plot of the force of adhesion (F_{ad}) versus applied potential for the cyclic voltammetric scan in Figure 6.24. (b) Plot of the fractional coverage (ϕ) of the reduced form of the ferrocenyl end group of the FcT monolayer versus applied potential. Reproduced with permission from Green et al. [161].

As an explanation of the interaction differences that give rise to the observed dependence in Figure 6.26, arguments based on miscibility differences between the methyl-terminated probe tip and the two different states of the ferrocenyl monolayer were presented [161]. This assertion, which built upon a parallel analysis for frictional data [159], is based on treatments of the work of cohesion (W_{co}) for chemically symmetric contacts and the work of adhesion (W_{ad}) for chemically asymmetric contacts. Thus, the inherent connection of W_{ad} and W_{co} to the surface free energy of each of the contacting surfaces, with the greater differences between W_{ad} and/or W_{co} diagnostic of lower miscibility. These studies demonstrate the utility of SFM-based adhesion measurements to probing electrochemical transformations in situ and in real time.

6.3.3 Polymer Films

6.3.3.1 Introduction As noted earlier, polymer films are critical to technologies such as optoelectronic devices, chemical sensors, rechargeable batteries, and related technologies [63]. As with STM, the SFM-based studies at polymer modified electrodes consist largely of investigations on three types of conductive polymers (polypyrrole, polythiophene, and polyaniline), and on dielectric films of electropolymerized polyphenyleneoxide. The main thrust of the work has concentrated on using SFM as a structural probe in attempts to connect structure to conduction mechanisms. For the movement of charge carriers (e.g., polarons, bipolarons, and exitons in these materials), there are two primary modes of transport: along the polymer backbone and "hopping" between polymer chains. The efficiency of both mechanisms are sensitive to changes in film structure and composition. The following subsections discuss insights into this issue gained by characterizations by SFM.

6.3.3.2 Polypyrrole (PPy) Polypyrrole has possible applications to fields ranging from microelectronic devices to chemical sensors. As with most conducting polymers, the conductivity of PPy is dramatically affected by its "oxidization state." In its reduced form, PPy is an insulator. In contrast, PPy is a strong conductor in its oxidized state and, depending upon the identity of the counter ion, has a conductivity approaching a metallic state. Thus, SFM is uniquely suited as an in situ characterization tool for unraveling key structural issues related to the effects of preparative conditions and electrolysis cycling on the conductivity of PPy [165]. In a recent in situ study, the evolution of the film morphology was monitored during both the deposition and electrochemical transformation of PPy [166]. Images at the micron-length scale revealed that the morphology of PPy changed markedly as the thickness of the deposited film increased. At early stages of growth, the morphology of the film was largely indistinguishable from the surface of the underlying gold electrode. The roughness of the film, however, increased with film growth, as evident from the appearance of small, island-like protrusions. The morphology was also found to change upon the repetitive cycling of a preformed PPy film between oxidized and reduced states. Images revealed that cracks and large pinholes irreversibly developed in the film structure as cycling increased, an observation attributed to the uptake/expulsion of counter ions by the film to maintain electronneutrality.

Studies using SFM have also investigated earlier observations that subtle changes in the formation conditions dramatically affect the charge transfer rates for the resulting polymer film [71]. For electropolymerizations at higher current densities (~3 mA/cm^2), PPy films exhibit a rougher microstructure, whereas preparations at lower current densities have a smoother microstructure. Images as a function of applied potential also confirmed that the film becomes rougher as PPy is increasingly oxidized, with the increase in roughness attributed to the difference in the tension at the surface and in the bulk of the film structure.

6.3.3.3 Polythiophene (PT) Polythiophenes and structurally related materials [167] have promise as optical switches and displays. As with PPy, most of the investigations of PT focus on structural observations of the film as a function of growth conditions and electrode material. Importantly, several of these studies have combined insights gained from vibrational spectroscopic characterizations of film structure [168] as well as semi-empirical structural calculations [169] to complement the observations by SFM. In one example [168], PT formed at a polycrystalline platinum electrode was examined by Raman spectroscopy to follow the electrochemically induced compositional changes in the film. Comparison of the intensities of vibrational modes diagnostic of doped or undoped PT suggested that the readily doped portions of the film grew preferentially as thickness increased. Results also argued that the intercalation of dopants into regions where the film is disordered led to the formation of large protrusions in roughness, as detected by SFM.

The studies of PT films have also explored the use of the *tapping mode* (similar to the noncontact AC mode in Fig. 6.16D) or the intermittent contact mode of SFM [169]. An advantage of this mode is that the extent of the interactions between a cantilevered tip and sample surface operative in contact imaging modes are greatly reduced, enhancing the acquisition of data not compromised by the imaging process. Using the tapping mode, images revealed the presence of coil-like PT strands at the surface of the film. These images, after accounting for the effects of the size and shape of the tip, indicated that these coils have a diameter between 1.0 and 1.5 nm. Geometric optimizations using semi-empirical calculations were used to refine further the interpretations of these images. We note that recent studies have extended the use of SFM to investigations of analogs of PT, including polyselenienyl thiophene [167].

6.3.3.4 Polyaniline (PAN) As with PPy and PT, interest in PAN is driven by technological opportunities in sensor, electrochromic display, and battery development [170]. In a study of sensor development [171], PAN was tested as a coating of an interdigitated electrode array for use in the detection of acetic acid. This scheme relies on the dependence of the conductivity of PAN to the extent of the protonation of its amine substituent. Investigations of long-term performance, however, indicated a decrease in detection capabilities over time. Images taken to delineate whether an evolution of the film morphology caused the observed degradation in performance found the gradual appearance of deep cracks in the film structure. Thus, the degradation in performance was viewed as a consequence of a gradual loss in the conductive pathways through the PAN connecting the electrodes of the interdigitated array, and therefore a decrease in the effective amount of responsive sensor material.

SFM-based characterizations were also used in an in situ mode to determine the alignment of small-sized polymer strands at the surface of the film [170] as well as to probe for subtle changes in the thickness of PAN as a function of the extent of its oxidation [172]. Of particular issue in the latter investigation was the influence of changes in the elasticity of the film in its different states of oxidation. Because such determinations are reliable only when a film exhibits a constant elasticity, changes in elasticity from electrochemical transformations can lead to a significant error in the apparent thickness. Thus, variations in the detected swelling of PAN at different locations across the film were attributed in part to a difference in the "softness" of the film at different locations.

6.3.3.5 Polyphenylene Oxide (PPO)   One of the most exciting applications of the scanning probe microscopies is their potential application to the manipulation of the molecular architecture at chemically modified electrode surfaces. In the case of STM, approaches using the probe tip to induce the local deposition in a predetermined pattern of electropolymerizable films of conductive polymers [15] or to remove a small area of different types of organic thin films from electrode surfaces have been reported [173]. Similar strategies have also been pursued using SFM. In an example of such a study, PPO was electropolymerized from phenolate ion to form thin, relatively smooth dielectric coatings on HOPG [174,175]. Images obtained at values of F_N below ~100 nN revealed a relatively smooth coating topography that followed the contour of the underlying HOPG. Raising the value of F_n above ~250 nN, however, was found to abrasively remove or "nanodoze" the film from the electrode surface. This concept was then extended to the nanodozing-based fabrication of microelectrodes and of spatially patterned, laterally heterogeneous polymeric films, setting precedence for the construction of a host of interesting material composites of possible utility to microelectronics and chemical and biochemical sensors design.

6.4 CONCLUSIONS

We have briefly overviewed in this chapter details related to the development and application of STM and SFM to problems of importance to the electrochemical interface. Efforts underway in several laboratories at an international level are finding new ways to apply the remarkable views of interfaces provided by these techniques. Indeed, some of the recent applications discussed herein give a glimpse of the potential of these techniques for unraveling long-standing issues in electrochemistry. The possibility of nondestructive chemical identification at a molecular-length scale afforded by some of the offspring techniques such as SFM adhesion- and friction-based measurements presents opportunities for mapping interfacial composition at previously inaccessible length scales. The challenge then is to harness the power of these techniques in addressing issues related to the structure of the electrical double layer and the nature (e.g., local structure and spatial distribution) of sites at electrode surface with high levels of reactivity (e.g., catalytic or corrosive sites), and to extend the resulting insights to the design of electrode surfaces with new or improved performance characteristics.

ACKNOWLEDGMENTS

The authors gratefully acknowledge the assistance of R. C. Brush in the conceptualization and design of some of the figures, and in assistance in preparing this review. The suggestions of the reviewers are also acknowledged. This work was supported by the Chemical Sciences Division of the U.S. Department of Energy, the Microanalytical Instrumentation Center of Iowa State University through Post Doctoral Fellowship (C.A.M. and M.T.M.), and an Energy Research Summer Fellowship (J.-B. G.) from the Electrochemical Society, Inc. The Ames Laboratory is operated for the U.S. Department Energy by Iowa State University under Contract No. W-7405-eng-82.

REFERENCES

1. R. W. Murray, in *Electroanalytical Chemistry: A Series of Advances*, A. J. Bard, ed. (Marcel Dekker, Inc., New York, 1984, vol. 13, pp. 191–368.
2. A. T. Hubbard, *Chemical Reviews*, **88**, 633–656 (1988).
3. M. P. Soriaga, *Chemical Reviews*, **90**, 771–793 (1990).
4. A. J. Bard, H. D. Abruña, C. E. Chidsey, L. R. Faulkner, S. W. Feldberg, K. Itaya, M. Majda, O. Melroy, R. Murray, M. D. Porter, M. P. Soriaga, H. S. White, *Journal of Physical Chemistry*, **97**, 7147–7173 (1993).
5. C.-J. Zhong, M. D. Porter, *Analytical Chemistry*, **67**, 709A–715A (1995).
6. H. O. Finklea, in *Electroanalytical Chemistry: A Series of Advances*, A. J. Bard, I. Rubenstein, eds. (Marcel Dekker, Inc., New York, 1996), vol. 19, pp. 110–337.
7. A. Ulman, *An Introduction to Ultrathin Organic Films: From Langmuir-Blodgett to Self-Assembly* (Academic Press, Inc., Boston, 1991).
8. L. H. Dubois, R. G. Nuzzo, *Annual Reviews of Physical Chemistry*, **43**, 437–463 (1992).
9. J. N. Israelachvili, *Intermolecular and Surface Forces: With Applications to Colloidal and Biological Systems* (Academic Press, Inc., San Diego, CA, 1985).

10. A. T. Hubbard, ed., *The Handbook of Surface Imaging and Visualization* (CRC Press, Inc., Boca Raton, 1995).
11. D. A. Bonnell, ed., *Scanning Tunneling Microscopy and Spectroscopy* (VCH, Inc., New York, NY, 1993).
12. L. A. Bottomley, J. E. Coury, P. N. First, *Analytical Chemistry*, **68,** 185R–230R (1996).
13. D. Sarid, V. Elings, *Journal of Vacuum Science and Technology B*, **9,** 431–437 (1991).
14. N. A. Burnham, R. J. Colton, in *Scanning Tunneling Microscopy and Spectroscopy*, D. A. Bonnell, ed. (VCH, Inc., New York, NY, 1993) pp. 191–249.
15. A. J. Bard, F. R. Fan, M. V. Mirkin, in *Electroanalytical Chemistry: A Series of Advances*, A. J. Bard, ed. (Marcel Dekker, Inc., New York, 1994), vol. 18, pp. 243–373.
16. C. J. Chen, *Introduction to Scanning Tunneling Microscopy*, Oxford Series in Optical and Imaging Sciences (Oxford University Press, New York, 1993), vol. 4.
17. D. Sarid, *Scanning Force Microscopy*, Oxford Series in Optical and Imaging Sciences (Oxford University Press, New York, 1991), vol. 2.
18. T. T.-T. Li, M. J. Weaver, *Journal of the American Chemical Society*, **106,** 6107–6108 (1984).
19. M. D. Porter, T. B. Bright, D. L. Allara, C. E. D. Chidsey, *Journal of the American Chemical Society*, **109,** 3559–3568 (1987).
20. A. Kumar, N. L. Abbott, E. Kim, H. A. Biebuyck, G. M. Whitesides, *Accounts of Chemical Research*, **28,** 219–226 (1995).
21. W. Mizutani, D. Anselmetti, B. Michel, *NATO ASI Ser., Ser. E*, **240,** 43–48 (1993).
22. L. Strong, G. M. Whitesides, *Langmuir*, **4,** 546–558 (1988).
23. C. E. D. Chidsey, G.-Y. Liu, P. Rowntree, G. Scoles, *Journal of Chemical Physics*, **91,** 4421–4423 (1989).
24. C. E. D. Chidsey, D. N. Loiacono, *Langmuir*, **6,** 682–691 (1990).
25. H. Sellers, A. Ulman, Y. Shnidman, J. E. Eilers, *Journal of the American Chemical Society*, **115,** 9389–9401 (1993).
26. L. Sun, R. M. Crooks, *Journal of the Electrochemical Society*, **138,** L23–L25 (1991).
27. L. Sun, R. M. Crooks, *Langmuir*, **9,** 1951–1954 (1993).
28. C. A. Widrig, C. A. Alves, M. D. Porter, *Journal of the American Chemical Society*, **113,** 2805–2810 (1991).
29. O. Chailapakul, L. Sun, C. Xu, R. M. Crooks, *Journal of the American Chemical Society*, **115,** 12459–12467 (1993).
30. Y. T. Kim, A. J. Bard, *Langmuir*, **8,** 1096–1102 (1992).
31. K. Edinger, A. Goelzhaeuser, K. Demota, C. Woell, M. Grunze, *Langmuir*, **9,** 4–8 (1993).
32. U. Durig, O. Zuger, B. Michel, L. Haussling, H. Ringsdorf, *Physical Review B*, **48,** 1711–1717 (1993).
33. C. Schonenberger, J. A. M. Sondag-Huethorst, J. Jorritsma, L. G. J. Fokkink, *Langmuir*, **10,** 611–614 (1994).

34. J.-P. Bucher, L. Santesson, K. Kern, *Langmuir*, **10**, 979–983 (1994).
35. J. A. M. Sondag-Huethorst, C. Schonenberger, L. G. J. Fokkink, *Journal of Physical Chemistry*, **98**, 6826–6834 (1994).
36. C. A. McDermott, M. T. McDermott, J.-B. Green, M. D. Porter, *Journal of Physical Chemistry*, **99**, 13257–13267 (1995).
37. T. Han, T. P. Beebe, *Langmuir*, **10**, 2705–2709 (1994).
38. S. J. Stranick, A. N. Parikh, Y. T. Tao, D. L. Allara, P. S. Weiss, *Journal of Physical Chemistry*, **98**, 7636–7646 (1994).
39. W. Mizutani, B. Michel, R. Schierle, H. Wolf, H. Rohrer, *Applied Physics Letters*, **63**, 147–149 (1993).
40. D. Anselmetti, C. Gerber, B. Michel, H. Wolf, H. J. Guentherodt, H. Rohrer, *Europhysics Letters*, **23**, 421–426 (1993).
41. R. L. McCarley, D. J. Dunaway, R. J. Willicut, *Langmuir*, **9**, 2775–2777 (1993).
42. R. L. McCarley, Y.-T. Kim, A. J. Bard, *Journal of Physical Chemistry*, **97**, 211–215 (1993).
43. Y. T. Kim, R. L. McCarley, A. J. Bard, *Journal of Physical Chemistry*, **96**, 7416–7421 (1992).
44. X. Gao, Y. Zhang, M. J. Weaver, *Journal of Physical Chemistry*, **96**, 4156–4159 (1992).
45. D. E. Weisshaar, M. M. Walczak, M. D. Porter, *Langmuir*, **9**, 323–329 (1993).
46. C.-J. Zhong, M. D. Porter, *Journal of the American Chemical Society*, **116**, 11616–11617 (1994).
47. G. E. Poirier, M. J. Tarlov, *Langmuir*, **10**, 2853–2856 (1994).
48. E. Delamarche, B. Michel, C. Gerber, D. Anselmetti, H.-J. Guntherodt, H. Wolf, H. Ringsdorf, *Langmuir*, **10**, 2869–2871 (1994).
49. G. E. Poirier, M. J. Tarlov, H. E. Rushmeier, *Langmuir*, **10**, 3383–3386 (1994).
50. E. Delamarche, B. Michel, H. Kang, C. Gerber, *Langmuir*, **10**, 4103–4108 (1994).
51. C. Schonenberger, J. Jorritsma, J. A. M. Sondag-Huethorst, L. G. J. Fokkink, *Journal of Physical Chemistry*, **99**, 3259–3271 (1995).
52. G. E. Poirier, M. J. Tarlov, *Journal of Physical Chemistry*, **99**, 10966–10970 (1995).
53. K. J. Tupper, R. J. Colton, D. W. Brenner, *Langmuir*, **10**, 2041–2043 (1994).
54. N. Camillone, C. E. D. Chidsey, G.-y. Liu, G. Scoles, *Journal of Chemical Physics*, **98**, 3503–3511 (1993).
55. P. Fenter, P. Eisenberger, K. S. Liang, *Physical Review Letters* **70**, 2447–2450 (1993).
56. R. G. Nuzzo, E. M. Korenic, L. H. Dubois, *Journal of Chemical Physics*, **93**, 767–773 (1990).
57. L. H. Dubois, B. R. Zegarski, R. G. Nuzzo, *Journal of Chemical Physics*, **98**, 678–688 (1993).
58. W. Mar, M. L. Klein, *Langmuir*, **10**, 188–196 (1994).
59. S. J. Stranick, A. N. Parikh, D. L. Allara, P. S. Weiss, *Journal of Physical Chemistry*, **98**, 11136–11142 (1994).

60. M. Sprik, E. Delamarche, B. Michel, U. Roethlisberger, M. L. Klein, H. Wolf, H. Ringsdorf, *Langmuir*, **10,** 4116–41130 (1994).
61. L. Haussling, B. Michel, H. Ringsdorf, H. Rohrer, *Angewandte Chemie (Int. Ed. Engl.)*, **30,** 569–572 (1991).
62. H. Wolf, H. Ringsdorf, E. Delamarche, T. Takami, H. Kang, B. Michel, C. Gerber, M. Jaschke, H. J. Butt, E. Bamberg, *Journal of Physical Chemistry*, **99,** 7102–7107 (1995).
63. J. W. Gardner, P. N. Bartlett, *Nanotechnology*, **2,** 19–32 (1991).
64. P. K. Hansma, J. Tersoff, *Journal of Applied Physics*, **61,** R1–R23 (1987).
65. R. Yang, D. F. Dalsin, D. F. Evans, L. Christensen, W. A. Hendrickson, *Journal of Physical Chemistry*, **93,** 511–512 (1989).
66. F. R. Fan, A. J. Bard, *Journal of the Electrochemical Society*, **136,** 3216–3222 (1989).
67. R. Yang, D. F. Evans, L. Christensen, W. A. Hendrickson, *Journal of Physical Chemistry*, **94,** 6117–6122 (1990).
68. R. Czajka, C. G. J. Koopal, M. C. Feiters, J. W. Gerritsen, R. J. M. Nolte, H. Van Kempen, *Bioelectrochem. and Bioenerg.*, **29,** 47–57 (1992).
69. M. P. Everson, J. H. Helms, *Synthetic Metals*, **40,** 97–109 (1991).
70. M. C. Montemayor, L. Vazquez, E. J. Fatas, *Applied Physics*, **75,** 1849–1851 (1994).
71. C. Froeck, A. Bartl, L. Dunsch, *Electrochemica Acta.*, **40,** 1421–1425 (1995).
72. L. L. Madsen, K. Carneiro, B. N. Zaba, A. E. Underhill, *Synthetic Metals*, **41–43,** 2931–2934 (1991).
73. R. Yang, D. F. Evans, W. A. Hendrickson, *Langmuir*, **11,** 211–213 (1995).
74. R. J. Willicut, R. L. McCarley, *Analytica Chimica Acta*, **307,** 269–276 (1995).
75. A. G. MacDiarmid, J. C. Chiang, A. F. Richter, A. J. Epstein, *Synthetic Metals*, **13,** 193–205 (1986).
76. D. Jeon, J. Kim, M. Gallagher, R. Willis, *Journal of Vacuum Science and Technology B*, **9,** 1154–1158 (1991).
77. T. L. Porter, T. R. Dillingham, C. Y. Lee, T. A. Jones, B. L. Wheeler, G. Caple, *Synthetic Metals*, **40,** 187–196 (1991).
78. T. L. Porter, C. Y. Lee, B. L. Wheeler, G. J. Caple, *Journal of Vacuum Science and Technology B*, **9,** 1452–1456 (1991).
79. G. E. Bowman, D. M. Cornelison, G. Caple, T. L. Porter, *Journal of Vacuum Science and Technology A*, **11,** 2266–2268 (1993).
80. S. P. Armes, M. Aldissi, M. Hawley, J. G. Beery, S. Gottesfeld, *Langmuir*, **7,** 1447–1452 (1991).
81. Y.-T. Kim, H. Yang, A. J. J. Bard, *Journal of the Electrochemical Society*, **138,** L71–74 (1991).
82. R. Nyffenegger, C. Gerber, H. Siegenthaler, *Synthetic Metals*, **55–57,** 402–407 (1993).
83. M. Wan, C. Zhu, J. Yang, C. Bai, *Synthetic Metals*, **69,** 157–158 (1995).
84. D. Jeon, J. Kim, M. Gallagher, R. Willis, *Science*, **256,** 1662–1664 (1992).

85. T. L. Porter, S. Jeffers, G. Caple, B. L. Wheeler, R. Swift, *Surface Science Letters*, **238**, L433–L438 (1990).

86. G. Caple, B. L. Wheeler, R. Swift, T. L. Porter, S. Jeffers, *Journal of Physical Chemistry*, **94**, 5639–5641 (1990).

87. J. Lukkari, L. Heikkilä, M. Alanko, J. Kankare, *Synthetic Metals*, **55–57**, 1311–1316 (1993).

88. F. Garnier, G. Tourillon, J. Y. Barraud, H. J. Dexpert, *Mater. Sci.*, **20**, 2687–2693 (1985).

89. S. R. Snyder, H. S. White, S. López, *Journal of the American Chemical Society*, **112**, 1333–1337 (1990).

90. H. Yang, F.-R. Fan, S.-L. Yau, A. G. Bard, *Journal of the Electrochemical Society*, **139**, 2182–2185 (1992).

91. P. Ocón, P. Herrasti, J. M. Vara, L. Vázquez, R. C. Salvarwzza, A. J. Arvia, *Journal of Physical Chemistry*, **98**, 2418–2425 (1994).

92. S. G. MacKay, M. Bakir, I. H. Musselman, T. J. Meyer, R. W. Linton, *Analytical Chemistry*, **63**, 60–65 (1991).

93. J. Wang, T. Martinez, D. R. Yaniv, L. D. McCormick, *Journal of Electroanalytical Chemistry*, **313**, 129–140 (1991).

94. K. M. Richard, A. A. Gewirth, *Journal of Physical Chemistry*, **99**, 12288–12293 (1995).

95. D. R. Yaniv, L. McCormick, J. Wang, N. J. Naser, *The Journal of Electroanalytical Chemistry*, **314**, 353–361 (1991).

96. J. A. DeRose, S. M. Lindsay, L. A. Nagahara, P. I. Oden, T. Thundat, R. L. Rill, *Journal of Vacuum Science and Technology B*, **9**, 1166–1170 (1991).

97. R. W. Keller, C. Bustamante, D. Bear, *Microbeam Analysis*, 365–366 (1991).

98. R. Srinivasan, J. C. Murphy, R. Fainchtein, N. Pattabiraman, *Journal of Electroanalytical Chemistry*, **312**, 293–300 (1992).

99. N. J. Tao, J. A. DeRose, S. M. Lindsay, *Journal of Physical Chemistry*, **97**, 910–919 (1993).

100. N. J. Tao, Z. Shi, *Journal of Physical Chemistry*, **98**, 7422–7426 (1994).

101. S. R. Snyder, H. S. White, *Journal of Physical Chemistry*, **99**, 5626–5632 (1995).

102. N. J. Tao, G. Cardenas, F. Cunha, Z. Shi, *Langmuir*, **11**, 4445–4448 (1995).

103. M. Kunitake, N. Batina, K. Itaya, *Langmuir*, **11**, 2337–2340 (1995).

104. F. Chuna, N. J. Tao, *Physical Review Letters*, **75**, 2376–2379 (1995).

105. E. Bunge, R. J. Nichols, B. Roelfs, H. Meyer, H. Baumgartel, *Langmuir*, **12**, 3060–3066 (1996).

106. A. P. Brown, C. Koval, F. C. Anson, *Journal of Electroanalytical Chemistry*, **72**, 379–387 (1976).

107. M. Radmacher, R. W. Tillmann, M. Fritz, H. E. Gaub, *Science*, **257**, 1900–1905 (1992).

108. G. Meyer, N. M. Amer, *Applied Physics Letters*, **53**, 1045–1047 (1988).

109. C. Schonenberger, S. F. Alvarado, *Review and Scientific Instruments*, **60**, 3131–3134 (1989).

110. P. Siedle, H.-J. Butt, *Langmuir*, **11**, 1065–1066 (1995).

111. G. Binnig, C. F. Quate, C. Gerber, *Physical Review Letters*, **56,** 930–933 (1986).
112. U. Durig, O. Zuger, A. Stalder, *Journal of Applied Physics*, **72,** 1778–1798 (1992).
113. R. J. Warmack, X.-Y. Zheng, T. Thundat, D. P. Allison, *Reviews of Scientific Instruments*, **65,** 394–399 (1994).
114. J. H. Hoh, A. Engel, *Langmuir*, **9,** 3310–3312 (1993).
115. S. Graftstrom, J. Ackermann, T. Hagan, R. Neumann, O. Probst, *Journal of Vacuum Science and Technology B*, **12,** 1559–1564 (1994).
116. N. A. Burnham, *Applied Physics Letters*, **63,** 114–116 (1993).
117. N. A. Burnham, R. J. Colton, H. M. Pollock, *Journal of Vacuum Science and Technology A*, **9,** 2548–2558 (1991).
118. M. Salmeron, A. Folch, G. Neubauer, M. Tomitori, D. F. Ogletree, W. Kolbe, *Langmuir*, **8,** 2832–2842 (1992).
119. M. Salmeron, G. Neubauer, A. Folch, M. Tomitori, D. F. Ogletree, P. Sautet, *Langmuir*, **9,** 3600–3611 (1993).
120. M. B. Salmeron, *MRS Bulletin*, **1993,** 20–25 (1993).
121. A. L. Weisenhorn, P. K. Hansma, T. R. Albrecht, C. F. Quate, *Applied Physics Letters*, **54,** 2651–2653 (1989).
122. W. A. Ducker, T. J. Senden, R. M. Pashley, *Nature*, **353,** 239–241 (1991).
123. W. A. Ducker, T. J. Senden, R. M. Pashley, *Langmuir*, **8,** 1831–1836 (1992).
124. N. A. Burnham, in *Computations for the Nano-Scale*, P. E. Blochl, ed. (Kluwer Academic Publishers, 1993), pp. 199–207.
125. N. A. Burnham, D. A. Dominguez, R. L. Mowery, R. J. Colton, *Physical Review Letters*, **64,** 1931–1934 (1990).
126. N. A. Burnham, R. J. Colton, *Journal of Vacuum Science and Technology A*, **7,** 2906–2913 (1989).
127. U. Durig, O. Zuger, D. W. Pohl, *Journal of Microscopy*, **152,** 259–267 (1988).
128. E.-L. Florin, V. T. Moy, H. E. Gaub, *Science*, **264,** 415–417 (1994).
129. V. T. Moy, E.-L. Florin, H. E. Gaub, *Science*, **266,** 257–259 (1994).
130. C. M. Mate, G. M. McClelland, R. Erlandsson, S. Chiang, *Physical Review Letters*, **59,** 1942–1945 (1987).
131. G. Meyer, N. M. Amer, *Applied Physics Letters*, **57,** 2089–2091 (1990).
132. D. R. Baselt, J. D. Baldeschwieler, *Journal of Vacuum Science and Technology B*, **10,** 2316–2322 (1992).
133. R. M. Overney, H. Takano, M. Fujihira, *Physical Review Letters*, **72,** 3546–3549 (1994).
134. T. J. Senden, C.J. Drummond, P. Kekicheff, *Langmuir*, **10,** 358–362 (1994).
135. Y.-H. Tsao, D. F. Evans, H. Wennerstrom, *Science*, **262,** 547–550 (1993).
136. Y. Martin, C. C. Williams, H. K. Wickramasinghe, *Journal of Applied Physics*, **61,** 4723–4729 (1987).
137. Y. Martin, H. K. Wickramasinghe, *Applied Physics Letters*, **50,** 1455–1457 (1987).
138. J. J. Saenz, N. Garcia, P. Grutter, E. Meyer, H. Heinzelmann, R. Weisendanger, L. Rosenthaler, H. R. Hidber, H.-J. Guntherodt, *Journal of Applied Physics*, **62,** 4293–4295 (1987).

139. R. Weisendanger, I. V. Shvets, D. Burgler, G. Tarrach, H.-J. Guntherodt, J. M. D. Coey, *Z. Phys. B Condensed Matter*, **86**, 1–2 (1992).
140. Y.-H. Tsao, S. X. Yang, D. F. Evans, H. Wennerstrom, *Langmuir*, **7**, 3154–3159.
141. Y.-H. Tsao, S. X. Yang, D. F. Evans, *Langmuir*, **8**, 1188–1194 (1992).
142. F. Lin, D. J. Meier, *Langmuir*, **10**, 1660–1662 (1994).
143. S. Manne, P. K. Hansma, J. Massie, V. B. Elings, A. A. Gewirth, *Science*, **251**, 183–186 (1991).
144. O. Marti, B. Drake, P. K. Hansma, *Applied Physics Letters*, **51**, 484–486 (1987).
145. F. Ohnesorge, G. Binnig, *Science* **260**, 1451–1456 (1993).
146. J. Yang, Z. Shao, *Ultramicroscopy*, **50**, 157–170 (1993).
147. W. Zhong, G. Overney, D. Tomanek, *Europhysics Letters*, **15**, 49–51 (1991).
148. F. J. Giessibl, *Science*, **267**, 68–71 (1995).
149. C. A. Alves, E. L. Smith, M. D. Porter, *Journal of the American Chemical Society*, **114**, 1222–1227 (1992).
150. C. A. Alves, M. D. Porter, *Langmuir*, **9**, 3507–3512 (1993).
151. H.-J. Butt, K. Seifert, E. Bamberg, *Journal of Physical Chemistry*, **97**, 7316–7320 (1993).
152. J. Pan, N. Tao, S. M. Lindsay, *Langmuir*, **9**, 1556–1560 (1993).
153. C. A. Widrig, C. Chung, M. D. Porter, *Journal of Electroanalytical Chemistry*, **310**, 335–359 (1991).
154. C. D. Bain, G. M. Whitesides, *Journal of the American Chemical Society*, **110**, 5897–5898 (1988).
155. W. A. Zisman, *Advances in Chemistry Series*, **43**, 1–51 (1964).
156. R. M. Overney, E. Meyer, J. Frommer, D. Brodbeck, R. Lüthi, L. Howald, H.-J. Güntherodt, M. Fujihira, H. Takano, Y. Gotoh, *Nature*, **359**, 133–135 (1992).
157. A. Noy, C. D. Frisbie, L. F. Rozsnyai, M. S. Wrighton, C. M. Lieber, *Journal of the American Chemical Society*, **117**, 7943–7951 (1995).
158. C. D. Frisbie, L. F. Rozsnyai, A. Noy, M. S. Wrighton, C. M. Lieber, *Science*, **265**, 2071–2074 (1994).
159. J.-B. D. Green, M. T. McDermott, M. D. Porter, *Journal of Physical Chemistry*, **99**, 10960–10965 (1995).
160. J. L. Wilbur, H. A. Biebuyck, J. C. MacDonald, G. M. Whitesides, *Langmuir*, **11**, 825–831 (1995).
161. J.-B. D. Green, M. T. McDermott, M. D. Porter, *Journal of Physical Chemistry*, **100**, 13342–13345 (1996).
162. M. Fujihira, H. Kawate, M. Yasutake, *Chemistry Letters*, **1992**, 2223–2226 (1992).
163. R. C. Thomas, P. Tangyunyong, J. E. Houston, T. A. Michalske, R. M. Crooks, *Journal of Physical Chemistry*, **98**, 4493–4494 (1994).
164. G. K. Rowe, S. E. Creager, *Langmuir*, **7**, 2307–2312 (1991).
165. G. M. Brown, T. Thundat, D. P. Allison, R. J. Warmack, *Journal of Vacuum Science and Technology A*, **10**, 3001–3006 (1992).
166. J. Li, E. Wang, M. Green, P. E. West, *Synthetic Metals*, **74**, 127–131 (1995).

167. V. Peulon, G. Barbey, J.-M. Valleton, S. Alexandre, *Synthetic Metals*, **74,** 15–19 (1995).
168. E. A. Bazzaoui, J. P. Marsault, S. Aeiyach, P. C. Lacaze, *Synthetic Metals*, **66,** 217–224 (1994).
169. T. L. Porter, D. Minore, D. Zhang, *Journal of Physical Chemistry*, **99,** 13213–13216 (1995).
170. A. G. Sykes, Y. Shi, R. Dillingham, T. L. Porter, G. Caple, *Journal of Electroanalytical Chemistry*, **380,** 139–145 (1995).
171. J. Y. Josefowicz, F. G. Yamagishi, C. I. v. Ast, *Materials Research Society Symposium Proceedings*, **338,** 605–611 (1994).
172. R. Nyffenegger, E. Ammann, H. Siegenthaler, R. Kotz, O. Haas, *Electrochemica Acta*, **40,** 1411–1415 (1995).
173. J. K. Schoer, C. B. Ross, R. M. Crooks, T. S. Corbitt, M. J. Hampden-Smith, *Langmuir*, **10,** 615–618 (1994).
174. C. A. Goss, J. C. Brumfield, E. A. Irene, R. W. Murray, *Langmuir*, **8,** 1459–1463 (1992).
175. J. C. Brumfield, C. A. Goss, E. A. Irene, R. W. Murray, *Langmuir*, **8,** 2810–2817 (1992).

7

THEORETICAL ASPECTS OF THE SCANNING TUNNELING MICROSCOPE OPERATING IN AN ELECTROLYTE SOLUTION

W. SCHMICKLER

Abteilung Elektrochemie, University of Ulm, D-89069 Ulm, Germany

7.1 INTRODUCTION

During the last ten years the scanning tunneling microscope (STM) has evolved into a versatile tool for the investigation of electrochemical interfaces. Images with blurred spots representing single atoms or molecules have become a familiar sight at meetings and in journals. However, in most electrochemical investigations the STM is used as an imaging tool only. The experimenter varies the system parameters, such as tunneling bias and current, until obtaining the desired picture. While this approach has made it possible to elucidate the interfacial structure in a number of cases, which is indeed a spectacular progress, it has not advanced our understanding of the operation of the STM in solutions. In addition, the potential of the STM as a spectroscopic tool has been neglected, even though the prospects of performing current/voltage spectroscopy are better in solutions than in the vacuum, because two potential differences between tip and substrate and between substrate and solution, can be varied independently.

In this article we will focus on these neglected aspects of the electrochemical STM, and consider two topics in detail: the tunneling of electrons through thin layers of water—a process of fundamental importance for the operation of the STM in solutions—and the investigation of electronic structure by STM spectroscopy. It is one of the few areas in electrochemistry in which theory is on a par with, or even ahead of, experiments, therefore our emphasis will be on the theory with relevant experimental results presented as well.

7.2 ELECTRON TUNNELING IN THE VACUUM

The exchange of electrons between a tip and a metal substrate is conceptually much easier in the vacuum than in a solution. The simplest reasonable model is shown in Figure 7.1. The substrate and the tip have work functions Φ_1 and Φ_2; their Fermi levels differ by the bias potential $e_0 V$. To a first approximation the potential experienced by the tunneling electron may be taken as a barrier of constant height W, which for small bias is equal to the average of the two work functions.

When the separation l between the tip and the substrate is sufficiently small, on the order of a few Ångstroms, electrons can traverse the gap by tunneling through the energy barrier. In general, they can tunnel from occupied states of one metal to the empty states of the other metal. To a good approximation the Fermi-Dirac distribution, which governs the occupation of the electronic states, can be taken as a step function, so that all the states below the Fermi level are considered as being occupied. In this case the electron will tunnel from the metal with the higher Fermi level (in Fig. 7.1 this is the tip) to states above the Fermi level of the other metal. The concomitant current i is proportional to

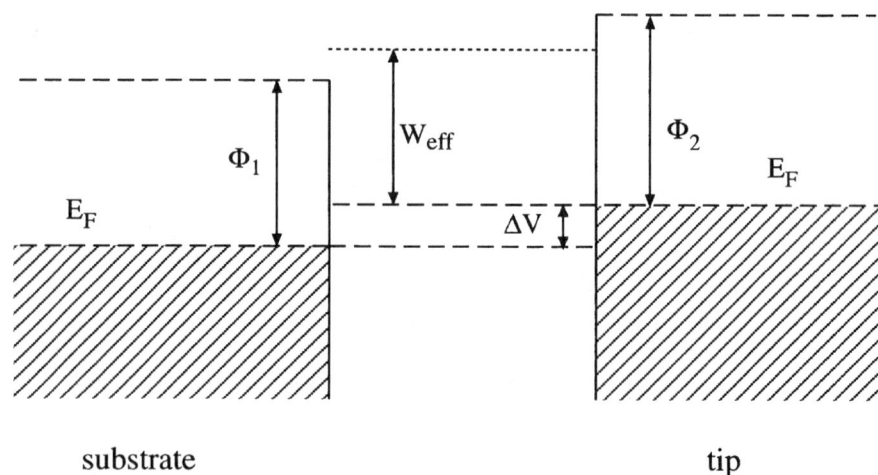

Figure 7.1. Schematic diagram of the effective barrier height for an STM in the vacuum.

7.2 ELECTRON TUNNELING IN THE VACUUM

the probability $T(\epsilon)$ for an electron of energy ϵ to tunnel from one metal to the other, integrated over the region between the two Fermi levels:

$$i \propto \int_0^{e_0 V} T(\epsilon)\, d\epsilon \tag{1}$$

where the Fermi level of the metal with the lower Fermi level has been taken as the energy zero. Equation (1) holds if the electron is transferred without change of energy (elastic tunneling); inelastic transitions usually make only a minor contribution to the current.

In a one-dimensional model the tunneling probability $T(\epsilon)$ decays roughly exponentially with the tip-substrate separation l, and can be estimated from the Gamov formula:

$$T(\epsilon) = \exp-\left\{ \frac{2}{\hbar} \sqrt{2m(W-\epsilon)} l \right\} \tag{2}$$

where m is the mass of the electron. Because the tunneling probability decreases strongly with $(W - \epsilon)$ the exchanged electrons come mostly from the Fermi level, and the average barrier height at small bias in this simple model is equal to the average work function.

In view of the Gamov formula it is natural to define an *effective barrier height* W_{eff} by the relation:

$$W_{\text{eff}} = \frac{\hbar^2}{8m} \left(\frac{\partial \ln i}{\partial l} \right)^2 \tag{3}$$

where i is the observed tunneling current.

Experimental values for the effective barrier height are usually somewhat lower than the work functions of the tip and the substrate, and tend to increase with growing separation. As an example, Figure 7.2 shows the results of Kuk and Silverman [1] for a reconstructed Au(111) surface. The barrier height is zero at contact, and rises continuously until it becomes constant at a separation of about 5 Å. The limiting value of about 3.2 eV is still substantially lower than the work function of the reconstructed Au(111) surface, which is about 5.4 eV.

Obviously, a rectangular barrier is a useful first approximation, but the true potential experienced by a tunneling electron is more complicated. Due to the long range of the exchange and correlation potentials, it does not rise abruptly to the value of the work function as the electron leaves the metal surface, but rises gradually, over a distance of several Ångstroms. This is illustrated in Figure 7.3, which shows a schematic diagram of the potential experienced by a single electron calculated from the jellium model. The work function is the limiting

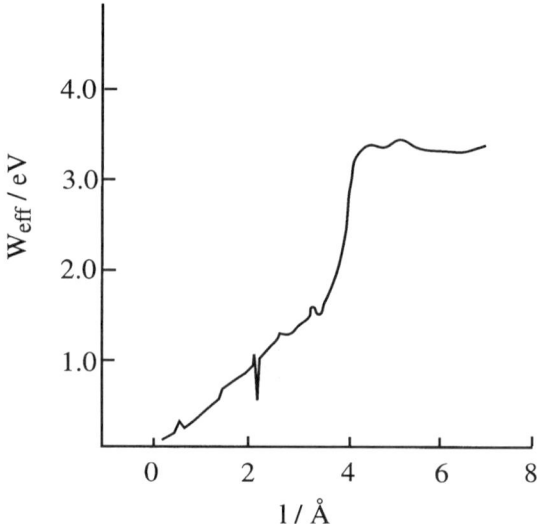

Figure 7.2. Effective barrier height as a function of the tip-sample separation [1]; the sample was a reconstructed Au(111) surface.

value of the potential, which is reached only at a distance of several Ångstroms from the surface.

For the STM the situation is complicated further by the electronic interaction between the tip and the substrate, which lowers the average barrier height even more. Figure 7.4 shows an example of the effective tunneling barrier, as

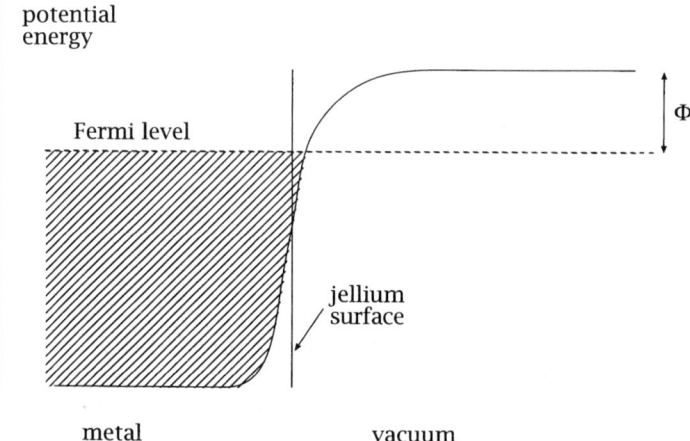

Figure 7.3. Potential energy experienced by an electron near the surface of jellium (schematic).

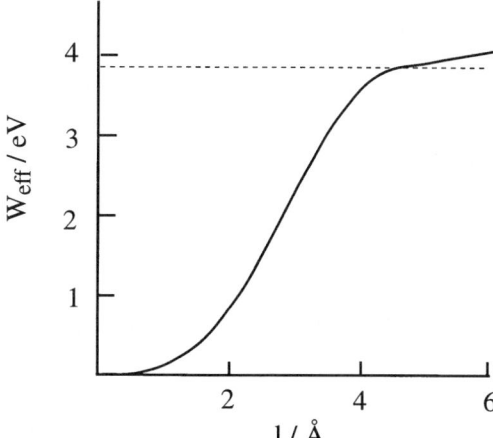

Figure 7.4. Effective barrier height according to the calculations of Lang [2]. The dashed line indicates the work function of the substrate.

defined by eq. (3), calculated from the jellium model. The theoretical curve agrees qualitatively quite well with experimental data: the calculated barrier height also vanishes at contact, and then rises towards the value of the work function. Note, however, that the experimental barrier height is always lower than the work function, while the theoretical curve rises slightly above it.

In summary: in vacuum the effective barrier heights are typically of the order of a few electron volts, somewhat lower than the work functions of the two metals involved.

7.3 EFFECTIVE BARRIER HEIGHTS IN SOLUTIONS

The first measurements of the effective barrier height for the STM in aqueous solutions gave surprisingly low values of a few tenths of an electron volt [3]. More recent measurements tend to give higher values of the order of about 1 eV [5,6,7]. It is not clear whether the low values are caused by a real physical effect as a few groups propose [4], or whether they are caused by contamination as others suggest [5]. We will return to this question later when we consider electron tunneling through intermediate states.

As an example of a recent investigation we consider the results of Vinzelberg and Lorenz [6] for a Au(111) surface in an aqueous solution of 0.01 M $HClO_4$ in greater detail. These authors obtained effective barrier heights by the usual method of measuring the tunneling current i as a function of distance and calculating W_{eff} from eq. (3). When the substrate potential and the tunneling current were kept constant the barrier height increased slightly with increasing bias V (see Fig. 7.5). Since increasing bias corresponds to increasing separation

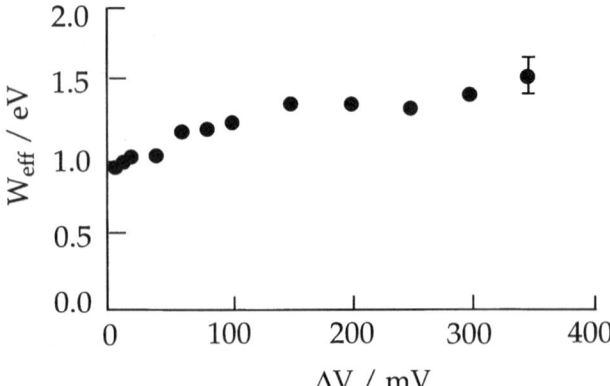

Figure 7.5. Effective barrier height for Au(111) electrode, kept at a potential of 200 mV versus NHE, as a function of the tunneling bias [6].

these data show the same trend as those observed in the vacuum (cf. Fig. 7.2). In contrast, when the bias was kept constant, the barrier height did not depend on the electrode potential (see Fig. 7.6). In any case, the limiting value for the barrier is substantially lower than in the vacuum and lies well below the work function. Therefore, early theoretical work focused on the effect of the solvent on electron tunneling, and looked for reasons why the presence of the solvent should lower the effective barrier.

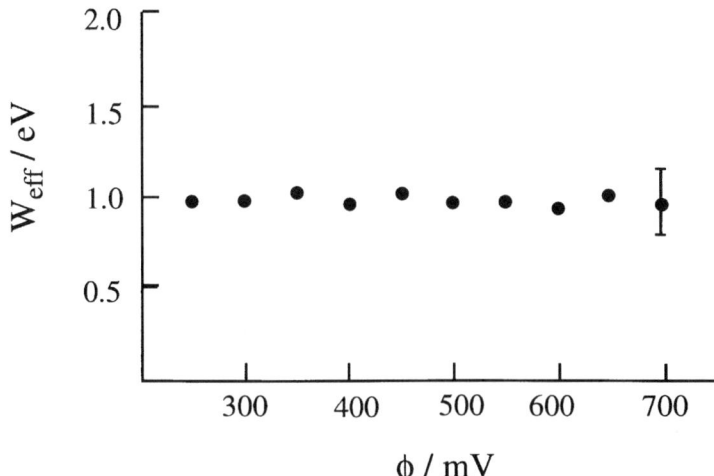

Figure 7.6. Effective barrier height for a Au(111) electrode as a function of the electrode potential [6].

7.4 CONTINUUM MODELS FOR THE EFFECT OF THE SOLVENT

The first explicit model for the STM in solutions was developed by Schmickler and Henderson [8]. Following the ideas of Lang [2] for the vacuum case, they represented the substrate as a semi-infinite jellium and the tip as a jellium sphere. The solution was treated on the Poisson-Boltzmann level; in particular, the solvent was modeled as a dielectric continuum.

An important feature of this model is the interaction of the tunneling electron with the polarizability of the solvent. When an electron escapes from the surface of a metal, it leaves behind an exchange-correlation hole, with which it interacts. At long distances this interaction becomes the classical image force. The polarizability of the solvent reduces this interaction, and consequently it also reduces the work function. Since the velocity of the tunneling electron is high, only the electronic polarizability can follow its movement and screen the interaction. For water, the corresponding optical dielectric constant has a value of $\epsilon_{opt} \approx 1.88$. The resulting change in the potential experienced by a tunneling electron is shown in Figure 7.7; in typical cases the reduction of the work function is of the order of 1 eV. This theoretical estimate agrees quite

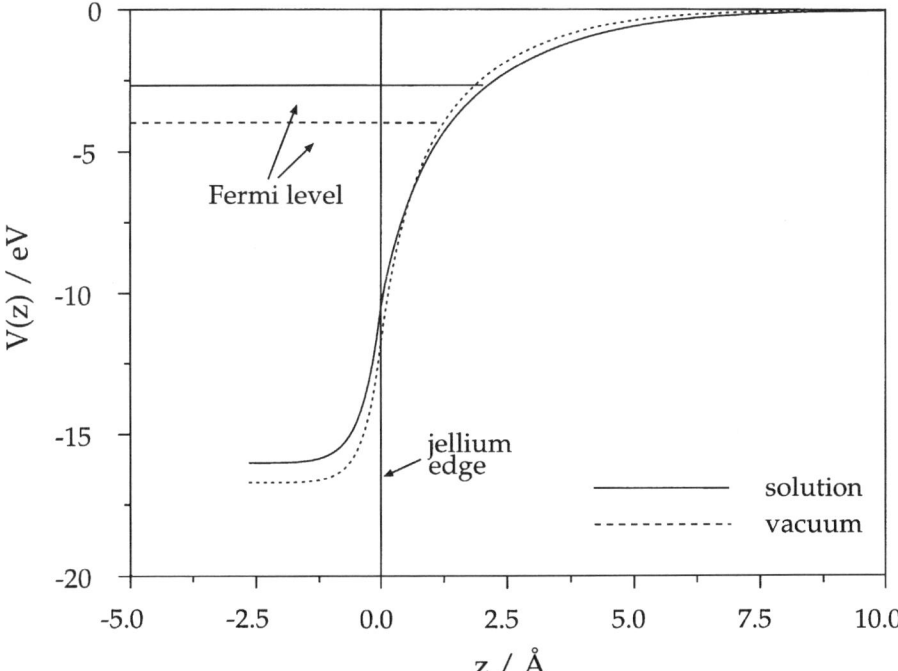

Figure 7.7. Electronic potential-energy profile for a semi-infinite jellium in contact with a solution of dielectric constant $\epsilon_{opt} = 1.88$ or with vacuum. The electronic density of the jellium corresponds to a Wigner-Seitz radius of $r_s = 2$.

well with experimental data: electronic work functions for photoemission into aqueous solutions are also about 1 eV lower than the vacuum values [9].

For small bias the tunneling probability $T(\epsilon)$ in eq. (1) can be taken as constant and can be evaluated at the Fermi levels of the two metals. This results in the formula [10]:

$$i \propto e_0 V \sum_{\mu,\nu} |M_{\mu\nu}|^2 \delta(\epsilon_\nu - E_F)\delta(\epsilon_\mu - E_F) \quad (4)$$

where $M_{\mu\nu}$ is the matrix element for tunneling between electronic states with energies ϵ_ν and ϵ_μ, and E_F denotes the Fermi level, which at small bias has approximately the same value for tip and substrate.

Schmickler and Henderson calculated the matrix element $M_{\mu\nu}$ from Oppenheimer's version of first order perturbation theory; the unperturbed wave functions are those for a semi-infinite jellium and for an isolated jellium sphere at the Fermi level. As expected, the resulting conductance $S = i/V$ decreases strongly with the separation l. Since the work function is lower in the presence of a solvent than in vacuum, the decrease of the conductance is less rapid (see Fig. 7.8). The same effect appears in the resulting effective barrier heights (Fig. 7.9). The values for the solution are consistently lower than those for the vac-

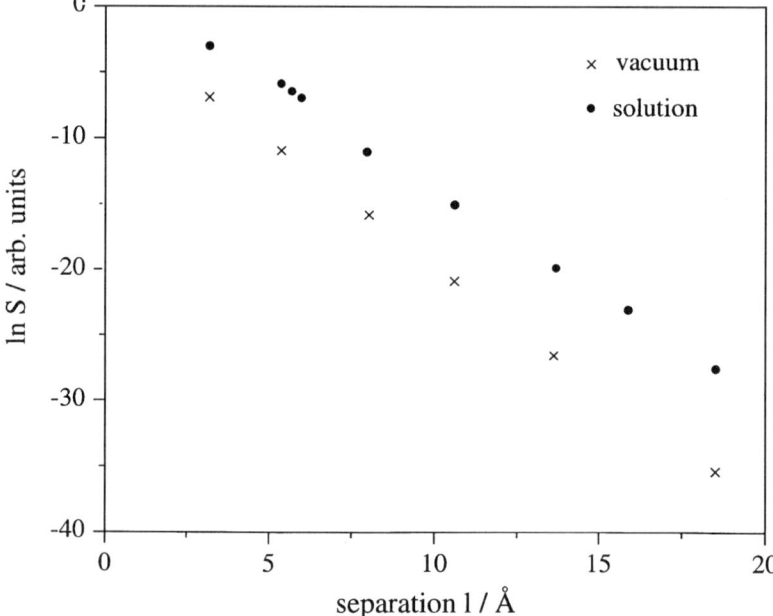

Figure 7.8. Conductance $S = i/\Delta V$ (in arbitrary units) as a function of the substrate-tip separation. Both substrate and tip have an electronic density corresponding to $r_s = 2$.

7.4 CONTINUUM MODELS FOR THE EFFECT OF THE SOLVENT

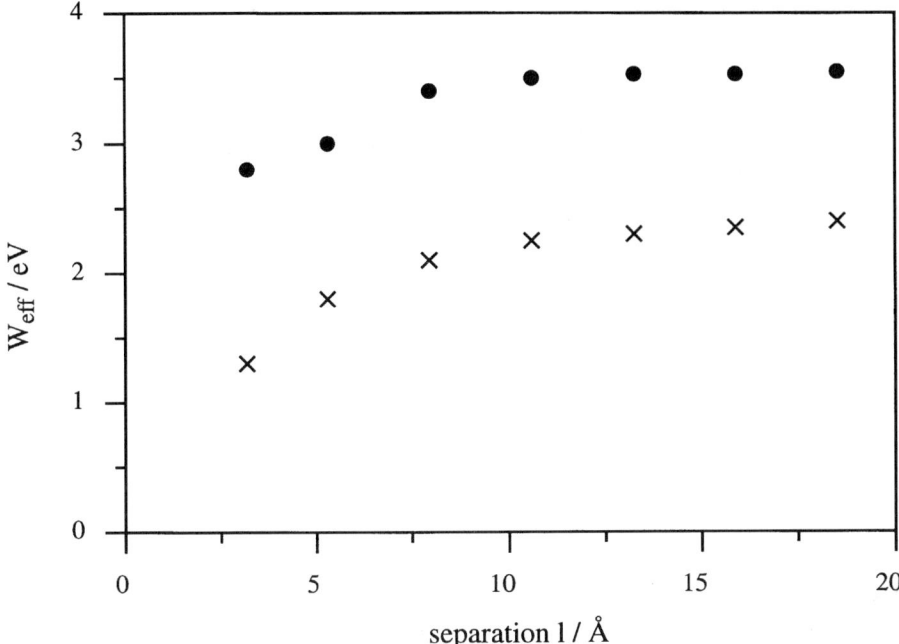

Figure 7.9. Effective barrier height as a function of the substrate-tip separation.

uum; both increase with the separation and tend toward the work function at large distances.

The use of perturbation theory limits the validity of these calculations to relatively large tunneling distances; in addition, they were restricted to the case of a small bias. To some extent these limitations were overcome in the work of Pecina et al. [11]. These authors used an approximate solution of the Poisson-Boltzmann equation [12] to calculate the electrostatic potential between a jellium substrate and a sphere. This potential was added to the electronic interaction between the transferring electron and the two metals involved. In order to simplify the calculations, the problem was reduced to one dimension by considering only the potential between the point on the sphere that is closest to the substrate and the point on the substrate that is closest to the sphere. The probability $T(\epsilon)$ that an electron of energy ϵ tunnels through this barrier was calculated exactly using Numerov's algorithm for solving the one-dimensional Schrödinger equation [13]. The current was then obtained from eq. (1) by integrating over the energy range between the two Fermi levels.

The resulting effective barrier heights are qualitatively similar to those calculated by Schmickler and Henderson (see Fig. 7.10). However, with increasing gap thickness the barrier rises more steeply, passes through a maximum and tends toward the work function at large separations. This overshoot of the work function can also be seen in the calculations by Lang in Figure 7.4. It is

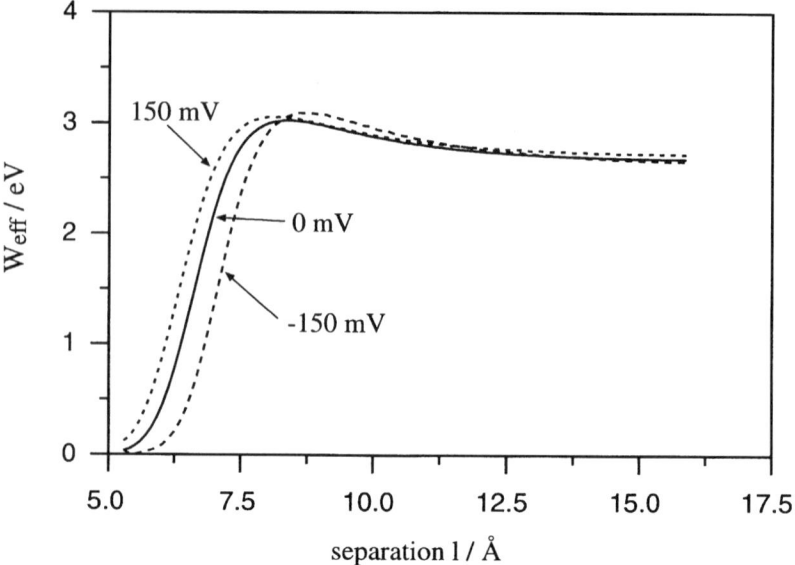

Figure 7.10. Effective barrier height for three different values of the bias potential; the tip (jellium sphere) was assumed to be uncharged.

caused by the following effect: at small separations both the barrier height and the barrier width increase as the gap is widened. Therefore the effective barrier height, which is a derivative quantity, seems larger than the real barrier.

At least within this model the bias potential, defined as the potential between the tip and the substrate, has only a minor effect on the effective barrier. A positive bias means that the substrate is negatively charged, so that the electrons spill out further, and the barrier begins to rise at larger distances from the substrate. If the tip remains uncharged, the effective barrier begins to rise at larger separations. Conversely, a negative bias implies a positive excess charge on the substrate, so that the electrons are pulled toward the interior, and the barrier starts to rise at shorter separations.

7.5 FLUCTUATIONS OF THE SOLVENT AND ADSORBATES

The preceding calculations ignored any effects of the orientational polarization on the tunneling probability. To some extent this neglect was justified by the work of Sebastian and Doyen [14]. Using the well-known results of Buettiker and Landauer [15] these authors estimated that the tunneling time for the electron is of the order of 10^{-15} s, while reorientation times for solvent molecules are of the order of 10^{-11}–10^{-13} s. Therefore, electron tunneling occurs at a fixed solvent configuration. However, fluctuations of the solvent, even if they are static on the timescale of tunneling, may still affect the transition proba-

7.5 FLUCTUATIONS OF THE SOLVENT AND ADSORBATES

bility. Using a continuum model for the solvent, Sebastian and Doyen showed that such fluctuations enhance the tunneling probability by a factor of:

$$A = \exp \frac{2\lambda mkTl^2}{\hbar^2 V_{barr}} \qquad (5)$$

where V_{barr} is the barrier in the absence of fluctuations, and λ measures the strength of the interaction between the solvent and the tunneling electron. It has the same meaning as the energy of reorganization in theories of electron transfer reactions. This enhancement effect can be understood from the following argument: Solvent fluctuations generate configurations that may be either more or less favorable than the average (i.e., they may present a higher or lower effective barrier). Since the barrier height enters exponentially into the tunneling rate the enhancement at favorable configurations will outweigh the reduction at unfavorable configurations, and the net effect will be an increase of the current.

However, the overall enhancement is quite small: typical values for λ lie in the range of 0.5–1 eV; the tunneling distances are on the order of a few Ångstroms, and the barrier heights V_{barr} at most a few electron volts, so that the resulting values for the enhancement differ little from unity. They do, however, increase with temperature since the fluctuations become larger (see Fig. 7.11).

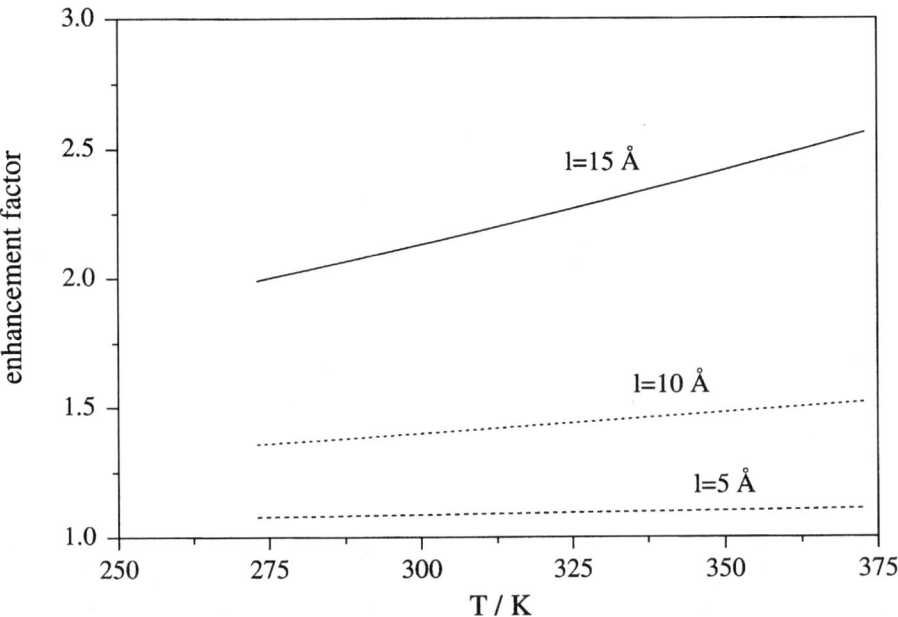

Figure 7.11. Enhancement factor due to solvent fluctuations as a function of temperature [14].

Many of the same conclusions were reached by Rostkier-Edelstein et al. [16], who studied the effect of solvent fluctuations by a different formalism. They also find a small enhancement factor, which increases with temperature and with tunneling distance. High frequency intramolecular modes are more effective in increasing the tunneling rate than the slower solvent reorientation modes, but their effect on the barrier height is still minor.

It was pointed out by Kornyshev [17] that solvent fluctuations should, in principle, be observable as noise, from which the autocorrelation function could be deduced. However, such measurements would require a much better time resolution than can be achieved by current instruments.

Similar fluctuations in the tunneling currents can be induced by the surface diffusion of adsorbates [18]; because this diffusion occurs on a much longer timescale than the solvent fluctuations, the associated noise has a much lower frequency, and should hence be easier to observe. In the vacuum, where experiments are not encumbered by the presence of the solution, the diffusion of oxygen on Si(111) has been studied by measuring the power spectrum of the current fluctuations [20]. The time resolution of such experiments can be extended to the picosecond range by combining the STM with ultrashort laser pulses [19]. It remains to be seen whether such techniques can be applied in solutions.

7.6 TUNNELING THROUGH INTERMEDIATE STATES IN WATER

As previously mentioned, early measurements gave surprisingly low values for the effective barrier heights in solutions. This result led to speculations that the tunneling mechanism in aqueous solutions could be quite different from that in the vacuum; several possibilities are discussed in a paper by Kuznetsov et al. [21]. A specific mechanism was first suggested by Sass and Gimzewski [22], who proposed that solvated electrons could serve as intermediate states (see Fig. 7.12). The presence of such states would diminish both the distance and the effective barrier height. In a simple approximation the tunneling probability via one intermediate state is given by:

$$T(\epsilon) \propto \frac{\exp-\left\{\frac{2m}{\hbar}\sqrt{2(W_{\text{eff}} - \epsilon)}l_1\right\} \times \exp-\left\{\frac{2m}{\hbar}\sqrt{2(W_{\text{eff}} - \epsilon)}l_2\right\}}{\exp-\left\{\frac{2m}{\hbar}\sqrt{2(W_{\text{eff}} - \epsilon)}l_1\right\} + \exp-\left\{\frac{2m}{\hbar}\sqrt{2(W_{\text{eff}} - \epsilon)}l_2\right\}} \quad (6)$$

where l_1 and l_2 are the two tunneling distances involved. This expression is dominated by the lower of the two tunneling probabilities, therefore the optimum position for an intermediate state is halfway between the substrate and the tip. However, it was pointed out both by Rostkier-Edelstein et al. [16] and by Schmickler [23] that the energy required for the formation of a solvated electron

7.6 TUNNELING THROUGH INTERMEDIATE STATES IN WATER

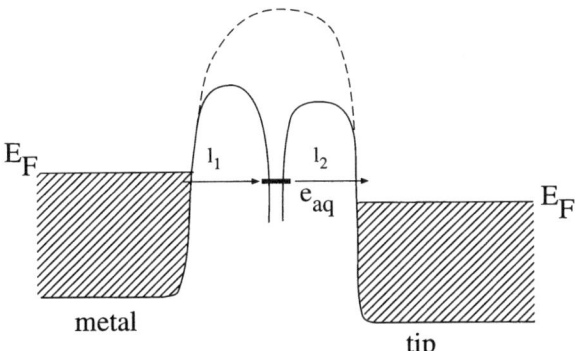

Figure 7.12. Tunneling with a solvated electron as an intermediate state. The dashed line gives the barrier for direct tunneling, the full line for indirect tunneling.

is of the order of 0.5–1.0 eV and thus too large for tunneling through these states to be effective. Nevertheless, other candidates for intermediate states, which will be discussed later, could be more effective, or certain solvent fluctuations, while not actually forming a cage for a solvated electron, could still enhance the tunneling mechanism, even though the crude estimates based on continuum theory suggest that they play only a minor role.

Another tunneling mechanism was proposed by Lindsay et al. [24]. They suggested that the intermediate states are provided by the water molecules themselves, and that the electrons tunnel along the hydrogen bonds. The tunnel current would mainly be carried by the shortest hydrogen-bonded path linking the tip and the substrate (see Fig. 7.13).

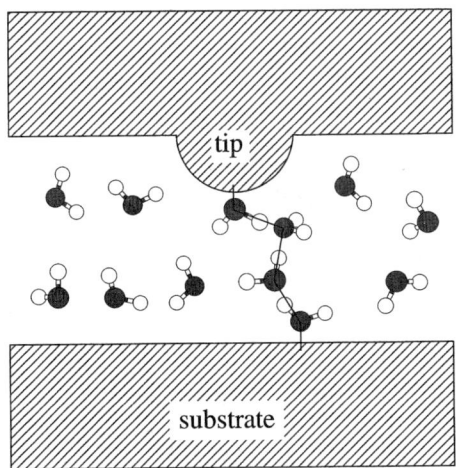

Figure 7.13. Water layer between a tip and a metal substrate; the line indicates a possible tunneling path along a hydrogen-bonded network (adapted from [24]).

The flux of electrons between a state i on one molecule and a state j on another can be calculated by Bardeen's formalism [25]. The relevant quantity is the mixed current:

$$\gamma_{ij} = \frac{\hbar^2}{2m} \left| \int_P dP \left(\phi_i \frac{\partial \phi_j^*}{\partial s} - \phi_j^* \frac{\partial \phi_i}{\partial s} \right) \right| \tag{7}$$

where ϕ_i and ϕ_j are the corresponding wave function, s is the coordinate along the most probable tunneling path, and P denotes any plane that traverses this path. Lindsay et al. [24] estimate that for water molecules the optimum overlap is between a 2p state on the oxygen atom and a 1s state on hydrogen. It is difficult to estimate the effective barrier height explicitly in this model since it is not clear how such tunneling paths would change when the gap width increases, but in any case it seems that such a network of hydrogen bonds provides an effective tunneling path.

A similar, but less detailed argument was advanced by Halbritter et al. [26]. If the gap offers a high density of possible intermediate states, the tunneling current can pass through those states that are at the optimum positions in space. If n such states, all optimally placed, participate, eq. (6), is replaced by:

$$T(\epsilon) \propto \exp - \left\{ \frac{2m}{\hbar} \sqrt{2(W_{\text{eff}} - \epsilon)} \frac{l}{n+1} \right\} \tag{8}$$

so that the effective barrier height is reduced by a factor $(n + 1)$. Obviously, this equation can be used to explain very low barrier heights by invoking a sufficient number of intermediate states. Possible candidates for such states are dipole resonances on water or on adsorbed hydroxide.

Halbritter et al. consider experimental data for the effective barrier height on Ag(111) obtained with a Pt-Ir tip [26] in greater detail. Figure 7.14 shows

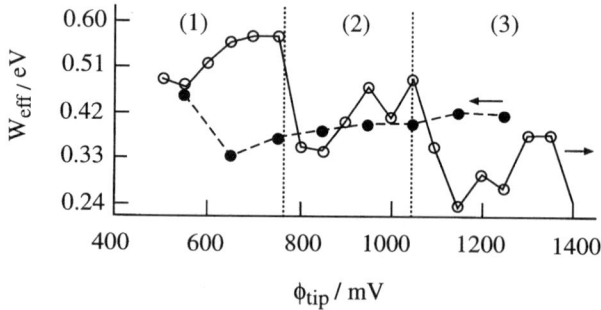

Figure 7.14. Effective barrier height on Ag(111) in aqueous solution as a function of the tip potential; the arrows indicate the scan direction. Adapted from [26].

the dependence on the potential of the tip keeping the electrode potential constant in the double layer region. The data show a substantial hysteresis, which is attributed to irreversible changes on the tip surface. The authors consider the anodic sweep in greater detail, and distinguish three regions: (1) a double layer region in which the tip surface is free of adsorbates, (2) a region in which the surface is covered by Pt-OH, and (3) an oxide region. The effective barrier height decreases from region (1) to region (3), which is attributed to an increasing number of intermediate states as the surface is covered with adsorbates.

The dipole resonances invoked by these authors have been observed in vacuum experiments by electron scattering through water layers at very small kinetic energies. Their energy is therefore close to the vacuum level (i.e., several electron volts above the Fermi level). They cannot serve as real intermediate states, but are virtual resonances, in spite of claims to the opposite by Repphuhn and Halbritter. In fact, there cannot be any real electronic states in water that lie near the Fermi level of electrodes, because such states could serve as redox centers. It should also be noted that the experimental data in Figure 7.14 contradict the data by Vinzelberg [6] obtained later in the same laboratory. Effective barrier heights are notoriously difficult to measure, and it is not unusual that laboratories publish conflicting data at different times. The fact that recent measurements agree on values of the order of 1 eV, or a little higher, may reflect increasing accuracy of the data.

In summary: it is quite possible that electronic intermediate states participate in the tunneling process. However, these must be virtual states with energies well away from the Fermi level. We will return to this point in the following section.

7.7 MOLECULAR MODELS FOR WATER

While the continuum models for the solution provide some physical insight into possible enhancement mechanisms a description at a molecular level is obviously preferable. This can be achieved by using effective potentials for the interaction of an electron with a water molecule. Suitable pseudopotentials have been devised by Barnett et al. [27] and by Schnitker and Rossky [28]. We consider the former in greater detail since it has been used to calculate the effective barrier experienced by a tunneling electron.

Figure 7.15 shows the potential energy of an electron interacting with a single water molecule. The most noticeable features are a strong repulsion from the oxygen atom, which is caused by the high electronic density on that atom, and an attraction from the hydrogen atoms, which carry a positive charge. In addition, there is a long-range attractive force due to the electronic polarizability of the molecule.

These pseudopotentials have been combined with the results of molecular dynamics simulations to construct the potential energy surface for a tunneling

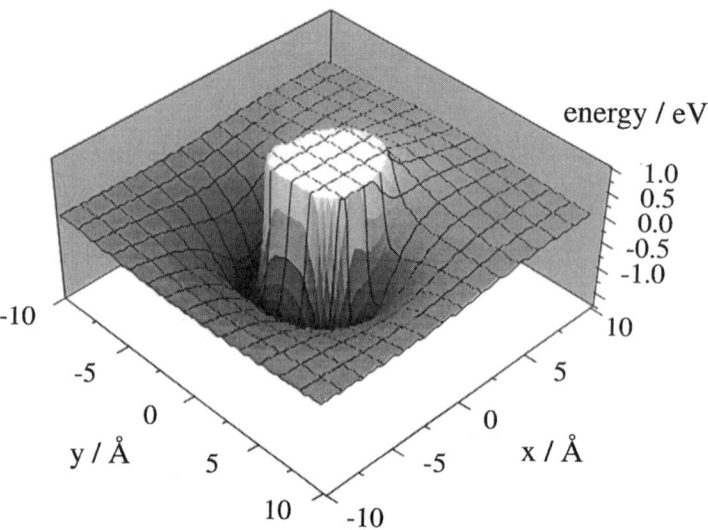

Figure 7.15. Potential energy experienced by an electron in the vicinity of a water molecule. The electron is situated within the plane spanned by the oxygen and the two hydrogens. The peak in the center gives the position of the oxygen atom; it has been cut off at a height of 1 eV for clarity. The hydrogen atoms sit in the well to the left of the oxygen atom.

electron. Explicit model calculations were performed by Schmickler [23] and by Mosyak, Nitzan, and Kosloff [29]. Both groups considered an ensemble of water molecules about three layers thick (9.6 Å) sandwiched between two metal plates (see Fig. 7.16). Molecular dynamics simulations were performed for the water ensemble. After the simulation had run for a certain time it was stopped, and the positions of the molecules recorded. The potential energy barrier experienced by a tunneling electron was calculated by adding the pseudopotential contributions from each water molecule. The resulting potential energy surface is three-dimensional; a typical two-dimensional cross section is shown in Figure 7.17. The oxygen atoms give rise to a high potential barrier, while the hydrogen atoms attract electrons.

Schmickler [23] treated the two metal plates as jellium, and added the interaction of the electron with the two jellium plates to the electron-water interaction. The resulting total interaction potential shows strong variations; in effect, the electron tunnels through a three-dimensional grid, where the hydrogen atoms provide the slits, and the oxygen atoms the barrier. In order to calculate the tunneling probability a window of 6 Å × 6 Å was chosen perpendicular to the z direction (see Fig. 7.16), and cyclic boundary conditions were imposed in the x and y directions. By performing a two-dimensional Fourier transform in the x and y directions the Schrödinger equation was transformed into a set

7.7 MOLECULAR MODELS FOR WATER   321

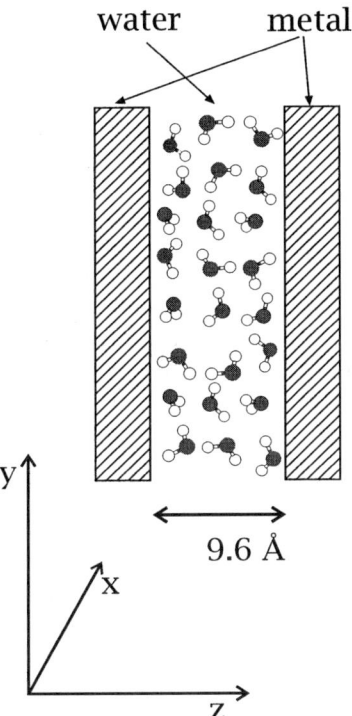

Figure 7.16. Model system for calculating the tunneling of electrons through a thin water layer.

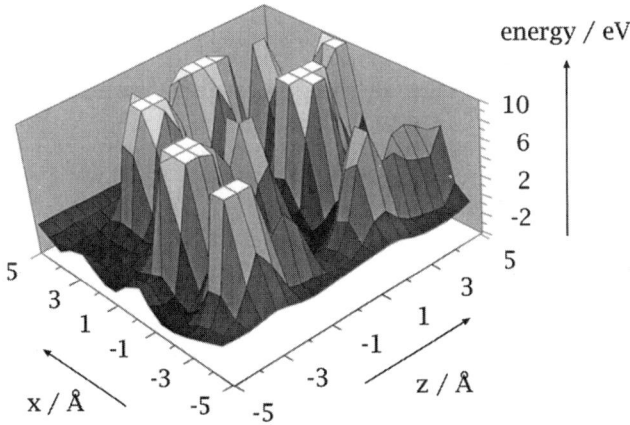

Figure 7.17. Cross section through the potential energy surface produced by the interaction with water. The y-coordinate was set to zero, and the potential is plotted as a function of x and z. The potential was cut off at 10 eV for clarity.

of coupled one-dimensional equations, which were solved numerically using the Numerov [13] method. The boundary conditions are those appropriate for a scattering problem:

at $z = -\infty$: incoming wave $\exp ikz$ + reflected wave
at $z = \infty$: outgoing wave

Obviously, the resulting tunneling probability depends on the particular water configuration employed. Therefore the calculations were repeated for many configurations obtained from the molecular dynamics simulations. A few typical results are shown in Figure 7.18. For the chosen window size the tunneling probability is of the order of 10^{-7}–10^{-6}, and fluctuates by a factor of ten. The calculations were repeated for the case in which the two metal plates carry charge densities of equal magnitude (10 μC cm^{-2}) and opposite sign, but the results were essentially the same.

The work of Mosyak et al. [29] is similar but differs in a few details. The two metal plates were not modeled as jellium but as hard planes. An additional barrier of 5 eV was superimposed onto the electron-water interaction to account for the work function of the metal. Tunneling probabilities were calculated by

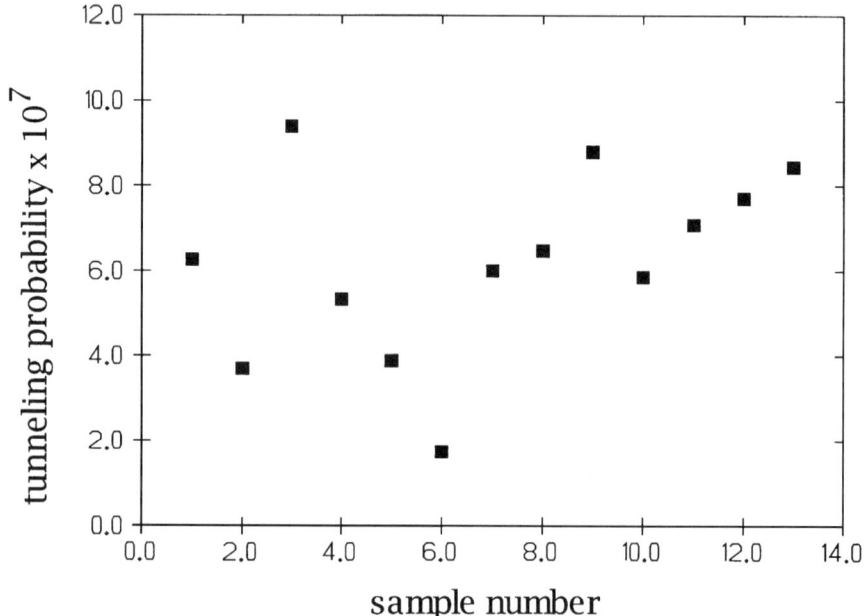

Figure 7.18. Tunneling probability through three layers of water for various configurations [23].

7.7 MOLECULAR MODELS FOR WATER

three different methods, which gave essentially the same results. The resulting values are of the same order of magnitude as those by Schmickler. The authors compare their calculations with those obtained by a one-dimensional treatment, in which a one-dimensional barrier is taken at an arbitrary position in the xy-plane. The tunneling probability through such a barrier fluctuates greatly: if the cross section is taken at a position in which the one-dimensional path passes through several protons, the barrier is low and the tunneling rate high; conversely, if the path traverses oxygen atoms, the tunneling rate is low. Averaging over such paths leads to an overeestimate of the tunneling rate since the high rates dominate in the averaging process (see Fig. 7.19). Conversely, if the average is first performed over one-dimensional cross sections of the barrier, the repulsive potential of the oxygen cores dominate, and the tunneling rate through this average barrier is much lower than the rate through the true three-dimensional barrier.

The same group has recently considered the effect of the electronic polarizability of water on the tunneling rate with greater accuracy. In the procedure

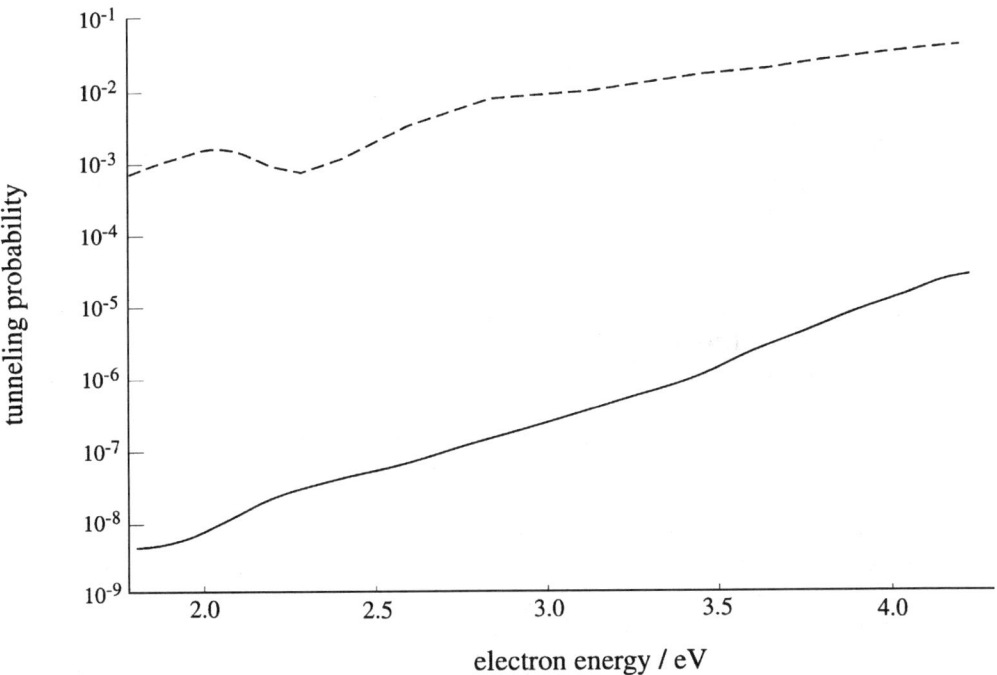

Figure 7.19. Tunneling probability as a function of the electronic energy; the full line is for a three-dimensional barrier calculated for a particular configuration of water; the dashed line was obtained by averaging the tunneling probabilities through one-dimensional cross sections; data taken from [29].

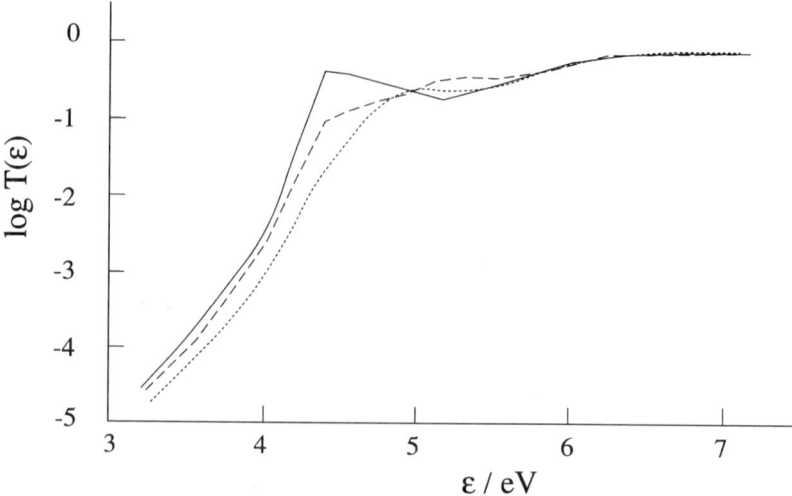

Figure 7.20. Electron tunneling probability through three layers of water as a function of the electron energy. The three lines correspond to different water configurations.

already outlined, the interaction between the tunneling electron and the water molecules was considered to be additive, the effect of the electric field generated by one water molecule on the electronic polarizability of the other water molecules was not accounted for. A consistent treatment of the polarizability enhances the tunneling rate by about two orders of magnitude. In addition, the tunneling probability $T(\epsilon)$ shows some structure at an electron energy of -0.5 eV below the barrier (i.e., 4.5 eV above the Fermi level), which may indicate a weak resonance caused by extended electronic states (see Fig. 7.20). By varying the number of water layers between the two metal plates, an estimate for the effective barrier height could be obtained [31]; for the system investigated, which models water between two slabs of Pt(100), the values are of the order of 4.5 eV (see Fig. 7.21), which is still substantially higher than experimental values obtained for gold surfaces, but lower than the work function of platinum.

To some extent these molecular models were confirmed by a study of Nagy [32], who measured the current-distance characteristics on a graphite (0001) electrode. These curves decay rapidly at a distance of a few Ångstroms, where the first water molecules would be expected to sit (see Fig. 7.22). From these data an average effective barrier height $V(z)$ can be estimated as a function of the separation. This barrier height shows a pronounced maximum at a distance of about 2.5 Å. The absolute magnitude obtained for the barrier profile depends on the value that is assumed for the kinetic energy of the tunneling electrons, but the overall shape of the barrier and the position of the maximum is always the same.

7.7 MOLECULAR MODELS FOR WATER   325

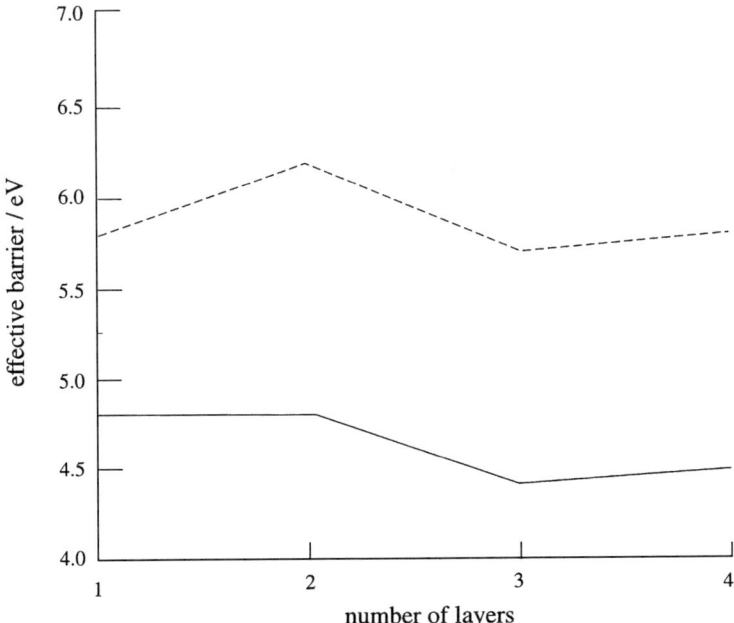

Figure 7.21. Effective barrier height, obtained by fitting the calculated tunneling rate to the expression for a rectangular barrier, for various numbers of water layers. The full line is for the polarizable water model, the dashed line for nonpolarizable water. In the absence of water the barrier height would be 5 eV.

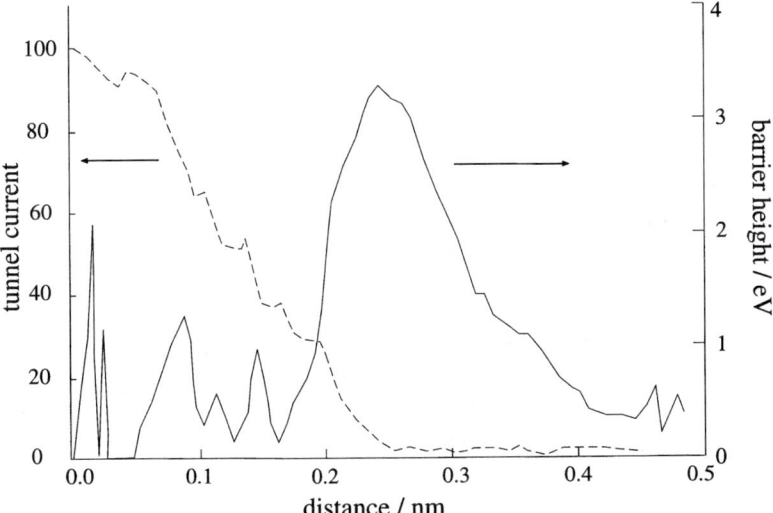

Figure 7.22. Tunneling current (dashed line) and effective barrier height as a function of the tip-substrate separation on graphite (0001); reprinted with permission from Nagy [32].

7.8 CURRENT-VOLTAGE SPECTROSCOPY

As discussed in Section 7.2 electrons tunnel from occupied states of the metal with the higher Fermi level to the empty states of the other metal. The current is therefore proportional to the electronic densities of states $\rho(\epsilon)$ at the surfaces of the substrate. On clean metals this density of states does not vary much near the Fermi level, and was therefore not considered explicitly in eq. (1) because it contributes only a constant factor. However, in the presence of adsorbates or on semiconductors it may depend strongly on the electronic energy, and should be considered explicitly:

$$i \propto \int_0^{e_0 V} \rho(\epsilon) T(\epsilon) \, d\epsilon \tag{9}$$

Information about the density of states may be obtained from current-potential curves at a fixed distance. Differentiation of eq. (9) gives:

$$\frac{di}{dV} = \rho(e_0 V) T(e_0 V) \tag{10}$$

So the derivative gives the product of the density of states and the tunneling probability. Since the latter depends exponentially on the electronic energy ϵ it may be eliminated to some extent by taking the logarithmic derivative:

$$\rho(\epsilon) \approx \frac{di}{dV} \frac{V}{i} \tag{11}$$

In this way, Stroscio et al. [33] obtained the density of states at the surface of Si(111)2×1 in the vacuum.

There have been only few applications of this technique in electrochemistry. An early example is the work of Robinson and Widrig [34], who examined the surface of a platinum film electrode by so-called *differential conductance tunneling spectroscopy* (DCTS). In this variant of tunneling spectroscopy the voltage is modulated as the tip scans the surface; Robinson and Widrig used modulation amplitudes of 30–75 mV at a frequency of 42 kHz. With their instrument, the tunneling conductance di/dV, which is obtained from the modulation, could be recorded simultaneously with the usual topographic images. For the investigated platinum film the two images look surprisingly similar (see Fig. 7.23). The DTCS signal was higher in valleys and pits and lower at the top of features, and thus follows the topography of the surface. The reason for this behavior is not understood; possibly the topography influences the local density of states and the work function.

Current-voltage spectroscopy should be particularly valuable for the investi-

7.9 INVESTIGATION OF ELECTRON-TRANSFER REACTIONS

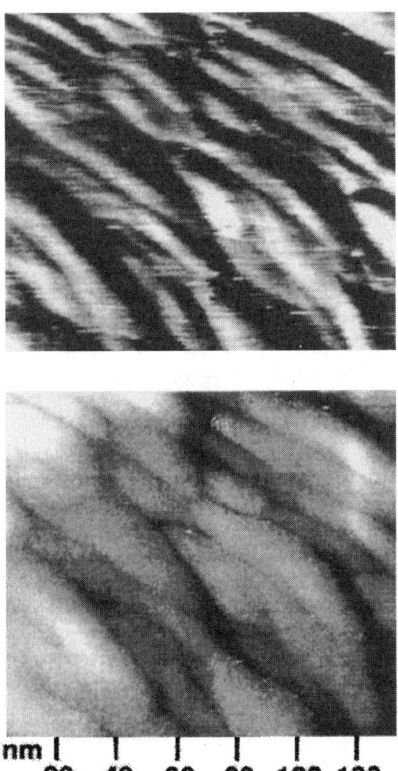

Figure 7.23. Images of a platinum film obtained by the conventional topographic STM technique (lower image) and by differential conductance tunneling spectroscopy (upper image). Reprinted with permission from Robinson and Widrig [34]; copyright (1992) American Chemical Society.

gation of adsorbates. The underlying processes are discussed in some detail by Kuznetsov et al. [21] and Andersen et al. [35], who distinguish a fair number of situations: strong and weak coupling of the adsorbate to the substrate and to the tip, associated reorganization of the solvent and of vibrational modes, participation of one or more electronic intermediate states. We refer the interested reader to the original publications.

7.9 INVESTIGATION OF ELECTRON-TRANSFER REACTIONS

7.9.1 Theoretical Considerations

An adsorbate with electronic levels near the Fermi level should be particularly easy to detect with an STM. Current-voltage spectroscopy of such states should yield valuable information about the properties of this level and its coupling to the solvent. Therefore, Schmickler and Widrig [36] suggested to investigate electron-transfer reactions of a monolayer of an adsorbed electroactive species by scanning tunneling spectroscopy.

Let us discuss briefly some of the conditions that a system must fulfill to be suitable for this purpose. If the redox center is to participate in tunneling, it is obviously necessary that it undergo fast and reversible electron transfer; this typically occurs when the reduced and oxidized forms of the molecule are similar, and are not strongly coupled to solvent modes; in other words, the concomitant energy of reorganization should be small. Furthermore, the identification of the active species is aided if it forms a regular lattice on the electrode surface. Under these conditions the electroactive species will appear as bright spots in a topographic STM picture, since a large part of the tunneling current will flow via these states. The tip can then be positioned immediately above a redox center, and current-voltage spectroscopy be performed (see Fig. 7.24).

If the tip is biased positively electrons will tunnel from the electrode to the tip with the active sites serving as virtual intermediate states (see Fig. 7.25). The energy range of the tunneling electrons is limited to the region between the Fermi levels of the two systems. Within a semi-classical framework, in which only the electrons are treated as quantum particles, the current can be written in the form:

$$i \propto \int_0^{e_0 V} D_{\text{ox}}(\epsilon) \, d\epsilon \qquad (12)$$

where $D_{\text{ox}}(\epsilon)$ is the probability of finding an empty state of energy ϵ on the redox couple, or the *density of oxidized states* in Gerischer's terminology [37]. In the Marcus theory it takes on the familiar form:

$$D_{\text{ox}}(\epsilon) = \sqrt{\frac{\pi}{kT\lambda}} \exp - \frac{(\epsilon - \epsilon_r)^2}{4\lambda kT} \qquad (13)$$

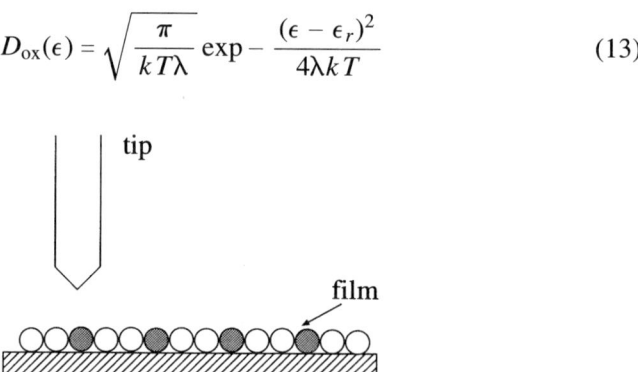

Figure 7.24. Investigation of a monomolecular film containing electroactive species (shaded) with an STM.

7.9 INVESTIGATION OF ELECTRON-TRANSFER REACTIONS 329

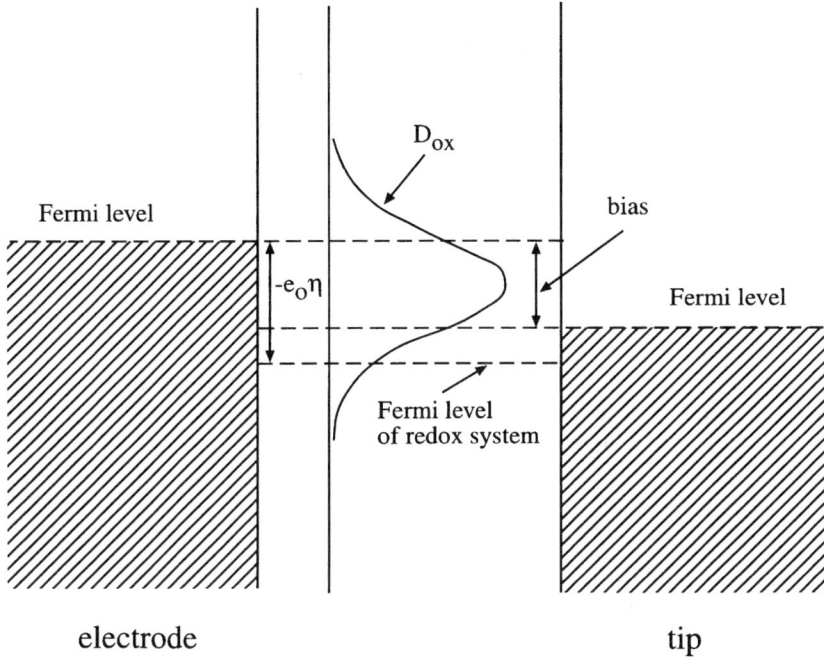

Figure 7.25. Resonant electron exchange via an adsorbed redox center.

where λ denotes the energy of reorganization for the electron-transfer reaction, and ϵ_r is the electronic energy of the electroactive level. When a redox species is adsorbed on a metal surface, its density of states can be broadened by the electronic interaction with the substrate, which induces an additional electronic width Δ:

$$D_{ox}(\epsilon) = \int_{-\infty}^{\infty} dt \, e^{-kT\lambda t^2} e^{-it(\epsilon - \epsilon_r)} e^{-\Delta|t|} \qquad (14)$$

The stronger the electronic interaction with the substrate, the wider is the redox density of states.

$D_{ox}(\epsilon)$ can be obtained from current-potential curves in the same way as the density of states of an adsorbate, if the energy difference between the intermediate level and the Fermi level of the substrate is kept fixed. A more elegant way consists of keeping the bias constant and shifting the redox level; this method will be discussed in the experimental section.

The classical treatment suffices if only classical modes are coupled to the redox center. If, however, quantum modes, which usually originate in the inner sphere of the reactant, are reorganized during the transfer the theory must be extended [38,39]. The main new effect is the possibility of inelastic transitions:

a tunneling electron arriving at the redox center with an energy ϵ may excite a quantum mode of frequency ν and arrive at the other metal with a diminished energy $\epsilon' = \epsilon - nh\nu$, where n is the number of vibrational quanta that has been excited (see Fig. 7.26). In this case the tunneling current passing through the redox level takes on the form:

$$i \propto \int d\epsilon \int d\epsilon' \, d\epsilon' \, T(\epsilon, \epsilon') f(\epsilon)[1 - f(\epsilon')] \qquad (15)$$

where $f(\epsilon)$ again denotes the Fermi-Dirac distribution. The transition matrix $T(\epsilon, \epsilon')$ exhibits a peak whenever $\epsilon - \epsilon' = nh\nu$.

An inelastic transition involving the excitation of n vibrational quanta can only occur if the bias is larger than $nh\nu$. This should make it possible to detect the participation of quantum modes through the following procedure [38]: Starting at a low bias, where $e_0 V < h\nu$, only elastic transitions can occur. As the bias is gradually increased a new inelastic channel opens whenever the bias passes a multiple of $h\nu$. The opening of a new channel leads to an increase in the current, which will appear as a step in the first derivative di/dV, and as a peak in the second derivative d^2i/dV^2; Figure 7.27 gives an example for

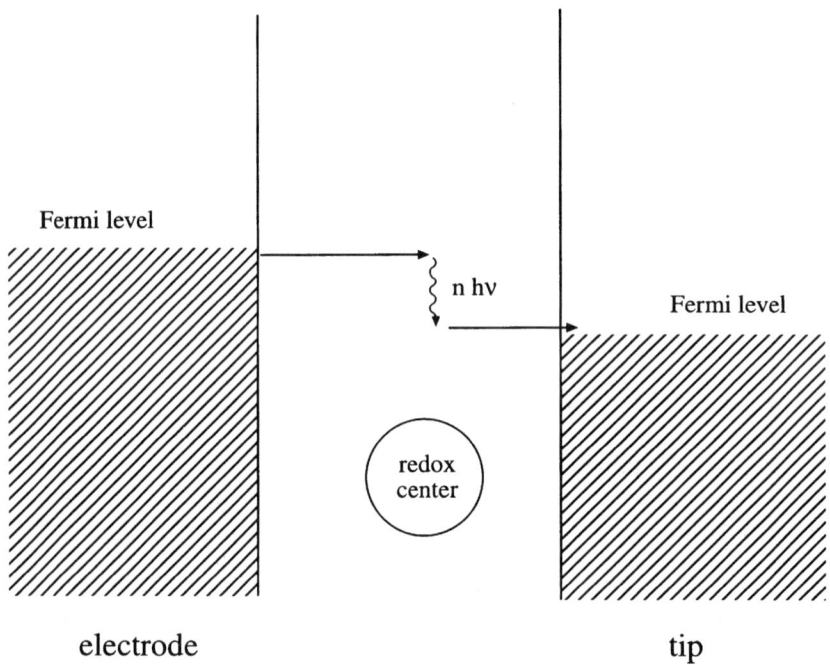

Figure 7.26. Electron exchange via a redox center with excitation of an inner-sphere mode.

7.9 INVESTIGATION OF ELECTRON-TRANSFER REACTIONS

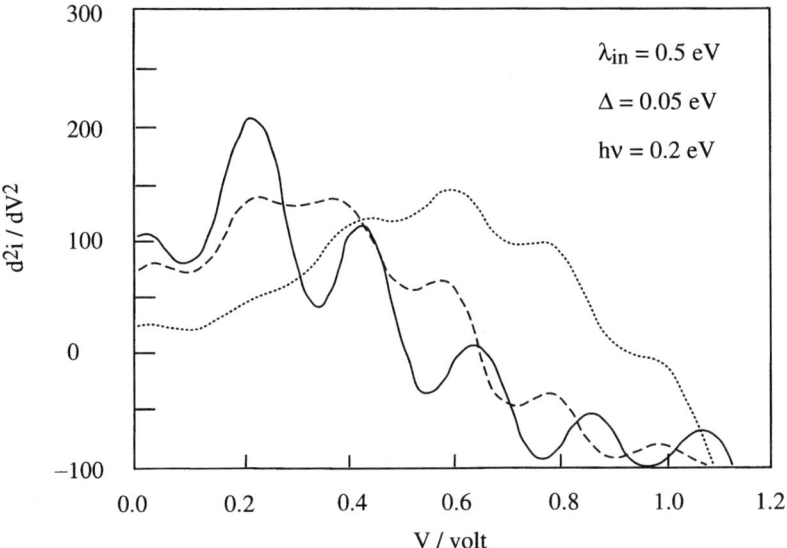

Figure 7.27. Second derivative d^2i/dV^2 for various energies of reorganization of the outer sphere; full line: $\lambda_{out} = 0.1$ eV; dashed line: $\lambda_{out} = 0.2$ eV; dotted line: $\lambda_{out} = 0.5$ eV.

a spectrum that was calculated for the case where one quantum mode is coupled to the electron transfer. The difference between adjacent peaks gives the frequency of the participating mode, and the peak height indicates the strength of the coupling: the higher the peaks are, the larger is the energy of organization λ_{in} of the quantum mode, and the smaller the energy of reorganization λ_{out} of the classical outer sphere modes. Depending on the system parameters, 10–30% of the current may pass through inelastic channels, so that the detection of these modes should be possible. Such experiments are, however, quite difficult to perform, and it is not surprising that none have been reported yet.

7.9.2 Experiments

So far, there has been only one convincing study of an electron-transfer reaction with an STM: the investigation of doped porphyrin films by Tao [40]. In these experiments a monomolecular film containing both Protoporphyrin IX (PP) and Fe(III)-Protoporphyrin IX (FePP) was adsorbed on a highly oriented pyrolytic graphite electrode. These films form regular two-dimensional lattices in which the molecules lie flat on the surface. The FePP molecules are electroactive and undergo a reversible electron transfer reaction, in which Fe(III) is reduced to Fe(II); the corresponding standard equilibrium potential is -0.48 V vs. SCE. These films were imaged with an STM in the usual constant current mode; the bias potential was set to the relatively small value of 0.1 V, and the tunneling current at 30 pA or less; under these conditions the tip sample separation is larger than 10 Å [41,42], and therefore the interaction between the tip and the sample is comparatively weak.

In the STM images both PP and FePP molecules appear as bright spots, but the apparent height of the FePP species depends on the electrode potential. This is clearly evident in Figure 7.28, which shows a series of images with a single FePP molecule surrounded by PP species at various electrode potentials. For a quantitative evaluation line scans were performed in the direction indicated in the top image. The FePP molecule is always brighter than the inactive PP species; obviously, the extra current is provided by resonant exchange with the Fe-ion at the center. Its apparent height changes by more than 2 Å (see Fig. 7.29), and reaches a maximum at potentials of -0.4 V to -0.5 V versus SCE, while that of the PP molecules remains constant. By subtracting the PP signal from that of the FePP, the extra apparent height Δz can be obtained as a function of the electrode potential. The resulting data are shown in Figure 7.30.

Obviously, the electronic levels of the electroactive molecules are shifted with respect to the Fermi level of the electrode as the potential is varied. The situation on graphite is particularly favorable, since it has a low density of states

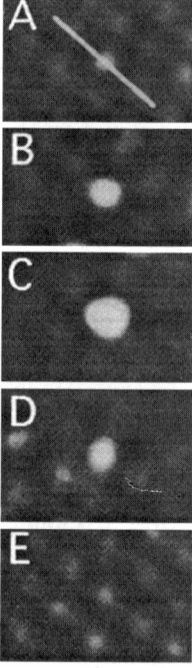

Figure 7.28. STM image of a single FePP molecules embedded in an array of PP when the substrate was held at a potential of: (a) -0.15 V, (b) -0.30 V, (c) -0.42 V, (d) -0.55 V, and (e) -0.65 V, respectively.

7.9 INVESTIGATION OF ELECTRON-TRANSFER REACTIONS 333

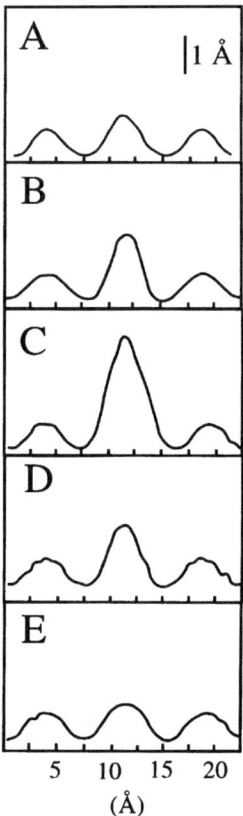

Figure 7.29. Line scan along the white line shown in Figure 7.27 (a); the electrode potentials are the same as in Figure 7.27.

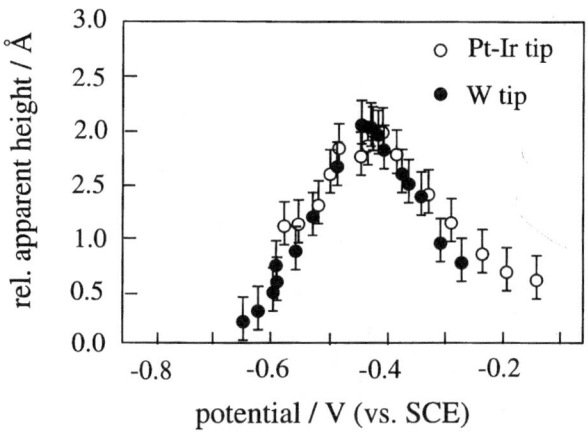

Figure 7.30. Apparent height of FePP relative to PP as a function of the electrode potential.

near the Fermi surface, and this results in the formation of a space charge layer at the surface, as is evidenced by a low double-layer capacity on the order of a few μF cm^{-2}. Therefore the relative shift in the energy levels between the electrode and the adsorbate should be approximately equal to the change in the electrode potential.

In order to extract the redox density of states from these data, the apparent height, which was measured as a function of the electrode potential at constant current, must be converted to the current that would have been measured at constant height. For this purpose the dependence of the STM current on the separation between the tip and the sample must be known. Since the distance between the tip and the adsorbate is much larger than the distance between the electrode and the adsorbate, the tunneling probability for resonant exchange is given by the tunneling probability T_{a-t} between the tip and the adsorbate. As discussed earlier, the latter depends exponentially on the separation: $T_{a-t} \propto \exp -\kappa z$, where κ is determined by the effective barrier height W_{eff}; with $W_{\text{eff}} \approx 1$ eV, $\kappa \approx 1$ Å^{-1}.

The resonant current through the FePP and the background current from the PP should have the same dependence on the separation. Therefore, within the semiclassical theory, the extra apparent height Δz is related to the redox density of states:

$$\exp(-\kappa \Delta z) - 1 \propto \int_0^V D_{\text{ox}}(\epsilon - e_0\eta + V)\, d\epsilon \tag{16}$$

Figure 7.31 shows the corresponding data obtained with $\kappa = 1$ Å^{-1}; they were normalized such that the average of the data near the maximum is unity. They can be fitted to a classical density of states with negligible electronic width and an energy of reorganization of $\lambda = 0.2$ eV. There is no indication that quantum modes participate in the tunneling process.

The energy of reorganization that is obtained from the data depends on the value assumed for the decay constant κ. For example, for $\kappa = 0.5$ Å^{-1} one obtains $\lambda = 0.25$ eV, while $\kappa = 1.5$ Å^{-1} gives $\lambda = 0.15$ eV. Obviously, an exact determination of the energy of reorganization would require a measurement of the decay constant in the same experiment. Therefore, the most significant result is the shape of the density of states, which is independent of the value assumed for κ. In particular, the data clearly show the decrease of the density of states at high energies, which is equivalent to a decrease of the rate constant in the inverted region postulated by Marcus [43].

In any case the energy of reorganization is small compared to those for electron-transfer reactions of nonadsorbed species. This is in line with a study by Smalley et al. [44], who showed that the reorganization energy of ferrocene attached to a monolayer of alkanethiols decreases from about 0.95 eV to 0.7 eV as the length of the chain is reduced from 16 to 5 members. In the latter case the redox center is estimated to be 7.4 Å away from the electrode surface,

7.10 CONCLUSIONS

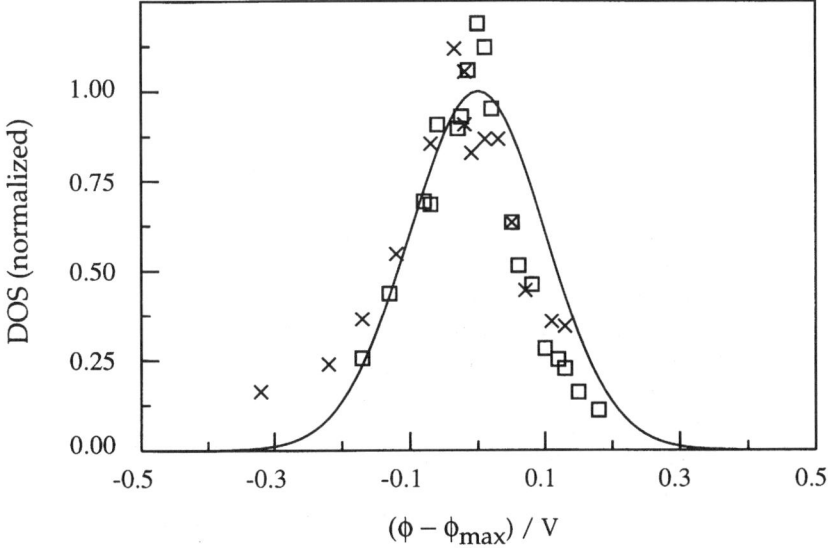

Figure 7.31. Apparent density of states of FePP. The squares and crosses denote the values derived from the experimental data for a tungsten and for a Pt-Ir tip, respectively.

while for FePP this distance is of the order of 3 Å, so that λ should be much smaller still. A simple model based on classical electrostatics [45] also predicts a strong decrease of λ in this situation.

This study by Tao is the first, and so far the only, direct measurement of the density of state of a redox couple; in particular, it constitutes a direct proof of the decrease at high energies [46]. A related investigation was performed by Snyder and White [47], who measured bias-current curves for a similar system. However, this work was performed ex situ, which makes an interpretation of the results difficult. Obviously, more experimental work in this area is highly desirable.

7.10 CONCLUSIONS

While the STM is nowadays routinely used to obtain topographic images of electrode surfaces, spectroscopic applications, such as the measurement of current-distance and current-voltage relations, have been rare. Indeed, there seem to be more theoretical than experimental works in this area. This is partially due to the difficulties encountered when the STM is used in anything but the standard imaging mode, but psychological effects may also play a role: The topographic images give the structural information that electrochemists had been lacking in the 1980s, and few experimentalists are tempted to go further and enter an area where success is less certain.

From the few current-distance measurements that have been performed it appears that the effective barrier height for electron tunneling is substantially lower in water than in the vacuum. There seem to be two reasons for this: (1) the electronic polarizability of water lowers the work function for electron emission into water; (2) the hydrogen atoms on the water molecules attract electrons. The latter effect can either be described in terms of valleys in the potential energy surface, or as virtual resonance states.

When the application of the STM to electrochemical interfaces was established, theoreticians were quick to point out that current-voltage spectroscopy could give chemical information of the imaged species, and that electrochemical systems offered more opportunities than surfaces in the vacuum, because two potential drops can be controlled independently. Practical applications have proved to be difficult, but the few successful studies, in particular the elegant work by Tao, have demonstrated the great potential of this method.

We hope that spectroscopic applications of the STM become widespread as experimental difficulties are overcome. So this review should be viewed as a snapshot of a growing area, which should spur further research, and thus catalyze its own obsolescence.

ACKNOWLEDGMENT

I would like to thank N. Tao from Florida International University for a fruitful scientific exchange. Financial support by the Deutsche Forschungsgemeinschaft is gratefully acknowledged. Thanks are also due to D. Kolb, University of Ulm, and to W. Lorenz and S. Vinzelberg, University of Karlsruhe, for allowing me to use their unpublished results.

REFERENCES

1. Y. Kuk and P. J. Silverman, *J. Vac. Sci. Technol.* **A8** (1990) 289.
2. N. D. Lang, *Imaging and Transfer of Single Atoms in the Scanning Tunneling Microscope*, in Electronic Processes at Solid Surfaces, E. Ilisca and K. Makoshi, eds., World Scientific, Singapore, 1996; N. D. Lang, *Phys. Rev.* **B37** (1988) 10395.
3. R. Christoph, H. Siegenthaler, H. Rohrer, and W. Wiese, *Electrochim. Acta* **34** (1989) 1011.
4. Repphuhn and Halbritter, *J. Vac. Sci. Technol.* **A13** (1995) 1693.
5. A. Vaught, T. W. Jing, and S. M. Lindsay, *Chem. Phys. Lett.* **236** (1995) 306.
6. S. Vinzelberg, Ph.D. thesis, Technical University of Karlsruhe, 1996; W. Lorenz and S. Vinzelberg, unpublished results.
7. D. Kolb, unpublished results, private communication.
8. W. Schmickler and D. J. Henderson, *J. Electroanal. Chem.* **290** (1990) 283.
9. Yu. V. Pleskov and Z. A. Rotenberg, in *Advances in Electrochemistry and Electro-*

chemical Engineering, Vol. 11, H. Gerischer and C. W. Tobias, eds., Wiley, New York, 1978.
10. J. Tersoff and D. R. Hamann, *Phys. Rev. B* **31** (1985) 805.
11. O. Pecina, W. Schmickler, K. Y. Chan, and D. J. Henderson, *J. Electroanal. Chem.* **396** (1995) 303.
12. D. Henderson, K. Y. Chan, M. Lozada-Cassou, and W. Schmickler, in *Lectures on Thermodynamics and Statistical Mechanics*, M. Lopez and C. Varea, eds., World Scientific, Singapore, 1992.
13. B. V. Numerov, *Mon. Not. R. Astron. Soc.* **84** (1924) 592.
14. K. L. Sebastian and G. Doyen, *Surf. Science* **290** (1993) L703; *J. Chem. Phys.* **99** (1993) 6677.
15. M. Buettiker and R. Landauer, *Phys. Rev. Lett.* **49** (1982) 1739.
16. D. Rostkier-Edelstein, M. Urbakh, and A. Nitzan, *J. Chem. Phys.* **101** (1994) 8224.
17. A. A. Kornyshev, *J. Electroanal. Chem.* **376** (1994) 9.
18. M. Sumetskii, A. Kornyshev, and U. Stimming, *Surf. Science* **307** (1994) 23.
19. S. Weiss, D. F. Ogletree, D. Botkin, M. Salmeron, and D. S. Chemla, *Appl. Phys. Lett.* **63** (1993) 2567.
20. M. L. Lozano and M. C. Tringides, *Europhys. Lett.* **30** (1995) 537.
21. A. M. Kuznetsov, P. Sommer-Larsen, and J. Ulstrup, *Surf. Science* **275** (1992) 52.
22. J. K. Sass and J. K. Gimzewski, *J. Electroanal. Chem.* **308** (1991) 333.
23. W. Schmickler, *Surf. Science* **335** (1995) 416.
24. S. M. Lindsay, T. W. Jing, J. Pan, D. Lamper, A. Vaught, J. P. Lewis, and O. F. Sankey, in *Nanoscale Probes of the Solid/Liquid Interface*, A. A. Gewirth and H. Siegenthaler, eds., NATO ASI Series E, Vol. 288, Kluwer, Dordrecht, 1995.
25. J. Bardeen, *Phys. Rev. Lett.* **6** (1961) 57.
26. J. Halbritter, G. Repphuhn, S. Vinzelberg, G. Staikov, and W. J. Lorenz, *Electrochim. Acta* **40** (1995) 1385; G. Repphuhn and J. Halbritter, *J. Vac. Sci. Technol.* **A13** (1995) 1693.
27. R. N. Barnett, U. Landmann, C. L. Cleveland, and J. Jortner, *J. Chem. Phys.* **88** (1988) 4429.
28. J. Schnittker and P. J. Rossky, *J. Chem. Phys.* **86** (1987) 3462.
29. A. Mosyak, A. Nitzan, and R. Kosloff, *J. Chem. Phys.* **104** (1996) 1549.
30. A. Mosyak, P. Graf, I. Benjamin, and A. Nitzan, *J. Chem. Phys.*, in press.
31. I. Benjamin, D. Evans, and A. Nitzan, *J. Chem. Phys.* **106** (1997) 6647.
32. G. Nagy, *J. Electroanal. Chem.* **409** (1996) 19.
33. J. A. Stroscio, R. M. Feenstra, and A. P. Fein, *Phys. Rev. Lett.* **57** (1986) 2579.
34. R. S. Robinson and C. A. Widrig, *Langmuir* **8** (1992) 2311.
35. J. E. T. Andersen, A. Kornyshev, A. M. Kuznetsov, L. L. Madsen, P. Møller, and J. Ulstrup, *Electrochim. Acta* **42** (1997) 819.
36. W. Schmickler and C. Widrig, *J. Electroanal. Chem.* **336** (1992) 213; **8** (1992) 2311.
37. H. Gerischer, *Z. Phys. Chem. NF* **26** (1969) 21.
38. W. Schmickler, *Surf. Science* **295** (1993) 43.

39. W. Schmickler, *Electrochim. Acta* **40** (1995) 1315; W. Schmickler, *The Metal-Solution Interface in the STM-Configuration*, Proceedings of the NATO Advanced Study Institute on Nanoprobes of the Solid/Liquid Interface, 1995.
40. N. Tao, *Phys. Rev. Lett.* **76** (1996) 4066.
41. A. Vaught, T. W. Jing, and S. M. Lindsay, *Chem. Phys. Lett.* **236** (1995) 306.
42. M. Bingelli, D. Carnal, R. Nyffenegger, and H. Siegenthaler, *J. Vac. Sci. Technol.* **B9** (1991) 1985.
43. R. A. Marcus, *J. Chem. Phys.* **43** (1965) 2654; **52** (1970) 2803.
44. J. F. Smalley, S. W. Feldberg, C. E. D. Chidsey, M. R. Lindford, M. D. Newton, and Yi-Ping Liu, *J. Phys. Chem.* **99** (1995) 13149.
45. Yi-Ping Liu and M. D. Newton, *J. Phys. Chem.* **98** (1994) 7162.
46. W. Schmickler and N. Tao, *Electrochim. Acta* **42** (1997) 2809.
47. S. R. Snyder and H. S. White, *J. Electroanal. Chem.* **394** (1995) 177.

INDEX

"Active" state, 140
Adenine, 213, 216
Adsorbate-induced etching process, 261
Adsorbed porphyrin films, 276
Adsorption capacitance, 17
Antiphase oscillation, 181
Atomic absorption spectroscopy (AAS), 261
Atomic force microscopy (AFM), 3, 100, 212
Attenuated total reflection (ATR), 182
Auger electron spectroscopy (AES), 108
Autocatalytic reaction, 140
Avrami equation, 18
Azobenzene-terminated alkanethiols at gold surfaces, 270

Benzene, 212
Benzene chemisorption on Rh(111), 234
2,2' Bipyridine(22Bpy), 213
22BPY adsorption on Au(111), 228
22BPY monolayer, 229
4,4' Bipyridine, 213
Biomolecular films, 276
Bipyridines, 212
Bistable complex, 201
Bistable regime, 201

Cantilever torsional response, 282
Characteristic length of patterns, 141
Charge transfer resistance, 17

Collective electron oscillations, 184
Colloidal solutions, 250
Commensurate or incommensurate 2D Me phases, 13
Concentration patterns, 141
Conductive polymers, 271
Continuously stirred tank reactor (CSTR), 166
CO oxidation in Pt(100), 142
Copper clusters, 117
Copper deposition on gold electrodes, 100
Coupling mechanisms, 150
Critical cluster (N_c), 112
"Critical" nucleus size, 113
Crystal violet, 123, 126
Current-voltage spectroscopy, 326
Cu underpotential deposition on Au (111), 69, 106, 109

Defect-induced techniques, 44
Density of states, 326, 329, 335
Diffraction-basic consideration of, 58
Diffusion coupled reaction fronts, 154
DNA bases, 213
Dodecanethiolate, 266
Dodecanethiolate monolayer at Au(111), 265
Double layer capacitance, 17
Dynamic instabilities in surface reactions, 145
Dynamics of low-dimensional Me phase formation, 15

339

ECALE systems, 128
Elastic scattering of monoenergetic electrons, 58
Elastic transitions, 330
Elastic tunneling, 307
Electrochemical impedance spectroscopy (EIS), 18
Electrochemical pattern formation, 176
Electrochemical sensors, 275
Electrochemical waves, 144
Electron diffraction (RHEED), 57, 69, 108
Electrocrystallization of Pb, 120
Electrodeposition of Cu on Ag (111), 115
Electronic configuration of the metal surface, 189
Electron microscopy, 57
Electron scattering quantum chemical procedure, 239
Electron-transfer reactions, 327, 331
Electron tunneling, 306
Electropolymerization of surface-confined pyrrole, 272
Electrolytic deposition of silver, 92
Ellipso-microscopy for surface imaging (EMSI), 143, 162
End-group derivatized monolayers, 269
Energy of reorganization, 329, 334
Epitaxial growth, 14
Evanescent wave, 186
Exchange-correlation hole, 311

Ferrocene-terminated alkanethiol, 290
Field electron microscopy (FEM), 144
Field-induced techniques, 44
Field ion microscopy (FIM), 144
First order phase transition, 10
Flame annealing method, 106
Flame-treatment, 69, 90
"Flat topped" crystallites, 118
Force measurements with SFM, 281
Fluctuations, 314
Fluorocarbon domains, 286
Frank-van der Merwe, 6
Frank-van-der-Merwe growth (layer-by-layer growth), 116
Free electron system, 181
Friction and adhesion-based measurements, 285, 286
Frictional parameters, 287

Gibbs energy of cluster formation, 10
Growing face of crystallite, 116

Guanine, 213
Guanine monolayer, 214, 215, 217

Helical structure of PT, 274
Highly oriented pyrolytic graphite (HOPG), 271
Hydrogen oxidation, 146

Image force, 311
Imaging in heterogeneous catalysis, 142
Imaging of organic molecules, 127
Imaging of reaction fronts, 139
Imaging patterns at electrode surfaces, 175
Imaging reaction fronts in heterogeneous catalysis, 145
Imaging the photoelectrons, 143
Indium-tin oxide (ITO), 274
Inelastic transitions, 329, 330
Infrared imaging, 169
Infrared reflection spectroscopic studies, 266
Inner Helmholtz plane (IHP), 191
"Insulating" organic molecules, 239
Intermediate states, 316
IR camera, 142
IR emissivity of Pt, 142

Jellium, 320

Kink sites, 113
Kossel-Stranski model, 112, 113
Kossel-Stranski theory, 111
Kretschmann configuration, 182, 186, 193

Langmuir-Blodgett, 286
Langmuir-Blodgett films, 211
Langmuir-Blodgett transfer techniques, 289
Langmuir-Hinshelwood mechanism, 146
Lateral force measurements with SFM friction loops, 282
Lead deposition on silver electrodes, 100
Leed and Rheed, 61
Local nanostructuring, 45
Local probe methods, 100
Low-dimensional metal phases, 1, 3
Low-energy electron microscopy (LEEM), 143

Massive gold, silver, platinum, and copper single crystals, 105
Me underpotential deposition, 2
Mercaptohexadecanol, 265
Mercury-water interface, 216
Metal adlayers, 105

INDEX **341**

Metal electrocrystallization, 2, 99, 100
Metal growth in electrocrystallization, 115
Metal overpotential deposition (OPD), 2
Metal-sulphur surface alloy formation, 3
Microlithography, 160
Mirror electron microscopy (MEM), 143
"Missing row" structure, 162
Modern nanotechnology, 3
Modes of operation for SFM, 280
Moire pattern, 110
Molecular dynamics, 320
Molecular motion and surface evolution, 268
Molecular packing structure, 230
Molecular packing structure, the role of hydrogen bonds, 216
Molecular-scale features, 259
Molecular-scale structures, 262
Monolayer 22BPY formation, 229
Monolayers from various alkanethiols, 265

Nano-clusters of Cu, 126
Nanostructuring and nanomodification of solid surfaces, 3
Nanostructuring of solid surfaces, 1, 41
Naphthalene, 235
Naphthoquinone, 235
Nernst potential for M/M^{n+}, 103
Nonlinear chemical reaction, 172
Nucleation and growth processes, 18
Nucleation and growth mechanism of PAN, 273
Nucleation preferential sites, 112
Nucleic acid bases, 232

One-dimensional electrode geometries, 176
Organic additives, 120, 126
Organic adsorbates, 119
Organic monolayers at electrode-electrolyte interfaces, 211, 212
Organic thin films at electrode surface, 249, 251, 278
Oscillation cycle for the CO oxidation, 149
Oscillation frequencies, 141
Oscillatory regime, 201
Oscillatory regime of Ni dissolution, 180
Otto-configuration, 182
Oxidation of ammonia and hydrocarbons on Pt, 146
Oxidative doping of emeraldine base, 273

Pattern formation in electrochemical systems, 181
"Passive" state, 140

Perfluorinated alkanethiols, 284
Phase difference, 162
Phenanthroline, 212, 213, 234
Phenanthroline monolayers, 234
Photoemission electron microscopy, 143, 153, 155
Photon scanning tunneling microscope, 183
Piezoelectric microactuators, 252
Polyaniline (PAN), 273, 294
Polymers, conducting, 274
Polymer dispersions, 250
Polymer films, 270, 293
Poly(phenylene oxide) (PPO), 275, 295
Polypyrrole(Ppy), 193, 271, 293
Polythiophene (Pt) STM, 274, 294
Porphyrins, 212
Potential drop across the double layer, 176
Potential-induced structural phase transition, 223
Potential probes, 176
Ppy/glucose oxidase (GOD) electrode, 276
Probe-induced techniques, 44
Probe microscopy (SPM), 3
Pseudopotentials, 319

Raman cross section, 183
Raman imaging with surface plasmons, 183
Raman scattering, 182
Reaction fronts in electrochemistry, 139, 144
Reciprocal lattices with Ewald spheres, 67
"Reconstruction model," 146
Reflectance anisotropy microscopy (RAM), 143, 162
Reflection electron microscopy images, 87
Reflection electron microscopy (REM), 143
Resonances, 319

Scanning force (SFM), 251
Scanning LEED, 142
Scanning photoemission microscope (SPM), 142, 151, 152
Scanning thermopower microscope, 214
Scanning tunneling microscope, principle of operation, 100, 212, 251, 252
Scanning tunneling optical microscopy, 183
Schwarzschild objective, 151
Second harmonic generation (SHG), 183
Self-assembly techniques, 211
SFM principles, 278
Single metal deposition systems, 101
Solitary waves, 160
Solvated electrons, 316
Spatio-temporal patterns, 172, 176

Spatio-temporal plot, 180
Spatio-temporal properties, 178
Specimen preparation for imaging, 68
Specimen treatment, 65
Spontaneously adsorbed monolayers, 253, 283
Stability conditions, 4
Standing waves, 156, 158
Step edges, 114
Sticking coefficient, 148
STM images of copper clusters, 111
STM modes of operation, 254
STM tip fabrication, 236
Stranski-Krastanov, 6
Stranski-Krastanov growth, 116, 120
STM, 3
STM imaging mechanism, 239
Structural imperfections of different dimensionality, 12
Structural phase transition in monolayers, 233
Structure of a H_2SO_4 monolayer on Au(111), 79
Subsurface oxygen formation, 148
Sum frequency (SFG), 183
Super- and undersaturation, 4, 6
Superperiodic structures, 220
Superperiodic structures in guanine monolayer, 221
Surface enhanced Raman scattering (SERS), 182
Surface faceting by fast potential cycling, 84
Surface imaging by diffracted electrons, 82
Surface inhomogeneities, 1, 3, 13
Surface microscopy, 67
Surface plasmon resonance, 176
Surface plasmons, 183
Surface plasmon (SP) microscopy, 144, 175, 181, 192, 193
Surface plasmon polarition fields, 183
Surface reconstruction of gold, 81, 82

Tapping mode, 294
Temperature front, 170
Temporal evolution of a population of spirals, 156
Temporal oscillations and spatial structures, 145
Tetramethylthiourea, 126, 127, 212
Thermography visual or infrared, 142
Tip induced local metal deposition, 46
Tip-molecule interaction in STM imaging, 240
Tribological properties, 283
Tunneling barrier height, 307, 309, 318, 324
Tunneling probability, 307, 309, 318, 324
Twin-electrode thin-layer (TTL), 25, 26
Two-dimensional nucleus, 116

Underpotential deposition, 106
Underpotential deposition of Cu on Pt(111) and Pt (100), 75
Underpotential deposition (UPD) of lead, 255
Undersaturation range, 2
Uric acid, 225

Vertical propagation of the copper crystallites, 123
Vibrational quanta, 330
Virtual intermediate states, 328
Virtual state, 319
Volmer-Weber growth, 116
Volume plasmons, 185

Waals radii, 218
Work functions, 306, 312

X-ray photoelectron spectroscopy (XPS), 250
Xanthine, 213, 216, 224, 225